# Brake Design and Safety

## Second Edition

Other SAE books on this topic:

**Electronic Braking, Traction, and Stability Control**
Edited by Ronald K. Jurgen
(Order No. PT-76)

**ABS - TCS - VDC Where Will the Technology Lead Us?**
Edited by Josef Mack
(Order No. PT-57)

For more information or to order this book, contact SAE at
400 Commonwealth Drive, Warrendale, PA 15096-0001
(724) 776-4970; fax (724) 776-0790
e-mail: publications@sae.org
web site: www.sae.org/BOOKSTORE

# Brake Design and Safety

## Second Edition

### Rudolf Limpert

Society of Automotive Engineers, Inc.
Warrendale, Pa.

**Library of Congress Cataloging-in-Publication Data**

Limpert, Rudolf
    Brake design and safety / Rudolf Limpert. — 2nd ed.
    p. cm.
  Includes bibliographical references and index.
  ISBN 1-56091-915-9
  1. Automobiles—Brakes—Design and construction. I. Title.

TL269.L56 1999                     98-53284
629.2'46—dc21                        CIP

SAE Order No. R-198

# Preface to the Second Edition

The Second Edition continues to provide a systems approach to designing safer brakes. Consulting experts will find it a single reference in determining the involvement of brakes in accident causation.

Brakes system technology has attained a high standard of quality over the last two decades. Nearly all automobiles are now equipped with antilock brakes. Federal braking standards require commercial vehicles to use antilock brakes. Revolutionary innovative brake designs are not expected. Improvements in brake systems will only be achieved through basic research, the application of sound engineering concepts, and testing resulting in small yet important design changes.

The objective of the Second Edition is to assist the brake engineer in accomplishing his task to design safer brakes that can be operated and maintained safely. The brake expert will find all the analytical tools to study and determine the potential causes of brake failures. The Second Edition is expanded to cover all essential subjects including the mechanical and thermal analysis of disk brakes. Mistakes found in the First Edition were corrected.

I thank all those who have made valuable suggestions and comments and helped me to understand brakes better, in particular the many individuals who attended my Brake Design and Safety seminars.

Rudy Limpert

# Preface to the First Edition

The purpose of this book is to provide a systems approach to designing safer brakes. Much of the material presented was developed during my work as a brake design engineer, conducting automotive research, consulting as a brake expert, and teaching brake design.

The book is written for automotive engineers, technical consultants, accident reconstruction experts, and lawyers involved with the design of brake systems, the analysis of braking performance, and product liability issues. Junior engineers will benefit from the book by finding in one single source all essential concepts, guidelines, and design checks required for designing safer brakes.

Chapter 1 reviews basic stopping distance performance, design rules, and product liability factors.

In Chapter 2, drum and disc brakes are discussed. Brake torque computations are shown for different drum and disc brake designs.

Temperature and thermal stresses are analyzed in Chapter 3. Practical temperature equations are shown whenever possible.

Chapter 4 briefly reviews basic concepts involved in analyzing mechanical brake systems.

The operation and design of hydraulic brakes are discussed in Chapter 5.

Air brake systems and their components are discussed and analyzed in Chapter 6.

Brake force distribution, braking efficiency, optimum brake force distribution, and vehicle stability during braking for the single vehicle are analyzed in Chapter 7.

Car-trailer and commercial truck-trailer braking is discussed in Chapter 8.

Important elements of anti-lock braking performance and design are introduced in Chapter 9.

Brake failures are discussed in Chapter 10.

# Table of Contents

Table of Contents

# CHAPTER 1

# Elements of Braking Performance, Design, and Safety

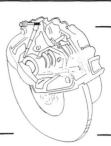

*In this chapter the basic brake functions and elements of brake systems are discussed. Operator pedal effort limits and braking performance are reviewed. Stopping distance equations are derived. The importance of driver reaction time is stressed. Brake design and product development guides are introduced along with a design selection procedure. Concepts of product liability are presented. Elements of U.S. and foreign brake safety standards are briefly reviewed.*

## 1.1 Functions of a Braking System

A vehicle is connected to the roadway by the traction forces produced by the tires. Any braking, steering, or accelerating forces must be generated by the small tire tread area contacting the road surface. Only forces equal to or less than the product of normal force and tire-roadway friction coefficient can be transmitted by the tire treads and wheels. Even the ideal braking system cannot utilize more traction than that provided by the tires and road.

The safe operation of a motor vehicle requires continuous adjusting of its speed to changing traffic conditions. The brakes and the tires along with the steering system are the most important safety-critical accident avoidance components of a motor vehicle. They must perform safely under a variety of operating conditions including slippery, wet, and dry roads; when a vehicle is lightly or fully laden; when braking straight or in a curve; with new or worn brake linings; with wet or dry brakes; when applied by the novice or experienced driver; when braking on smooth or rough roads; or when pulling a trailer.

1

These general uses of the brakes can be formulated in terms of three basic functions a braking system must provide:

1. Decelerate a vehicle including stopping.

2. Maintain vehicle speed during downhill operation.

3. Hold a vehicle stationary on a grade.

Deceleration involves the change of the kinetic and potential energy (if any) of a vehicle into thermal energy. Important factors a brake design engineer must consider include braking stability, brake force distribution, tire/road friction utilization, braking while turning, pedal force modulation, stopping distance, in-stop fade, and brake wear.

Maintaining vehicle speed on a hill involves the transfer of potential into thermal energy. Important considerations are brake temperature, lining fade, brake fluid vaporization in hydraulic brakes, and brake adjustment of air brakes.

Holding a vehicle stationary on a grade with the parking brake is mainly a problem of force transmission between the application lever and the tire. However, since a parking brake may be used for vehicle deceleration in an emergency, both thermal and vehicle dynamic factors must be considered by the design engineer.

## 1.2  Brake System Overview

### 1.2.1  Purpose of Brake System

The basic functions of a brake system are to slow the speed of the vehicle, to maintain its speed during downhill operation, and to hold the vehicle stationary after it has come to a complete stop.

These basic functions have to be performed during normal operation of the brakes, and to a lesser degree of braking effectiveness, during a brake system failure.

Consequently, brakes can be classified as *service brakes*, used for normal braking, *secondary or emergency brakes*, used during partial brake system failure, and *parking brakes* In current design practice, some components of the service brake are used for the secondary system and parking brake system.

## 1.2.2 Brake System Components

A typical hydraulic brake system is illustrated in Figure 1-1. All brake systems can be divided into four basic subsystems discussed next.

1. Energy source: This includes the components of a brake system that produce, store, and make available the energy required for braking. The energy source subsystem ends where driver-controlled modulation of the energy supply begins.

2. Apply system: This includes all components that are used to modulate the level of braking. The apply subsystem ends where the energy required for applying the brakes enters the energy transmission system.

3. Energy transmission system: This includes all components through which the energy required for applying the brakes travels from the apply system to the wheel brakes. Energy accumulators

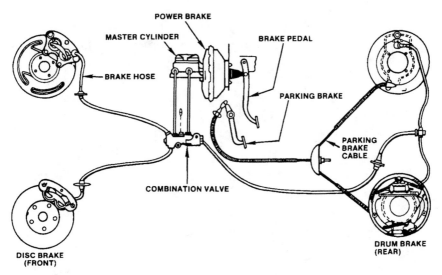

*Figure 1-1. Hydraulic brake schematic with parking brake (Bendix).*

3

located within this system are part of the energy transmission system. The energy transmission system ends at the component where the retarding brake torque is developed. For a hydraulic brake it ends where the wheel cylinder piston contacts the brake shoe. Brake tubes required for carrying hydraulic or air medium are part of the energy transmission system. Metallic rigid brake tubes commonly are called *brake lines*, whereas flexible tubes are called *brake hoses*.

4. <u>Wheel or foundation brakes</u>: These are the components where the forces are produced which oppose the existing or intended vehicle motion.

## 1.2.3 Type of Energy Source

The energy source is the medium that provides for the shoes to be pressed against the drums, or pads against the discs. Brake systems can be grouped by energy source as follows:

1. <u>Muscular driver pedal effort</u> is the basic system, often referred to as manual or standard brakes. It involves a brake system in which the shoe/drum pressing force and, hence, vehicle braking effectiveness is only related to the driver's pedal force and pedal displacement.

2. <u>Brake boost assist systems</u> are brake systems in which the shoe/drum pressing force is determined by the driver pedal force aided by one or more energy sources. Included under brake boost assist systems are the vacuum booster and hydro-boost system. A basic characteristic of brake boost assist systems is that they use a normal master cylinder, and that the driver can apply the brakes by muscular effort in the event of an energy source failure.

3. <u>Power brake systems</u> use one or more energy sources. The driver pedal effort is used only to modulate the energy source and not to apply force to press the shoes against the drum. No braking force is available in the event the energy source is depleted. Examples are air brakes and the hydraulic brakes of the Rolls Royce passenger cars.

4. <u>Surge brakes</u> for trailers use the motion energy of the trailer relative to the tow vehicle as the energy source to apply the brakes of the trailer.

5. <u>Drop weight brakes</u> use the potential energy of the trailer tongue, which when dropped to the ground applies the trailer brakes.

6. <u>Electric brakes</u> use magnetic force or electric motors to apply the brakes, most commonly in trailers.

7. <u>Spring brakes</u> use the force of a compressed spring as an energy source to apply the shoes against the drum.

## 1.2.4 Energy Transmission Medium

Brake systems can be grouped according to the medium by which the shoe/drum apply energy is transmitted from the energy source to the energy user or wheel brake as follows:

1. <u>Mechanical brakes</u> involve designs where only mechanical devices such as rods, levers, cables, or cams are used to transmit energy to the wheel or foundation brakes. In many cases parking brakes are mechanical brakes.

2. <u>Hydraulic brakes</u> use a fluid as the medium to transmit energy to the wheel brakes.

3. <u>Air brakes</u> use air to transmit energy to the foundation brakes. In automotive applications the air is pressurized. Vacuum brakes are used on trains as the medium to transmit energy to the wheel brakes.

4. <u>Electric brakes</u> use electrical current as the medium to transmit energy to the wheel brakes.

5. <u>Mixed brakes</u> use two or more of the means by which energy can be transmitted to the brakes. For example, air brakes use compressed air to transmit energy from the reservoir storing compressed air to the brake chamber near the foundation brake, and mechanical means such as pushrod, shaft, cam, and rollers to transmit energy from the brake chamber to the brake shoes.

*Air-over-hydraulic brake systems* use compressed air to transmit energy from the air reservoir to a converter, and hydraulic fluid to transmit energy from the converter to the wheel brakes.

## 1.2.5 Type of Dual Split System

Brake systems for passenger vehicles and trucks are designed so that a partial braking effectiveness is provided in the event of subsystem circuit failure. Accordingly, brake systems can be grouped according to their different split designs as follows:

1. Single-circuit systems use only one circuit to transmit braking energy to the wheel brakes of the service brake system. No braking is provided in the event of a circuit failure that is so severe that no brake line pressure can be produced. Since 1968, federal motor vehicle safety standards prohibit single-circuit systems for hydraulic brakes on passenger cars and pickup trucks, and since the mid-70s for air brakes. Since 1983, dual brake systems are required on all trucks with hydraulic brakes.

2. Dual-circuit systems use two or more circuits to transmit braking energy to the wheel brakes. In the event of a circuit failure, partial braking effectiveness is provided. Depending on the vehicle weight distribution, rear-front, diagonal, or a combination of splits are in use. A detailed discussion of the various dual system designs is presented in Chapter 10.

## 1.2.6 Type of Friction Brake

Automotive friction brakes are grouped according to their basic designs into two classes:

1. Drum brakes use brake shoes that are pushed out in a radial direction against a brake drum.

2. Disc brakes use pads that are pressed axially against a rotor or disc. Advantages of disc brakes over drum brakes have led to their universal use on passenger-car and light-truck front axles, many rear axles, and medium-weight trucks on both axles. See Chapter 2 for a detailed discussion of drum and disc brakes.

## 1.3  Pedal Force and Pedal Travel

Safety standards provide for certain limitations on pedal force. Ergonomic considerations and driver acceptance limit pedal force and pedal travel to a particular range established over the years. The maximum force exerted with the right foot for the 5th percentile female is approximately 445 N (100 lb); for the male approximately 823 N (185 lb) (Ref. 1). Both pedal force and pedal travel are important parameters for the human operator to safely modulate braking effectiveness. Brake systems without sufficient pedal travel feedback, particularly on slippery roads, may cause loss of vehicle control due to inadvertent brake lockup.

The pedal apply speed of skilled drivers is approximately 1 m/s (3 to 3.5 ft/s); of normal drivers around 0.5 ft/s. For a pedal travel of 100 mm pedal apply times are between 100 and 200 ms. A "soft" pedal does not only cause unsafe driver response but also increased stopping distances.

### 1.3.1  Manual or Standard Brakes

For brakes without a booster, the brake system should be designed so that for a maximum pedal force of 445 to 489 N (100 to 110 lb), a theoretical deceleration of 1 g is achieved when the vehicle is loaded at GVW (Gross Vehicle Weight; maximum allowable weight designated by the manufacturer). Maximum pedal travel between fully released and where the master cylinder piston bottoms out should not exceed 150 mm (6 in.). Drivers generally rate pedal force/deceleration ratios of 267 to 445 N/g (60 to 100 lb/g) as very good, and 445 to 668 N/g (100 to 150 lb/g) as good.

### 1.3.2  Brake Systems with Booster

A maximum pedal force of approximately 223 to 334 N (50 to 75 lb) should provide a deceleration of 0.9 to 1 g. The associated pedal travel should not exceed 75 to 90 mm (3 to 3.5 in.) for "cold" (less than 366 K or 200°F) brakes. The booster characteristic should increase linearly with pedal force and pedal travel. The booster run-out point should be reached for decelerations greater than 0.9 to 1 g. In order to ensure proper brake force modulation, a pedal force not greater than 13 to 22 N (3 to 5 lb) should be required to start boost assist. The boost ratio or gain should not be greater than approximately 4 to 6 in order to ensure safe vehicle deceleration in the event of a boost failure. Hydraulic brake line pressure rise time delays of 100 ms for single, and up to 180 ms for double diaphragm boosters must be considered.

### 1.3.3 Partial Failure Performance

Federal Motor Vehicle Safety Standard (FMVSS) 105 provides certain limits on pedal force and stopping distances in the event of a partial brake failure. These are considered minimum requirements. In general, brake systems are capable of achieving higher braking effectiveness at lower pedal forces than those required by the safety standard.

A maximum pedal force of approximately 445 N (100 lb) should achieve a deceleration of approximately 0.3 g for the vehicle loaded at GVW in the event of a booster failure.

In the case of a hydraulic circuit failure, a maximum pedal force of approximately 445 N (100 lb) should slow the vehicle laden at GVW at a deceleration of approximately 0.3 g.

In the event of repeated or continued braking with increased brake temperatures, a pedal travel of approximately 115 to 130 mm (4.5 to 5 in.) out of 150 mm (6 in.) available should not be exceeded for a maximum pedal force of approximately 445 N (100 lb).

### 1.3.4 Parking Brake

The parking brake should hold the vehicle stationary when laden at GVW on a 30% slope (16.7 degrees) with a hand force of not more than 356 N (80 lb) or a foot force of less than 445 N (100 lb). With the apply force limitations stated, the parking brake should be able to slow a vehicle laden at GVW at approximately 0.3 g.

## 1.4 Vehicle Deceleration and Stopping Distance

### 1.4.1 Basic Measures of Motion

The motion of a decelerating vehicle can be described by four measures of physics, namely *distance, time, velocity,* and *deceleration*. Distance and time are fundamental measures, that is, they cannot be broken down into submeasures. Velocity and deceleration are measures derived from distance and time, i.e., they can be broken down into fundamental measures.

The velocity V of a vehicle is computed by the ratio of distance S and time t:

$$V = S / t \quad , \quad m/s \ (ft/s) \tag{1-1}$$

where   S = distance, m (ft)

t = time, s

The term "speed," often used to describe velocity, only refers to the magnitude of the velocity, and does not indicate angular orientation and direction of the moving vehicle.

The velocity of a vehicle is constant or uniform when it travels the same distances in equal time intervals. The velocity of a vehicle is changing when it travels different distances in equal time intervals.

The deceleration of vehicle, a, is computed by dividing the velocity decrease by the time interval during which the velocity change has occurred:

$$a = \frac{\Delta V}{\Delta t} = \frac{V_2 - V_1}{t_2 - t_1} \quad , \quad m/s^2 \ (ft/s^2) \tag{1-2}$$

where   $t_1$ = time at start of deceleration, s

$t_2$ = time at end of deceleration, s

$V_1$ = velocity at start of deceleration, m/s (ft/s)

$V_2$ = velocity at end of deceleration, m/s (ft/s)

With the basic motion parameters defined, we can now compute the stopping distance and other related factors of a moving vehicle.

## 1.4.2 Simplified Stopping Distance Analysis

The motion of a vehicle as it changes over time can be shown graphically in the *velocity-time diagram* (V-t diagram). In the case of constant velocity, the velocity curve is a straight line as illustrated in Figure 1-2. The rectangular area under the V-line is given by the product of height and length, or velocity multiplied by time. Inspection of Eq. (1-1) reveals that the distance S traveled is also equal to velocity multiplied by time, i.e., equal to the area under the V-curve.

This observation can be expressed as:

> *The distance traveled by a vehicle is equal to the area under the velocity-time curve.*

In a simple analysis, the vehicle motion for an emergency braking maneuver with constant deceleration can be approximated as show in Figure 1-3. After the driver's reaction time $t_r$ and the brakes are applied, the vehicle begins to slow at constant deceleration from its travel speed $V_{tr}$ and the vehicle stops after the braking time $t_s$.

The V-t diagram shown in Fig. 1-3 consists of a rectangle under the constant speed portion, and a triangle under the decreasing speed portion of the

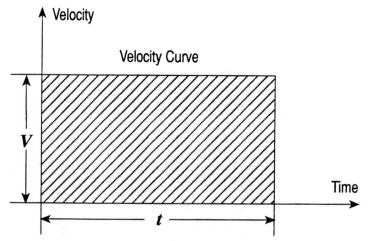

*Figure 1-2. Constant velocity-time diagram.*

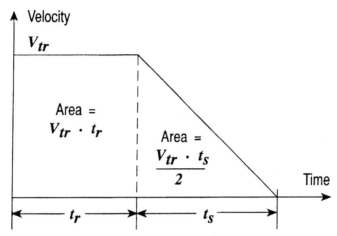

*Figure 1-3. Velocity-time diagram for stopping process.*

maneuver. The total distance $S_{total}$ is equal to the area of the rectangle plus the area of the triangle, or

$$S_{total} = V_{tr}t_r + V_{tr}t_s / 2 \quad , \quad \text{m (ft)} \tag{1-3}$$

The last term of Eq. (1-3) can be rewritten by using $t_2 - t_1 = t_s$ or $t_s = V_{tr}/a$ in Eq. (1-2) as:

$$S_{total} = V_{tr}t_r + V_{tr}^2 / 2a \quad , \quad \text{m (ft)} \tag{1-4}$$

Eq. (1-4) is the basic equation used for simple speed and stopping distance calculations in accident reconstruction.

## 1.4.3 Expanded Stopping Distance Analysis

In braking maneuvers where the maximum sustained vehicle deceleration is not achieved quickly and the deceleration rise cannot be ignored, a more detailed stopping distance analysis must be carried out.

Consider the basic braking parameters illustrated in Figure 1-4. The idealized pedal force as a function of time is shown in Fig. 1-4a. At time zero (0) the driver recognizes the danger. After the reaction time $t_r$ has elapsed, the

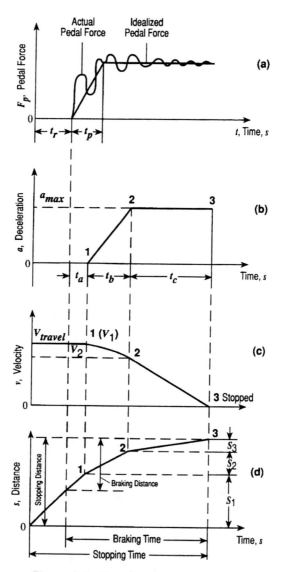

*Figure 1-4. Stopping distance analysis.*

driver begins to apply pedal force. After the brake system application time $t_a$ has passed, the brake shoes contact the drums and vehicle deceleration begins. The linear rise of pedal force is an approximation and occurs over the time $t_p$. In critical emergency situations, unskilled drivers tend to reduce their pedal forces somewhat after 0.1 to 0.2 s of brake initiation in an attempt to modulate the braking process (Ref. 2). When the obstacle comes closer, pedal forces rise again. Skilled drivers generally have pedal forces that more closely resemble the idealization. At higher speeds the pedal force rise characteristics actually present may be of lesser importance since their influence on overall stopping distance is small.

The idealized deceleration is shown in Fig. 1-4b. The deceleration begins to rise linearly at point 1 when brake torque development starts, and stops at point 2, either because the pedal force is constant or all brakes are locked and no further increase in tire-road braking forces is possible.

The velocity change as a function of time is shown in Fig. 1-4c. Prior to any deceleration, the travel velocity remains constant. Deceleration forces other than those produced by the wheel brakes themselves may slow the vehicle before the brake pedal is applied. Depending on the vehicle and braking process involved, these retarding forces may come from engine drag, retarders, aerodynamic drag, or gravity when braking on an incline. If they are significant relative to the decrease in travel speed prior to brake application, then they must be included by having a bilinear deceleration rise characteristic. In most emergency situations with rapid pedal force applications, the single linear deceleration rise idealization provides excellent correlation with actual stopping distance tests. The stopping distances predicted with a linear deceleration rise generally are only 0.5 to 1% longer than those obtained with bilinear rise.

In the velocity diagram shown in Fig. 1-4c, the velocity is a curved line between points 1 and 2. The deceleration remains constant when it has reached its maximum value. The velocity line between points 2 and 3 is straight. The vehicle stops at point 3.

The total stopping distance is the summation of the individual distances associated with the different time intervals, i.e., begin of reaction until deceleration begins to rise, deceleration rise time, and when the deceleration is constant until the vehicle stops.

a. Distance $S_1$ traveled during reaction and application time, $t_r$ and $t_a$, respectively:

$$S_1 = V_1(t_r + t_a) \quad , \quad \text{m (ft)} \tag{1-5}$$

where $V_1$ = velocity at point 1, m/s (ft/s)

b. Distance traveled during deceleration rise time $t_b$: The deceleration is given by the ratio of maximum deceleration and rise time, or

$$a(t) = a_{max}t / t_b \quad , \quad \text{m/s}^2 \text{ (ft/s}^2) \tag{1-6}$$

where $a_{max}$ = maximum deceleration, m/s² (ft/s²)

$a(t)$ = deceleration as function of time, m/s² (ft/s²)

$t$ = time, s

The velocity as a function of time during the deceleration rise is equal to the initial velocity minus the change in velocity, or

$$V(t) = V_1 - \int a_{max}t / t_b dt = V_1 - a_{max}t^2 / 2t_b \quad , \quad \text{m/s (ft/s)} \tag{1-7}$$

The distance $S_2$ traveled during the deceleration rise time $t_b$ is computed by integrating Eq. (1-7) between time zero and $t_b$, or

$$S_2 = \int_0^{t_b} V dt = V_1 t_b - a_{max}t_b^2 / 6 \quad , \quad \text{m (ft)} \tag{1-8}$$

where $V$ = velocity as function of time, m/s (ft/s)

c. Distance traveled during constant deceleration time interval: With the deceleration $a_{max}$ constant, the velocity as a function of time is computed by

$$V(t) = V_2 - a_{max} \int dt = V_2 - a_{max}t \quad , \quad \text{m/s (ft/s)} \tag{1-9}$$

where $V_2$ = velocity at point 2, m/s (ft/s)

The velocity at point 2 is computed by Eq. (1-7):

$$V_2 = V_1 - a_{max} t_b / 2 \quad , \quad \text{m/s (ft/s)} \tag{1-10}$$

The time required for the vehicle to stop, or for velocity $V(t)$ to be equal to zero, is computed by setting $V(t) = 0$ in Eq. (1-9), solving for $V_2$ and substituting into Eq. (1-10), and solving for time $t_c$:

$$t_c = V_2 / a_{max} = V_1 / a_{max} - t_b / 2 \quad , \quad s \tag{1-11}$$

The distance $S_3$ traveled during the constant deceleration interval is computed by

$$S_3 = \int_0^{t_c} V dt = V_2 t_c - a_{max} t_c^2 = V_2^2 / 2a_{max} \quad , \quad \text{m (ft)}$$

$$S_3 = 1 / 2a_{max} (V_1^2 + a_{max}^2 t_b^2 / 4 - V_1 a_{max} t_b) \quad , \quad \text{m (ft)} \tag{1-12}$$

The total stopping distance $S_t$ is computed by the sum of all individual distances, or

$$S_t = S_1 + S_2 + S_3 = V_1(t_r + t_a + t_b / 2)$$
$$+ V_1^2 / 2a_{max} - a_{max} t_b^2 / 24 \quad , \quad \text{m (ft)} \tag{1-13}$$

In most cases the third term in Eq. (1-13) is small for short deceleration rise times $t_b$ as compared to the other terms and, consequently, is neglected, yielding for the total stopping distance $S_t$:

$$S_t = V_1(t_r + t_a + t_b / 2) + V_1^2 / 2a_{max} \quad , \quad \text{m (ft)} \tag{1-14}$$

The total time $t_t$ from the driver reaction to vehicle stopping is given by

$$t_t = t_r + t_a + t_b / 2 + V_1 / a_{max} \quad , \quad s \tag{1-15}$$

The following example illustrates the influence of the different factors on stopping distance. A vehicle travels at a speed of 96 km/h or 26.7 m/s (88 ft/s), when the driver recognizes a hazard. After one second reaction time ($t_r = 1$ s), 0.25 s brake system application time ($t_a = 0.25$ s), and 0.3 s deceleration rise time ($t_b = 0.3$ s), the vehicle slows at a constant deceleration of 0.6 g or 5.9 m/s² (19.3 ft/s²).

Substitution into Eq. (1-13) yields:

$$S_t = (26.7)(1.55) + (26.7)^2 / 2(5.9) - (5.9)(0.3)^2 / 24$$

$$= 41.3 + 60.3 - 0.02 = 101.58 \text{ m}$$

$$\left[ \begin{array}{l} S_t = (87.5)(1.55) + (87.5)^2 / 2(19.3) - (19.3)(0.3)^2 / 24 \\ = 135.6 + 198.3 - 0.07 = 333.8 \text{ ft} \end{array} \right]$$

The total time is computed by Eq. (1-15) as

$$t_t = 1 + 0.55 + 0.3 / 2 + 26.7 / 5.9 = 5.93 \text{ s}$$

Inspection of the numerical values reveals the third term with 0.02 m (0.07 ft) to be insignificant compared to the others.

### 1.4.4 Driver Reaction Time

The overall stopping distance is strongly affected by driver reaction time. Driver reaction times used in accident reconstruction generally cover a time period from the perception of the hazard until some or all brakes are locked or the ABS brakes produce control tire marks (if any).

In general, driver reaction consists of four phases, namely, perception, judgment, reaction initiation, and reaction execution (Ref. 1). In certain cases such as a panic-type brake application, the judgment time may be at a minimum. Results of a large number of reaction time tests measured in simulated emergency braking maneuvers show that typical values of 0.75 to 1.5 s are generally acceptable (Refs. 1, 2). Statistical analyses of a large body of test data suggest that differences may exist for reaction times used in accident reconstruction (Ref. 3). A brief review of the findings is presented next.

In general, an object or hazard will first appear in the driver's peripheral vision. Only after the driver has focused the eyes on the object can an intended and planned human reaction begin. It is important to recognize that the first appearance of an object in the driver's peripheral vision, such as a pedestrian stepping off the curb on the left side of the highway, is not the beginning of the driver's reaction time. Experimental psychology has also determined that human reaction times are shorter for an expected signal than for less-observed unexpected ones. Drivers use *distributive* attention as they drive to scan the entire scene around them for signal gathering and possible conflicts. Only after they change to *concentrative* attention and focused on the hazard can a controlled reaction begin.

Prior to focusing the eyes they may have to be moved to bring the object into direct vision. Test results show that between 0.32 and 0.55 s will elapse from the time an object has entered a driver's peripheral vision to when the eyes are focused on the object.

The basic reaction time follows and runs from the moment the eyes are focused until the driver begins to lift the foot off the gas pedal. Test results show a basic reaction time range of 0.22 to 0.58 s. It is noted again that no general judgment time or actual accident threats were associated with the tests.

The pedal switchover time covers the period from the moment the right foot lifts off the gas pedal and begins to displace the brake pedal. Measurements show a range of 0.15 to 0.21 s.

For hydraulic brakes, a brake system response or application time of 0.03 to 0.06 s was measured, indicating that only a small amount of time is required to bring the shoes or pads in contact with the drums or discs.

The deceleration rise or buildup time is the time during which the wheel brake torque increases from zero to its maximum value until brakes are locked or ABS control tire marks appear. Measurements indicate a range of 0.14 to 0.18 s. These time values are a function of vehicle speed and tire/road friction levels. Details for computing brake lockup times are found in Section 9.2.

The total time from the moment the object entered the driver's peripheral vision until the brakes are locked ranges from 0.86 to 1.58 s.

In certain accident situations not requiring head movement, the driver may not claim the extended reaction time. For example, when a driver follows a truck too closely and is focusing on the tail lights of the truck, the hazard signal indicated by the brake lights coming on does not enter through the driver's peripheral vision. Under these circumstances a reaction of only 0.54 to 1.03 s should be used.

## 1.5  Elements of Engineering Design

### 1.5.1  Basic Design Objectives

Design engineers satisfy human needs problems. Whether a device is a complicated machine consisting of many parts or a simple item such as a paper clip, it was planned and designed before it was manufactured.

Most design errors and malfunctioning of devices are caused by insufficient planning and lack of proper identification of requirements and constraints. To achieve a certain design objective, several alternative solutions are generally available. Automotive braking systems are no exception. The brake design engineer must be able to rate the significance of a host of influence factors including braking stability, stopping distance, response time, reliability, safety, cost, maintainability, wear, noise, or human factors. The engineer must decide whether to use disc or drum brakes, vacuum or hydro-boost, deceleration or load-sensitive proportioning valves, standard or anti-locking brakes, parking brakes using in-hub or disc application, diagonal or front-to-rear dual split, wedge or S-cam brakes, and many more. In addition, proper sizing of component parts is essential for an effective and safe brake system operation.

The first design solution is generally not the best one. Alternative designs must be considered and evaluated by a rational process frequently employing a design selection table, resulting in a prototype final design. Only when the designer has found the best compromise among the different constraints will the design be judged as best. The prototype final design is then optimized relative to several critical and important influence factors. With the prototype final design completed in most respects, a prototype braking system is tested and evaluated. Questions answered include: Does it work and function properly? Are all critical design and operational objectives met? Are safety standards and industry practices satisfied? Will it last? Is the customer happy with it?

The prototype is followed by the production model. This is the brake system sold to the customer. Future design improvements are made based on simplifications, cost reductions, and hopefully, infrequent customer complaints and safety recalls. Modifications may be introduced based on different applications and markets. Standardization, limitations to certain models, sizes or performance levels, different materials or manufacturing methods may be considered as running production changes to optimize cost-benefit ratios.

## 1.5.2  Product Design and Development Guides

For an engineer to accomplish the design task, he or she must consider a number of engineering design concepts, guides, standards, and practices. In the search for the "best" brake system design, the basic design and product development rules that follow must be considered:

1. Reliability takes precedence over such considerations as efficiency or cost. An unreliable brake system will create safety hazards. Reliability is achieved through proper design based on sound engineering principles, use of proven machine elements, testing, and other factors.

2. System-Based Design Methods will ensure that a safe functioning of the brake system is obtained. Guard against making singular changes that accomplish a specific objective but cause system performance to suffer. For example, increasing the brake drum diameter on the rear axle to improve lining life without an appropriate change on the front brakes will shift brake balance to the rear, thus increasing the potential for premature rear brake lockup.

3. Safety and Product Liability requires that the brake system absolutely does not exhibit any unreasonable safety hazards or develop any during the operation of the vehicle. The design

engineer must know basic human ergonomics, not only based on what a driver can do in a laboratory experiment, but also based on what typical drivers will do during an emergency. During foreseeable operations the vehicle should remain stable and controllable by the driver. Any unexpected vehicle behavior, especially when uncontrollable, will create critical situations and may cause accidents.

4. Material Selection is based on cost-efficiency with respect to strength, weight, wear, life, and performance.

5. Surface Finish should be the least expensive one in terms of machining or surface treatment required to ensure proper and safe functioning of components and subsystems.

6. Economics is considered by including prefabricated materials or subcomponents and proven in-house parts.

7. Production Methods are based on a consideration of all possible methods such as machining, casting, welding, forging, or gluing, in connection with the number of pieces to be produced.

8. Assembly is considered during the design in terms of cost-effective manufacturing, maintenance, repair, and inspection. Brakes that are difficult to maintain tend to be unsafe.

9. Warnings are part of the responsibility of the designer when he or she knows of an inherent design hazard but is unable to design the hazard out or otherwise guard against it.

10. New-Versus-Used conditions must be considered by the design engineer. For example, new brakes tend to produce lining friction different from that of broken-in or burnished linings.

11. Failure Analyses show the effect of critical component failure on performance, reduced safety, and potential for driver error.

12. Safety Standards for federal, state, and industry level are satisfied and/or exceeded.

13. Accelerated Testing is used to reveal any in-use conditions that may show problems.

14. Inspection and Maintenance procedures are established which ensure a safe and efficient operation of the brakes.

15. <u>Advertisement Guidelines</u> are provided which guard against misleading claims.

16. <u>Production Approval</u> is given only after the design is reviewed by persons experienced in the use, inspection, repair, maintenance, safety, and manufacturing of the brake system.

17. <u>Packaging, Labeling, and Shipping</u> may be of lesser importance than other considerations. However, labels must clearly identify parts and state use limitations (if any).

18. <u>Customer Complaints and Accident Data</u> are analyzed relative to potential input data for design modifications.

The design engineer considers most if not all of the points mentioned when designing a braking system or evaluating alternative design solutions. Frequently, additional specific factors are included that have a direct bearing on the particular design analyzed.

## 1.5.3 Design Solution Selection Process

In the design solution selection process, ranking points are assigned to each different solution relative to the various influence factors or constraints (Ref. 4). A point spread of zero to five has worked well, with the ideal solution receiving five points. It should be recognized that design experience and personal inclinations may affect certain rankings. Since the final result, however, is based on many rank entries, reasonable objectivity is ensured. The ranking matrix or design selection table has the different design solutions written across the top, and the influence factors or constraints in the left-hand column. Frequently, cost and safety are ranked individually to assign more weight to their respective contribution in the ranking process and, hence, final design.

The use of the design selection process and table is demonstrated in the example that follows.

The braking system of a trailer for use with passenger cars and pickup trucks must be designed. For the purpose of this demonstration we will evaluate three different design solutions, namely, mechanical, hydraulic surge, and electrical brakes. No claims are made that each ranking entry reflects the latest information on research, testing, or usage.

## DESIGN SELECTION TABLE

| Influence Factor | Design Solutions | | | |
| --- | --- | --- | --- | --- |
| | Mechanical | Hydraulic | Electric | Ideal |
| 1.  Reliability | 3 | 4 | 3 | 5 |
| 2.  Complexity | 4 | 3 | 3 | 5 |
| 3.  Maintenance | 3 | 4 | 4 | 5 |
| 4.  Versatility | 3 | 3 | 2 | 5 |
| 5.  Design Choice | 2 | 4 | 4 | 5 |
| 6.  Repairability | 4 | 3 | 3 | 5 |
| 7.  Wear | 3 | 4 | 3 | 5 |
| 8.  Materials | 4 | 4 | 4 | 5 |
| 9.  Inspection | 4 | 4 | 3 | 5 |
| 10.  Mechanic Skill | 5 | 4 | 4 | 5 |
| 11.  Durability | 4 | 4 | 3 | 5 |
| 12.  Efficiency | 2 | 4 | 4 | 5 |
| 13.  Effectiveness | 3 | 4 | 3 | 5 |
| 14.  Corrosion/Water | 3 | 3 | 2 | 5 |
| 15.  Rental Use | 3 | 5 | 1 | 5 |
| 16.  Tow Vehicle Hookup | 4 | 4 | 4 | 5 |
| 17.  Panic Braking | 2 | 3 | 4 | 5 |
| 18.  Trailer Stability | 2 | 2 | 4 | 5 |
| 19.  Brake Failure | 1 | 0 | 3 | 5 |
| 20.  L/R Brake Balance | 1 | 5 | 3 | 5 |
| 21.  Braking on Grade | 2 | 2 | 5 | 5 |
| 22.  Trailer Weight | 2 | 4 | 5 | 5 |
| 23.  Parking Brake | 3 | 3 | 2 | 5 |
| 24.  Brake Fade | 2 | 2 | 3 | 5 |
| 25.  Driver Control | 2 | 2 | 4 | 5 |
| Total Points: | 71 | 84 | 83 | 125 |
| Technical Value $TV = Z/Z(i)$ | 0.568 | 0.672 | 0.664 | 1.00 |
| Production Cost Value $PV = Cost/Cost(i)$ | 1.7 | 2.00 | 2.2 | 1.00 |
| Comparison Value $CV = TV/PV$ | 0.334 | 0.336 | 0.301 | 1.00 |

It is apparent that no brake systems can be rated superior to the other in all respects. Furthermore, the actual cost involved in the production may be different from the ratios assumed in the example.

The design selection can be extended into more detail by isolating those influence factors that have a direct relationship to safety. Although any factor listed in the selection table may be a key safety factor, depending on a particular set of circumstances or accident specifics, points 17 through 25 imply a fairly direct safety meaning.

Grouping points 17 through 25 into a separate safety value analysis yields for each design solution: mechanical - 17 points, hydraulic - 23 points, electrical - 33 points. The safety value ratios are 0.37, 0.51, and 0.73, respectively. It must be said that these safety ratios are based on maintenance practices that keep the brakes in good mechanical condition. Mechanical trailer brakes use cables and levers that may exhibit excessive friction which limits the application force. The same may be true for hydraulic surge brakes. For more details see Chapter 8. However, if it turned out, for example, that electrical brakes, in general, have unsafe brakes due to poor maintenance, then the design engineer must take that influence factor into consideration by including it in the safety value analysis.

The design engineer must use the design selection process to find rational design solutions for most automotive systems including parking brake (drum or disc), parking brake apply mechanism (hand or foot), manual or booster assisted brakes, plastic or metal wheel cylinder piston, soft or firm pedal, standard or ABS brakes, solid or ventilated rotors, standard or two-slope master cylinder, and many more.

## 1.6 Basic Safety Considerations

The safety of a braking system is affected by many factors. Brake component and vehicle manufacturers are responsible for the inherently sound design, manufacture, and reliability of the brakes. Users are responsible for continued safety of their brakes by ensuring proper maintenance and repair. Governmental agencies are responsible for meaningful and safety-oriented standards and regulations.

Safety standards are continuously updated and improved to address newly emerging safety problems, yet significant issues still remain unanswered. For example, Federal Motor Vehicle Safety Standard FMVSS 105, since its inception in 1968, does not contain any requirements for braking on low-friction road surfaces, nor does it consider vehicle directional stability when brakes are locked. The National Highway Traffic Safety Administration (NHTSA) has recognized this limitation of the existing standard by introducing FMVSS 135, which addresses vehicle stability while braking on a full range of road surface conditions, basically requiring front wheels to lock first for decelerations between 0.15 and 0.8 g. FMVSS 135 applies to passenger cars, and becomes effective September 1, 2000. See Section 1.8 for a brief review of safety standards.

Reliability is an important safety consideration for a design engineer. It is defined as the probability that a component or subassembly will not fail within a specified time period. The change of failure probability with time, or failure rate, is the probability that a given component will fail after a specified period of time. Failures occurring early in the life of a vehicle are generally caused by manufacturing defects, whereas late failures are caused by aging and wear. Time-independent failures are caused by such events as accidental rock impact, improper repairs, or misuse. We must also recognize that the relationship between failure probability and accident probability cannot be established easily.

Accident statistics show that slightly less than 2% of all highway accidents involve brake malfunctioning as a contributing accident causation factor (Ref. 1). Of these, nearly 90% are related to brake system defects caused by improper maintenance, whereas the remaining 10% involve directional braking instability. Approximately 30% of heavy-truck accidents are caused by air S-cam brakes being out of adjustment, when only manually adjusted air brakes are considered.

Increased use of anti-lock brake systems (ABS) is expected to improve braking safety. Early accident statistics do not show sufficient detail to provide any specific trend data. German accident studies conducted after the introduction of ABS brakes on Mercedes-Benz vehicles in 1978 indicated a reduction of accidents for ABS-equipped vehicles. Recent German accident data appear to indicate that ABS-equipped vehicles may be over-involved in certain accidents due to drivers overestimating the safety contribution of the brakes, especially on ice or when following too closely. A German study evaluating taxicabs with and without ABS showed nearly the same number of accidents for either group (Ref. 5). These initial, and admittedly spotty, data appear to indicate that brake and vehicle manufacturers must follow a carefully developed program in educating the public about the potential differences between objective ABS safety and subjective safety perceived by drivers. Whenever subjective safety exceeds the objective safety actually available, an accident is preprogrammed to happen, regardless of how advanced an ABS system is. Studies published by NHTSA in the mid-'90s indicate that ABS equipped passenger cars are over-involved in single-vehicle rollover accidents. Advertisements must not contribute to this potential difference in objective and subjective safety.

## 1.7 Elements of Product Liability

### 1.7.1 Basic Product Liability Concepts

Being liable relative to a product means that a company or person is held responsible for the harm the product may have caused. The relationship between defendant and product alleged to be defective may be direct or remote, and may include vehicle and component manufacturers, dealers and distributors, advertisers, raw-material producers, aftermarket component manufacturers, and others.

### 1.7.2 Product Liability Terms and Definitions

Product liability analyses frequently involve terms such as "danger," "hazard," "risk," unreasonably defective, and many more. It appears helpful to define these terms for the limited scope of this book. According to Webster's Dictionary, risk is the chance of injury, damage or loss; a hazard exists when a risky or a dangerous condition is present; and danger is the likelihood of injury, or a thing that may cause injury.

For our purposes we will define the terms as follows: *Hazard* is the potential for causing injury or loss; *danger* is the likelihood that a hazard will be involved in causing injury; *risk* is a person's planned or inadvertent operation of a vehicle in such a manner that injury or harm may occur; and *safety* is a measure of the probability that a hazard or danger does not exist. We should be aware that the terms are often used interchangeably in the literature and by attorneys.

An example may help to illustrate the use of the terms. Consider an empty pickup without anti-lock brakes whose rear brakes may lock first when sufficient pedal force is applied. Since locking of the rear brakes first may cause directional instability and loss of control and, consequently, is a potential for causing an accident and injury, it is a hazard. How dangerous is the vehicle with this brake system in the empty condition? Obviously it depends on how often the pickup truck is operated in the empty or driver-only condition relative to the loaded one, and how often brakes are locked. Other factors of importance are road conditions, speed, and general usage of the pickup truck. For example, on a wet or slippery road surface it is more likely for wheels to lock, while vehicle miles driven on wet and slippery roads are fewer.

The dangerous nature associated with a particular design may be expressed as the product of hazard consequences and frequency of hazard occurrence, or (Refs. 6, 7):

$$Danger = Hazard\ Consequences \times Hazard\ Frequency.$$

We see that products including braking systems will have a high degree of danger, which is often expressed as being unreasonably dangerous and, hence, defectively designed, when there is a great level of hazard consequence associated with the use, and when the hazardous condition has a high likelihood of occurring. For example, heavy commercial vehicles are extremely hazardous when their S-cam manual slack adjusters are at a critical adjustment level that may render the vehicle virtually without brakes under certain operating conditions. The fact that manual S-cam air brakes frequently are not adjusted near their optimum level is well known, resulting in a high degree of hazard occurrence. Consequently, the danger associated with manually adjusted air brakes is high, since the product of hazard consequences and hazard frequency is high.

An example of a low danger level and, hence, safe braking system notwithstanding a high hazard is that associated with the safety analysis of brake fluid vaporization of diagonal split dual brake systems. When the brake temperatures of both front brakes reach a critical level sufficient to boil and vaporize brake fluid, the entire service or foot brake will fail since no brake line pressure can be produced in either brake circuit. There is no question that this is a very hazardous condition with a high potential for doing harm. However, the likelihood for brakes to reach temperatures sufficiently high for brake fluid to vaporize is extremely low. Under normally foreseeable conditions this may never occur. Therefore, the danger given by the product of hazard consequences and frequency is low relative to diagonal split brake systems and brake fluid vaporization. We should, however, recognize that under abnormal yet somewhat foreseeable conditions, diagonal split systems may fail due to brake fluid vaporization. These abnormalities may result from improper parking brake release for both drum and disc brakes, dragging brake pads, excessive braking on extended down grades, and lowering of the brake-fluid boiling-point temperature through a high water content. Changing brake fluid every one to two years will minimize brake fluid boil.

## 1.7.3  Concepts of Recovery

There are three basic ways through which a plaintiff may try to recover in a product liability lawsuit (Ref. 8).

1. <u>Negligence</u> means that the manufacturer did not use reasonable care in the design and/or manufacture of the product. Except in rare cases, a claim of negligence is generally difficult to prove.

2. <u>Breach of Implied Warranty</u> means that the manufacturer implied a warranty that the product and its design are fit for reasonable and foreseeable use; however, the product failed to perform as warranted.

3. <u>Strict Liability</u> means that the product and its design are defective if it is unreasonably dangerous.

Design liabilities may be shown when the designer used an inadequate design process or incomplete testing, or the product failed to fulfill one of its functions or stated criteria. Since all accidents are preceded by one or more critical conditions relating to either the environment, the vehicle, or the driver, the designer must minimize the potential for design-induced driver error. The designer should also understand that the term "state-of-the-art" does not refer to how everyone else is building a vehicle or designing a brake system, but what is technically feasible at the time of the design of the product.

Investigations by plaintiffs generally try to collect information on the following in an attempt to support liability claims:

1. Circumstances under which the product was designed.

2. State of the art, both domestic and foreign.

3. Existing technology.

4. Governmental standards and industry practices.

5. Who controlled and established industry practices.

6. Documents showing that manufacturer knew of existing design flaws.

7. Improved technology sold in foreign markets with different safety requirements.

8. Inadequate design and testing due to marketing pressure.

9. Documents showing that business decisions overruled safety.

# 1.8 Elements of Braking Safety Standards

No attempt is made to fully discuss presently existing or proposed federal or foreign safety standards. Readers interested must contact the appropriate Department of Transportation agencies. A number of SAE publications address the issues involved in the different safety standards. Besides regulatory standards, a large number of recommended industry practices have been formulated concerning brakes and braking performance. Requests should be directed to the Society of Automotive Engineers in Warrendale, Pennsylvania.

## 1.8.1 Federal Motor Vehicle Safety Standard 105

The first federal safety standard regulating the braking performance of new passenger vehicles and light trucks using hydraulic brakes became law in 1968. Major portions of FMVSS 105, the hydraulic brake system standard, were based initially on then-existing SAE recommended practices.

A hydraulic brake system is defined as a system that uses hydraulic fluid as a medium for transmitting force from a service brake control (brake pedal) to the service brake (wheel brakes). It may incorporate a brake power assist (vacuum booster) or power unit (full pump power or hydro-boost).

The standard specifies requirements for the hydraulic service brake and associated parking brake system of passenger cars, multipurpose vehicles, trucks and buses.

The braking effectiveness of the brake system of a vehicle is measured in terms of stopping distance under a variety of operating conditions, including green or unburnished brakes, burnished brakes, lightly and fully laden, and different speeds. For example, in the lightly laden condition, the vehicle must stop within 60 m (196 ft) from a speed of 96 km/h (60 mph). The brakes must produce a certain vehicle deceleration when the brakes are heated through repeated brake applications and recover in a prescribed manner. The parking brake must hold the fully laden and empty vehicle stationary on a slope of 30% for standard transmission vehicles, or 20% for vehicles with automatic transmission.

The major shortcoming of FMVSS 105 is the lack of any performance requirements on low friction road surfaces and any significant stability specifications. Since lockup of more than one wheel is prohibited, basic vehicle stability under real-life emergency braking conditions faced by the motoring public is not addressed.

## 1.8.2 Federal Motor Vehicle Safety Standard 135

After September 1, 2000, safety standard FMVSS 135 replaces FMVSS 105 for vehicles with a GVW of less than 10,000 lb. FMVSS 135 is similar to the European safety standard as it requires stringent stability performance under a variety of braking conditions. The most significant stability specification is expressed by not allowing rear brakes to lock before the front under any loading and nearly all road friction conditions.

As discussed in Chapter 7, locking front brakes first will result in a stable braking process by avoiding vehicle spinning associated with premature rear brake lockup.

### 1.8.3 Federal Motor Vehicle Safety Standard 121

Trucks, trailers, and buses equipped with air brakes are regulated by FMVSS 121. Not included in air brake systems are systems that use compressed air only to assist the driver in applying muscular force to hydraulic or mechanical components. In other words, if the assist energy source has failed and the driver has the "push-through" ability to apply the brake, then it is not an air brake system according to FMVSS 121 definitions.

The standard addresses equipment specifications, stopping distances, fade performance and parking brakes.

FMVSS 121 specifies stopping distances for low- and high-friction surfaces for different speeds. Brake actuation and release times are specified. The brake retardation for towed vehicles is specified in terms of percentage or fraction of the gross axle weight rating for different brake line pressures. For example, at a pressure of $5.5 \times 10^5$ Pa or 5.5 bar (80 psi), the brake retarding force must be 0.41 times GAWR. Brake power or fade performance is specified in terms of inertia dynamometer consecutive decelerations.

The parking brake with all other brakes rendered inoperative must have a static retarding force equal to 28% of the GAWR. The parking brake must also hold the loaded vehicle on a 20% slope.

ABS brakes and automatic slack adjusters are required by FMVSS 121 for air brake vehicles. By 1999, ABS brakes are required on hydraulic brake trucks.

### 1.8.4 Federal Motor Vehicle Safety Standard 122

FMVSS 122 specifies performance requirements for motorcycle brake systems. The standard applies to two-wheeled and three-wheeled motorcycles. Each motorcycle must have either a split hydraulic service brake system or two independently actuated service brake systems, i.e., front and rear brakes. Performance requirements are specified in terms of stopping distances.

### 1.8.5 European Safety Standard ECE 13

The standard specifies a number of different requirements including friction utilization or braking efficiency. Several mathematical relationships are provided which are used to compute friction utilization. The basic requirements include that for tire-road friction coefficients between 0.2 and 0.8, the front brakes must lock before the rear brakes. Optional exceptions are provided for passenger vehicles when the rear brakes may lock first and the friction utilization curve of the rear axle does not depart from the optimum curve by more than 0.05. Friction utilizations are specified for commercial vehicles as well.

In addition, performance requirements for braking effectiveness for the normal and partially failed service brake system, fade performance, and others are specified.

## 1.9 Basic Brake System Design Considerations

In most cases the brake engineer has the following data available when designing the brakes of a vehicle. In some cases, certain data such as maximum weight may change as an entirely new vehicle is developed.

1. Empty and loaded vehicle weight.

2. Static weight distribution lightly and fully laden.

3. Wheelbase.

4. Center of gravity height lightly and fully laden.

5. Intended vehicle function.

6. Tire and rim size.

7. Maximum speed.

8. Braking standards.

The design of a new brake system begins with the selection of the brake force distribution, that is, how much braking force is produced by the front rakes in relationship to the rear brakes. The optimum brake force distribution is only a function of the basic vehicle dimensions and weight distribution.

In the next step, the dual circuit system is designed by selecting the proper sizes of wheel cylinders front and rear, and master cylinder.

In the third step, the wheel or foundation brakes are designed in terms of their basic size to ensure sufficient wear life, thermal performance, and low noise. The maximum allowable brake diameter is limited by rim size and, as such, is determined by vehicle weight.

In the last step, the pedal assembly and power boost system is designed.

The design of a braking system must always be based on a systems approach. A small change in one area may adversely affect the overall performance of the braking system in a safety critical area. For example, increasing the drum radius on the rear brakes to improve lining life will increase the rear brake force, and hence potential for premature rear brake lockup and vehicle instability during braking.

The design of the braking system must include the following design check-points:

1. Braking effectiveness:

    a. Maximum straight-line wheels-unlocked deceleration.

    b. Braking effectiveness, that is, brake line pressure-deceleration characteristic.

    c. Pedal force-deceleration characteristic.

    d. If appropriate, vacuum-assist characteristic.

    e. If appropriate, full-power characteristic.

    f. If appropriate, retarder characteristic.

2. Braking efficiency:

    a. Maximum straight-line wheels-unlocked deceleration for low and high roadway friction coefficient, both lightly and fully laden.

   b. Maximum curved-line wheels-unlocked deceleration for low and high roadway friction coefficient, both lightly and fully laden.

3. Stopping distance, lightly and fully laden:

   a. Minimum stopping distance without wheel-lockup.

   b. Minimum stopping distance without loss of directional control with wheel-lockup, for wet and dry brakes, and for "cold" and heated brakes.

   c. Minimum stopping distance without wheel-lockup while turning.

4. Response time:

   a. For air brakes, application and release time lags.

   b. For hydraulic brakes, pedal force-boost lag.

5. Partial failure:

   a. Braking effectiveness with service-system circuit failure.

   b. Braking effectiveness with partial or complete loss of power assist.

   c. Braking effectiveness with brakes in thermal fade condition.

   d. Directional stability with diagonal split failure.

   e. Increased pedal travel with service-system circuit failure.

   f. Increased pedal force with service-system circuit failure.

6. Brake fluid volume analysis:

   a. Master cylinder bore and piston travel for each brake circuit.

   b. Wheel-cylinder piston travel.

7.  Thermal analysis:

    a.  Heat-transfer coefficient for drum or rotor.

    b.  Brake temperature during continued and repeated braking, and maximum effectiveness stop.

    c.  Reduced braking effectiveness during faded conditions.

    d.  Thermal stresses to avoid rotor cracking and heat checking.

    e.  Brake fluid temperatures in wheel cylinders to avoid brake fluid vaporization.

8.  Emergency or parking brake:

    a.  Maximum deceleration by application of emergency brake lever on level and sloped roadway.

    b.  Maximum grade-holding capacity.

    c.  Determination under what conditions an automatic emergency brake application should occur.

9.  Specific design measures:

    a.  Heat flux into drum or rotor surface.

    b.  Horsepower absorbed by brake lining or pad.

    c.  Wear measure in form of product of lining friction coefficient and mechanical pressure.

10. In-use factors:

    a.  Determination of whether certain maintenance practices or lack of maintenance by particular user groups may require redesign to ensure adequate component performance and life.

    b. Determination of whether the operating environment is adverse to parts of the brake system (corrosion, dust, mud, water, etc.).

    c. Determination of whether wear or use affects brake force distribution and, hence, braking stability due to premature rear brake lockup ("green" versus burnished brakes).

11. Component sizing:

    a. Based on fatigue loading.

    b. Based on overload.

12. Safety regulations:

    a. Federal standards.

    b. Foreign standards.

    c. Industry standards.

    d. Consumer expectations and limitations.

# CHAPTER 2

# Design and Analysis of Friction Brakes

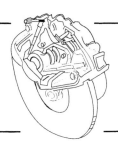

*In this chapter basic brake designs used for hydraulic and air brakes and their advantages and disadvantages are presented. Brake shoe travel, wear and adjustment, self-energizing, self-locking, and brake torque production are discussed. The general makeup of brake linings and lining friction are briefly reviewed. Practical engineering equations for computing brake torque of different disc and drum brakes are presented. The problems involved in computing brake torque developed by a nonrigid shoe are discussed. Engine brakes and retarders and their analysis are reviewed.*

## 2.1 Different Brake Designs

### 2.1.1 Drum Brakes

Friction brakes used in automotive applications can be divided into radial or drum and axial or disc brakes. Drum brakes subdivide into external band and internal shoe brakes. Typical shoe brakes subdivide further according to the shoe arrangement into leading-trailing, two-leading, or duo-servo brakes. Drum brakes may be further divided according to the shoe abutment or anchorage into shoes supported by parallel or inclined sliding abutment, or pivoted shoes. A sliding abutment supports the tip of the shoe but permits a sliding of the shoe relative to the fixed abutment. The brake shoe actuation may be grouped into hydraulic wheel cylinder, wedge, cam, screw, and mechanical linkage actuation.

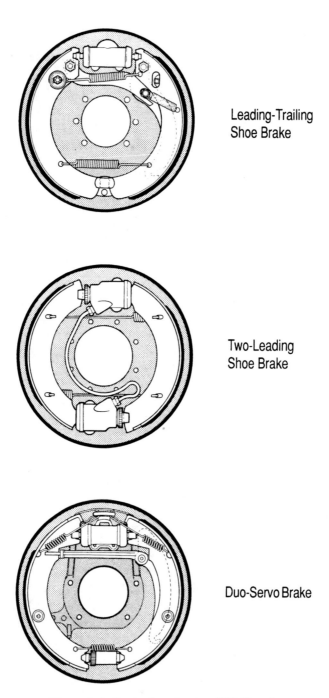

Leading-Trailing
Shoe Brake

Two-Leading
Shoe Brake

Duo-Servo Brake

*Figure 2-1. Basic drum brakes (ITT-Teves).*

The basic shoe arrangements for hydraulic drum brakes are illustrated in Figure 2-1. In the case of the leading-trailing and two-leading shoe brakes each shoe has its own abutment or anchorage to the backing plate. With the duo-servo brake only the secondary shoe is anchored to the backing plate, in most cases by pivot. The primary shoe pushes against the bottom of the secondary shoe thereby increasing the torque effectiveness of the duo-servo brake.

The basic components including shoes, wheel cylinder, automatic adjuster, and parking brake mechanism of a leading-trailing shoe brake are shown in Figure 2-2. The leading-trailing shoe design is used extensively as rear brake on passenger cars and light pickup trucks not using rear disc brakes. With a few exceptions, front-wheel-driven vehicles use rear leading-trailing shoe brakes. The advantage of this arrangement is a low sensitivity to lining friction changes and, hence, stable brake torque production.

The basic component parts of a duo-servo brake design are illustrated in Figure 2-3. The primary shoe reaction force at the bottom of the shoe is used as application force of the secondary shoe by pushing through the adjustment mechanism. The main advantage of the duo-servo brake is its high brake torque or brake factor for a given input force from the wheel cylinder pushing

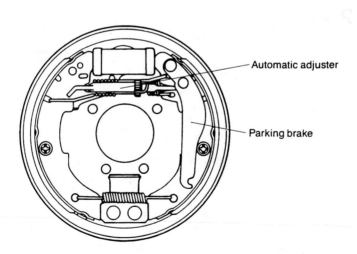

*Figure 2-2. Leading-trailing shoe brake (ITT-Teves).*

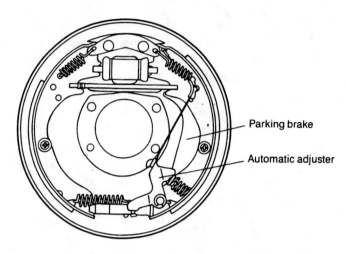

Parking brake

Automatic adjuster

*Figure 2-3. Duo-servo brake (ITT-Teves).*

the shoes apart. The major disadvantage of the duo-servo brake design is its high variation in brake torque for small changes in lining friction coefficient. For example, a lining friction coefficient increase of 15% due to moisture, thermal conditions, or other operational factors may result in a torque increase of 40 to 50%. This drastic unintended increase in rear brake torque may cause premature rear brake lockup and, hence, loss of vehicle stability during braking.

Drum brakes are the predominant foundation air brake on medium- and heavy-duty trucks, tractors, and trailers in North America. Over 90% of air brake-equipped heavy vehicles use either the S-cam or wedge actuated foundation designs. In some cases flat-cam brakes are used, and then primarily on front axles.

The *S-cam brake* uses the leading-trailing shoe design. The shoes are applied mechanically by rotation of a cam shaped in an S-form, hence, the name S-cam brake. A typical S-cam brake design for use on a trailer axle is shown in Figure 2-4. The main parts of this design are: leading (top) and trailing (bottom) shoes, S-cam, automatic slack adjuster, and air brake chamber. Rotation of the cam pushes the rollers and tips of the shoes apart. Due to cam geometry the application force against the leading shoe will have a

*Figure 2-4. S-cam brake with automatic slack adjuster (Rockwell International).*

smaller lever arm relative to the pivot anchor of the leading shoe than that of the trailing shoe, resulting in nearly uniform wear of both the leading and trailing shoe and, therefore, long lining life. As discussed in Section 2.8.2.j, this is also the reason that the standard leading-trailing shoe brake torque equation must be modified for S-cam brakes. S-cam brakes are simple and rugged. They can be inspected and maintained easily. Their major disadvantage is in stop fade, a limited brake factor, and the need for "tight" adjustment.

When the adjustment is at a critical level, often not detectable by the operator, thermal drum expansion and brake lining fade may cause ineffective truck braking. The thermal conditions do not have to involve excessive brake temperatures associated with extensive downhill operation of the truck. Simply exiting a freeway at 50 or 60 mph may be sufficient to cause the truck to "run out of brakes" or, more specifically, out of the remaining pushrod travel.

Obtaining adequate pushrod travel can be ensured by automatic slack adjusters. Automatic slack adjusters are standard equipment on all S-cam-equipped trucks and trailers. Automatic slack adjusters are required by FMVSS 121.

*Wedge brakes* use either the leading-trailing or two-leading shoe design. A dual-chamber two-leading shoe-type foundation wedge brake is shown in Figure 2-5. At present, approximately five percent of the air-braked trucks in North America are equipped with wedge brakes. The usage in Europe is considerably higher and increasing. In the wedge brake, a wedge is forced between the tips of the shoes, forcing the linings against the drums. The leading-trailing shoe brake uses one brake chamber, the two-leading shoe brake two. One benefit of wedge brakes is the integral automatic adjuster which ensures optimum drum-to-lining clearance. Another advantage over S-cam brakes is the higher brake factor and, hence, more compact size and lower weight.

*Figure 2-5. Dual-chamber wedge brake (Rockwell International).*

## 2.1.2 Disc Brakes

A typical disc brake is illustrated in Figure 2-6. The rotor or disc rotates through the caliper. The wheel cylinder pistons force the pads against the rotor and produce brake torque.

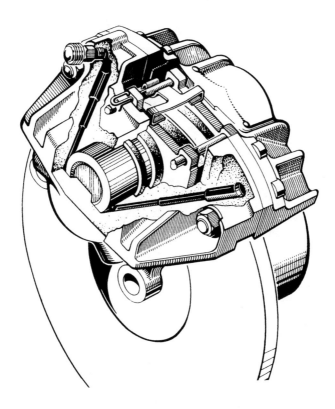

*Figure 2-6. Caliper disc brake (ITT-Teves).*

A *fixed-caliper* design is illustrated in Figure 2-7. The caliper, bolted solidly to the flange, has either two or four pistons which push the pads out. Fixed-caliper disc brakes have more balanced inner and outer pad wear with less pad taper than floating caliper designs. They require no anchor or integral knuckle for shoe support. They attach with standard fasteners, have no sleeves, grommets or hold-down springs, and require fewer service parts.

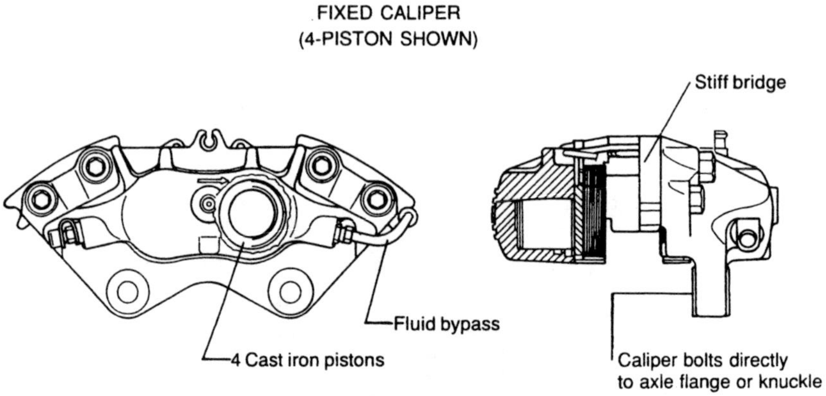

FIXED CALIPER
(4-PISTON SHOWN)

Stiff bridge

Fluid bypass

4 Cast iron pistons

Caliper bolts directly
to axle flange or knuckle

*Figure 2-7. Fixed-caliper disc brake (ITT-Teves).*

A typical *floating caliper* disc brake is shown in Figure 2-8. One or two pistons are used on the inboard side only. The hydraulic pressure forcing the piston and pad toward the rotor also forces the piston housing (wheel cylinder) in the opposite direction to apply the outboard pad against the rotor. Floating caliper brakes offer a number of advantages over fixed caliper designs. They are easier to package in the wheel since they do not have a piston on the outboard or wheel side. They have a lower brake fluid operating temperature than the fixed caliper and, hence, lower brake fluid vaporization potential. They also have fewer leak points, and are easier to bleed in service.

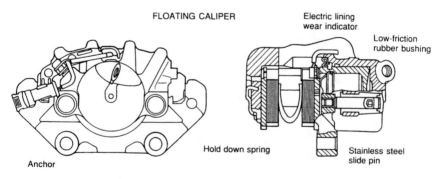

*Figure 2-8. Floating caliper disc brake (ITT-Teves).*

A major disadvantage of floating caliper disc brakes is the potential for pad dragging due to insufficient pad return since one piston seal must provide the clearance for both pads. There is also an increased potential for squeal due to a larger number of degrees of freedom when compared with the fixed caliper designs, and on some earlier designs reduced effectiveness due to corrosion of sliding surfaces. Modern calipers all have sealed-lubricated sliding pins as guiding members. The problems mentioned tend to be more pronounced for floating caliper disc brakes with wheel cylinder sizes greater than 2 to 2.25 inches in diameter.

Medium-weight trucks in the U.S. generally use disc brakes on all four wheels in connection with a full hydraulic pressurized system.

An air disc brake using the floating caliper design is illustrated in Figure 2-9. Shown in this figure are the ventilated rotor, sealed actuation mechanism, automatic slack adjuster, and air chamber. The rotation of the slack adjuster turns a screw, which forces the inboard and outboard pads against the disc. The swing-away caliper provides for easy lining changes.

*Figure 2-9. Air disc brake (Rockwell International).*

45

## 2.2 Brake Shoe Adjustment

To keep the clearance between brake lining and drum or pad and rotor at an optimum, adjustment becomes necessary as the linings wear. To accomplish this, either manual or automatic adjustment mechanisms are provided.

### 2.2.1 Drum Brake Adjustment

Since the return springs pull the shoes against a stop to their fully retracted position, the clearance between shoe lining and drum increases as linings wear. By adjusting the brake shoes out, the stops are moved toward the drum, thus preventing excessive return movement of the shoes.

*Manual adjusters* should be adjusted only when the brakes are cold and the parking brake is released. The adjustment mechanism may be located on the shoe, at the wheel cylinder, or at the fixed or floating abutment. Frequently, a screw is turned in or out to move the ends of a tappet relative to the brake shoe as shown in Figure 2-10. Manual brakes are rarely used on today's passenger cars and light to medium trucks. The use of manually and automatically adjusted brakes on different axles of the same vehicle may create a safety problem by confusing maintenance personnel.

*Automatic adjusters* are designed to keep the lining-to-drum clearance at an optimum value. The designs illustrated in Figs. 2-2 and 2-3 are used most frequently on domestic drum brakes. Some drum brakes use automatic adjusters employing bimetallic temperature sensors to prevent inadvertent over-adjustment when the drum diameter is expanded temporarily due to temperature.

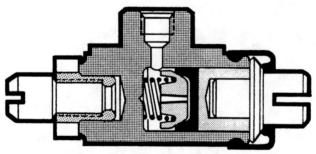

*Figure 2-10. Shoe adjustment (ITT-Teves).*

An automatic slack adjuster for S-cam brakes is shown in Figure 2-11. An internal worm gear rotates the S-cam shaft relative to the slack adjuster lever as the adjustment nut is turned. Several different mechanisms including the simple ratchet type have been designed to ensure a minimum clearance to prevent drum drag when the brakes are released. As many as 50 adjustments may be necessary to maintain the proper lining-to-drum clearance during the life of the brake lining. With proper maintenance every 30,000 to 50,000 miles, automatic slack adjusters minimize the need for routine manual adjustments, and vehicle downtime, and greatly improve brake safety.

## 2.2.2 Disc Brake Adjustment

Hydraulic disc brake adjustment is accomplished automatically by the wheel cylinder piston seal. The seal is designed so that in the event of a piston displacement, it distorts elastically for about 0.152 mm (0.006 in.). Provided no pad wear has occurred, the piston seal pulls the piston back on releasing the brake line pressure, as shown in Figure 2-12. If the clearance between pad and rotor becomes greater due to wear, the piston travels in excess of 0.006 in., and the piston seal preload is overcome, forcing the piston closer to the rotor. The return movement of the piston is determined by how much the seal can deflect during application. In the floating caliper design, one seal must return the pad on the piston side as well as the outboard caliper pad. Low-drag caliper designs use a special chamfer to increase seal deflection. Since the resulting pad clearance is greater than normal, special quick take-up master cylinders may be used to provide the extra brake fluid to bring the pads against the rotor (see Section 5.4.3).

In some disc brake designs with integral parking brakes, the rear brake pads are automatically adjusted when the parking brake is applied. Since some drivers rarely use the parking brake, safety problems have arisen where rear disc brakes were found to be severely out of adjustment or corroded.

Adjustment of air disc brakes is accomplished by a regular automatic slack adjuster as illustrated in Fig. 2-9.

*Figure 2-11. Automatic slack adjuster (Rockwell International).*

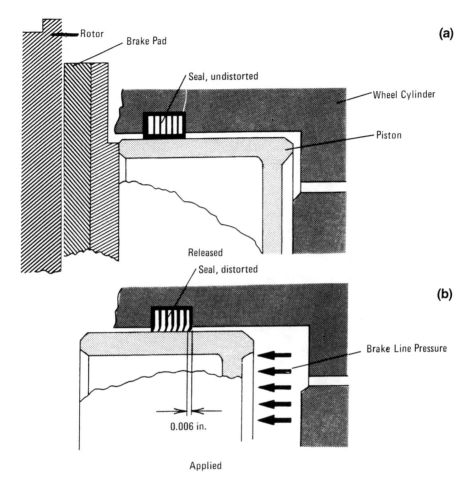

Figure 2-12. Disc brake clearance adjustment. (a) Released brakes.
(b) Applied brakes.

## 2.3 Lining Wear and Pressure Distribution

As a rule of thumb, a drum or rotor should last two to three sets of linings or pads before they are worn sufficiently to be replaced (see Section 2.7.3).

The lining material is in all practical applications the wear component of the brake, i.e., the wear of the drum or disc is negligible compared with the lining or pad wear.

## 2.3.1 Lining Pressure and Wear in Drum Brakes

With the assumption that the brake drum and brake shoes are rigid and all deformation occurs within the lining material, the compression of the lining as a result of the shoe displacement against the drum, measured by the angle rotated by the shoe about its pivot, is related to the strain and the original lining thickness $d_{Lo}$ by

$$\varepsilon = d_L \, / \, d_{Lo} \qquad\qquad (2\text{-}1)$$

where  $d_L$ = lining compression, mm (in.)

$d_{Lo}$ = original lining thickness, mm (in.)

$\varepsilon$ = strain of lining material

Tests have shown that the pressure p is approximately proportional to strain, i.e., Hooke's Law is valid provided excessive mean pressures are avoided. The actual pressure distribution between lining and drum is bound by functional relationships of the form

$$p = E\varepsilon = E(a\varphi \, / \, d_{Lo})\sin\alpha \quad , \quad N/m^2 \ (psi) \qquad\qquad (2\text{-}2)$$

and

$$p = c(e^{ka\varphi \, \sin\alpha / d_{Lo}} - 1) \quad , \quad N/m^2 \ (psi) \qquad\qquad (2\text{-}3)$$

where   a = brake dimension, mm (in.)

c = constant for determining pressure distribution between lining and drum, $N/m^2$ (psi)

E = Elastic modulus, $N/m^2$ (psi)

k = constant for determining pressure distribution between lining and drum

$\varepsilon$ = strain of lining material

$\alpha$ = lining angle, deg (see Fig. 2-33)

$\phi$ = shoe rotation, rad

The results obtained for several lining materials with different elastic behaviors are presented in Figure 2-13 where the pressure distribution over the lining angle is shown (Ref. 10). Inspection of Fig. 2-13 reveals that the constant c varies between 0.2 and $5 \times 10^5$ N/m$^2$ (2.94 and 73.5 psi) for the linings tested. The information contained in Fig. 2-13 may be used to compute the approximate strain values. At a lining angle of 50 deg, the strain $\varepsilon$ of the soft lining is approximately 0.05, that of the hard lining 0.005. The corresponding values of the elastic modulus are 165 to $1200 \times 10^5$ N/m$^2$ (2400 to 17,500 psi) for the soft and hard lining, respectively.

When the wear behavior of the lining material is known, the pressure distribution along the lining arc can be determined. A detailed analysis is complicated. Only some basic observations are presented.

For a pivoted leading shoe a wear relationship of the form

$$w_1 = k_1 \mu_L p v_1 \quad , \quad m^3 \ (in^3) \tag{2-4}$$

is assumed, where

$k_1$ = wear constant, s m$^4$/N (s in$^4$/lb)

$p$ = pressure, N/m$^2$ (psi)

$v_1$ = sliding speed, m/s (in/s)

$w_1$ = lining wear, m$^3$ (in.$^3$)

$\mu_L$ = lining drum friction coefficient

With Eq. (2-4) a sinusoidal pressure distribution may be found to exist along the brake lining. The pressure distribution obtained analytically after successive brake applications and, thus, wear are presented in Figure 2-14. Inspection of Fig. 2-14 reveals that a sinusoidal distribution $p = 9.1 \times 10^5 \sqrt{\sin \alpha}$, N/m$^2$ ($p = 132.2 \sqrt{\sin \alpha}$, psi) is developed after 11 brake applications.

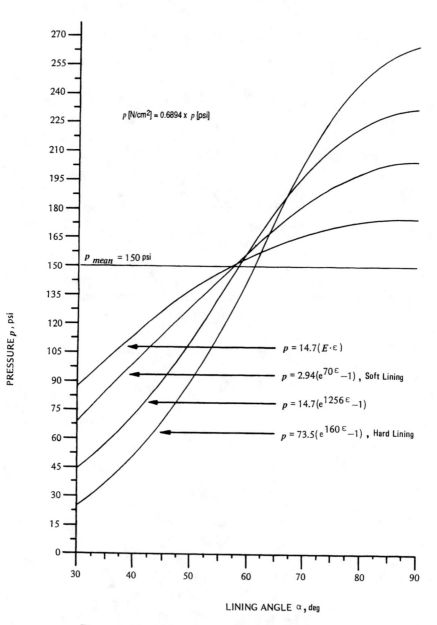

Figure 2-13. Measured pressure distribution over
lining angle α for different linings.

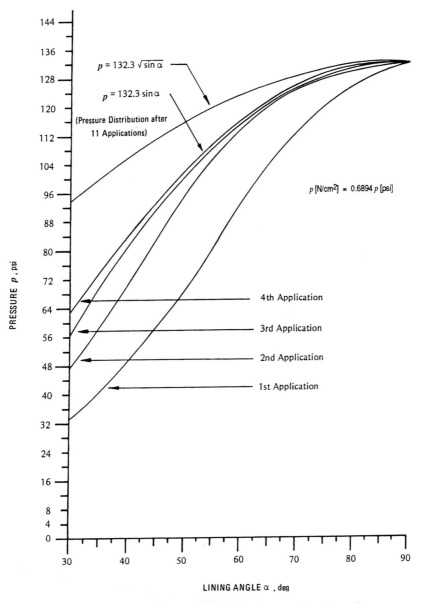

*Figure 2-14. Computed pressure distribution as a function of wear after successive brake applications.*

For a wear relationship of the form

$$w = k_2 \mu_L p^2 v_1^2 \quad , \quad \text{m3 (in.}^3) \tag{2-5}$$

where $k_2$ = wear constant, $s^2$ $m^5/N^2$ ($s^2$ in.$^5$/lb$^2$)

$w$ = lining wear, $m^3$ (in.$^3$)

$v_1$ = sliding speed, m/s (ft/s)

a pressure distribution of the form $p$ = constant $\times \sin\alpha$ is obtained. This pressure distribution is shown in Fig. 2-14.

Inspection of the curves in Fig. 2-14 reveals that new brakes will have a different pressure distribution than brakes in service. For an exact prediction of pressure distribution, hence brake torque, a knowledge of both the wear relationship and the elastic behavior of the lining material is essential. It is an established fact that the pressure distribution changes during the run-in periods. Burnishing procedures subject the vehicle brake system to a series of brake applications during which the pressure distribution along the lining tends to approach run-in conditions.

New or unburnished—sometimes called green—brakes can have a drastic effect on vehicle braking stability, particularly in connection with rear duo-servo drum brakes where variations in pressure distribution with wear may increase brake torque and, hence, the potential for premature rear brake lockup. Green or not fully burnished drum brakes often exhibit higher brake torque than those in the burnished condition.

## 2.3.2 Pad Pressure and Wear in Disc Brakes

One of the major requirements of the disc brake caliper is to press the pads against the rotor as uniformly as possible. Uniform pressure between pad and rotor results in uniform pad wear and brake temperatures, and more stable pad/rotor friction coefficients. Nonuniform pressure distribution wears the brake pads unevenly, particularly during severe brake applications from high speeds. Pad wear increases rapidly for brake temperatures in excess of approximately 573 to 623 K (600°F) resulting in tapered pad wear when ineffective caliper designs are used. Uniform pad wear is a major indicator of a quality caliper design.

Worn disc brake pads may show significantly more wear on the leading end (rotor entrance) as compared to the trailing end (rotor exit). This nonuniform wear is caused by higher pressure between pad and rotor at the leading end compared to that at the trailing end. The nonuniform pressure distribution is caused by the lever arm between pad drag force and abutment force. For a symmetrical wheel cylinder piston and pad design, the drag/abutment force moment results in pad pressures at the leading end that are approximately one-third greater than the average pressure. The corresponding pressure at the trailing end is approximately two-thirds of the average pad pressure.

Solutions to minimize or eliminate tapered pad wear involve an off-center pad application force produced by an asymmetrical caliper piston contact edge, effectively moving the piston force more in the direction of the trailing end of the pad, which creates a counter moment balancing the pad friction moment.

Other solutions have the piston located closer to the trailing end of the pad, again producing a counter moment.

A design patented by ITT-Teves, called "hammerhead" design because of its shape, is illustrated in Figure 2-15. In this particular caliper/pad design, the pad is pulled by the drag force rather than pushed. This design solution has proven to be reliable for both fixed and floating caliper disc brakes. Other advantages of this pad anchor system include lack of pad vibrations and,

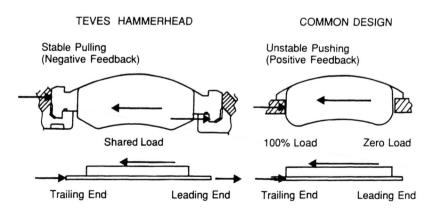

*Figure 2-15. ITT-Teves "hammerhead" design.*

hence, low potential for brake noise or brake judder, less weight, since the reaction load is carried by each end of the pad; minimum deflection; and uniform pad temperatures.

Expensive designs minimizing tapered pad wear use four pistons per caliper. The pads are pushed with two pistons of different diameter with the smaller piston located at the leading end of the pad.

Some basic design analyses to minimize non-uniform pad wear are presented next.

### 2.3.2.a Non-Uniform Pad Pressure Distribution

The non-uniform pressure distribution with a linear pressure change is illustrated in Figure 2-16. The average force pressing the pad against the rotor is indicated by $F_{av}$. The pressure change at the leading and trailing edge of the pad is indicated by $\Delta F$. When a linear pressure variation is assumed, a triangular pressure distribution results as shown in Fig. 2-16. The resultant force of the pressure triangle is located $2/3\ell_p$ from the tip of the triangle.

Application of moment balance about point A yields:

$$F_{av}\mu_p t_p + F_{av}\mu_p \mu_f \frac{\ell_p}{2} = \frac{\Delta F \ell_p}{6}$$

Solving for pressure change $\Delta F$ results in:

$$\Delta F = \frac{F_{av}6}{\ell_p}\left(\mu_p t_p + \mu_p \mu_f \frac{\ell_p}{2}\right)$$

With $F_{max} = F_{av} + \Delta F$ the maximum pad pressure is:

$$F_{max} = F_{av}\left[1 + \frac{6}{\ell_p}\left(\mu_p t_p + \mu_p \mu_f \frac{\ell_p}{2}\right)\right], \quad \text{N (lb)} \qquad (2\text{-}6)$$

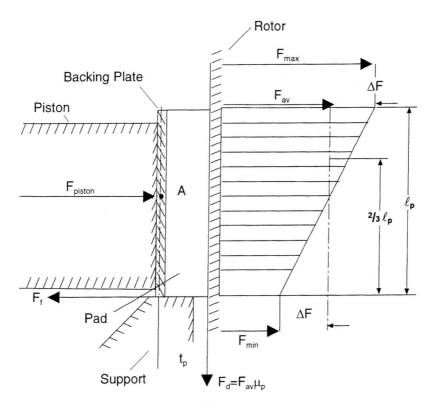

*Figure 2-16. Non-uniform pad pressure.*

where  $\ell_p$ = pad length, mm (in.)

  $t_p$ = pad thickness/support distance, mm (in.)

  $\mu_f$ = pad support friction coefficient

  $\mu_p$ = pad/rotor friction coefficient

Substitution of typical values for a disc brake yields $F_{max} = 1.33\, F_{av}$, indicating that pressure at the rotor entrance will be as much as one-third greater than the average pressure, and only two-thirds of the average pressure at the rotor exit.

### 2.3.2.b Offset Piston Design

The basic layout and dimensions are illustrated in Figure 2-17. The offset distance is designated by c. If the distance c is computed properly, the pressure distribution will be uniform over the length of the pad.

$F_d$ is the friction drag force between rotor and pad. $F_{av}$ is the piston application force. $F_f$ is the friction force at the pad backing plate support.

Moment balance about point A yields:

$$F_{av}c = F_d t_p + F_f \frac{\ell_p}{2}$$

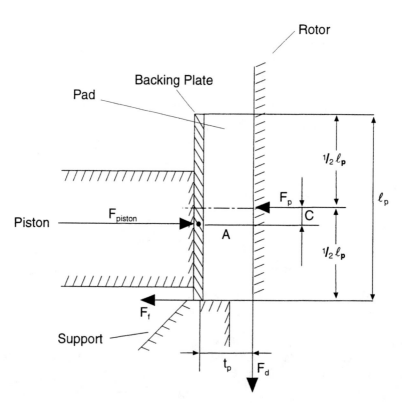

*Figure 2-17. Uniform pad pressure with offset piston.*

Force balance yields:

$$F_{av} = F_p + F_p \mu_p \mu_f$$

Combining both equations and solving for offset c results in:

$$c = \frac{\mu_p t_p + \mu_p \mu_f \dfrac{\ell_p}{2}}{1 + \mu_p \mu_f} \quad , \quad mm \ (in.) \tag{2-7}$$

The offset computed by Eq. (2-7) produces a uniform pressure distribution for the data used. As the pad thickness decreases, uneven pad wear will result. A somewhat smaller distance $t_p$ may be used to adjust for pad wear with use of the vehicle. Manufacturing costs due to non-symmetry may be excessive, particularly for smaller production numbers.

### 2.3.2.c  Pulled or "Hammerhead" Pad Design

The basic schematic of the pad design is illustrated in Figure 2-18. The forces are identified in Section 2.3.2.b.

Moment balance about point A yields:

$$F_{av} \mu_p t_p - F_f b = \Delta F \frac{\ell_p}{6}$$

Solving for pressure change $\Delta F$ results in:

$$\Delta F = F_{av} \frac{6}{\ell_p} \left( \mu_p t_p - \mu_p \mu_f b \right)$$

where   b = distance from piston center to pad support, mm (in.)

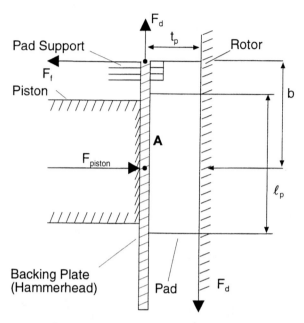

*Figure 2-18. Pulled or hammerhead pad design.*

With $F_{max} = F_{av} + \Delta F$ the maximum pressure becomes:

$$F_{max} = F_{av}\left[1 + \frac{6}{\ell_p}\left(\mu_p t_p - \mu_p \mu_f b\right)\right] \quad , \quad N \text{ (lb)} \quad (2\text{-}8)$$

Substitution of typical data yields a maximum pressure of $F_{max} = 1.033\, F_{av}$. The result shows a significant improvement with the pulled pads versus the pushed pads by providing a nearly uniform pressure distribution. Pulled pads can carry heavier specific loadings and are used increasingly in high-performance vehicles.

### 2.3.2.d  Four-Piston Fixed Caliper Design

A more expensive solution to achieve uniform pad wear is the four-piston caliper design as illustrated in Figure 2-19. The pads are pushed together by two opposing pistons of different diameters.

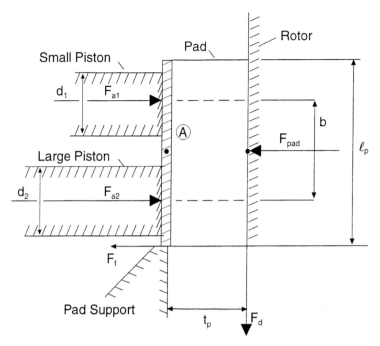

*Figure 2-19. Four-piston fixed caliper design.*

Uniform pad pressure distribution is achieved when all moments balance about point A.

Hence:

$$F_d t_p + F_f \frac{\ell}{2} + F_{a1} \frac{b}{2} - F_{a2} \frac{b}{2} = 0$$

where $F_{a1}$ = force of piston 1, N (lb)

$F_{a2}$ = force of piston 2, N (lb)

With $F_d = \mu_p F_{pad}$ and $F_f = F_d \mu_f = F_{pad} \mu_p \mu_f$ and $F_{pad} = F_{a1} + F_{a2} - F_f$ we have:

$$F_{a1} = F_{pad} \left( \frac{1}{2} - \frac{\mu_p t_p}{b} - \frac{\mu_p \mu_f \ell}{2b} + \frac{\mu_p \mu_f}{2} \right)$$

61

$$F_{a2} = F_{pad}\left(\frac{1}{2} + \frac{\mu_p t_p}{b} + \frac{\mu_p \mu_f \ell}{2b} + \frac{\mu_p \mu_f}{2b}\right)$$

Finally, the piston application force, and hence diameter, ratio is:

$$\frac{F_{a1}}{F_{a2}} = \frac{d_1^2}{d_2^2} = \frac{\dfrac{1}{2} - \dfrac{\mu_p t_p}{b} - \dfrac{\mu_p \mu_f \ell}{2b} + \dfrac{\mu_p \mu_f}{2}}{\dfrac{1}{2} + \dfrac{\mu_p t_p}{b} + \dfrac{\mu_p \mu_f \ell}{2b} + \dfrac{\mu_p \mu_f}{2}}$$

where  $d_1$ = diameter of piston 1, mm (in.)

$d_2$ = diameter of piston 2, mm (in.)

The individual diameters are obtained from the fictitious single wheel cylinder diameter $d_3$ by

$$d_3^2 = d_1^2 + d_2^2$$

The single diameter $d_3$ is obtained from a brake balance analysis employed for a single piston caliper (see Chapter 7). Solving for the individual diameters results in:

$$d_1 = \sqrt{\frac{d_3^2}{1 + \dfrac{1}{F_{a1}/F_{a2}}}} \quad , \quad \text{mm (in.)} \tag{2-9a}$$

$$d_2 = \frac{d_1}{\sqrt{F_{a1}/F_{a2}}} \quad , \quad \text{mm (in.)} \tag{2-9b}$$

Typical brake data and $d_3$ = 57 mm used in Eq. 2-9 may result in $d_1$ = 36 mm and $d_2$ = 44 mm. For vehicles not requiring extreme braking performance, a smaller diameter difference such as 38 and 42 mm may be used to account for the fact that increased pad wear reduces $t_p$. High performance sports and race cars would use a 36/44 diameter ratio since generally pads are not worn to minimum levels.

All previous uniform pad wear analyses apply fully only to fixed caliper designs. For floating calipers limitations exist including play tolerances and deformations. Design solutions involving two pistons per pad, piston offset, and others have not proven fully successful in practical applications. Only pulled pad solutions are advantageous for floating caliper disc brakes.

## 2.4　Parking Brake Design

### 2.4.1　Drum Parking Brakes

Parking brakes generally use the same drum and lining components as the service brake, but have different components for brake shoe application. Drum brakes are ideally suited for parking brake application as illustrated in Figs. 2-2 and 2-3. Due to its high torque output the duo-servo brake is well suited for heavier vehicles or when wheel packaging is restricted.

In the past, medium-weight trucks used driveshaft-mounted internal shoe or external band brakes.

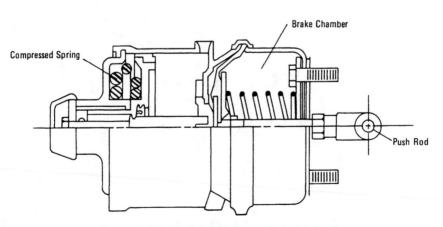

*Figure 2-20. Brake chamber with spring brake.*

Air S-cam and wedge brakes use compressed mechanical springs as shoe application force for parking brake purposes, as illustrated in Figure 2-20. The dual chamber consists of the regular service brake chamber and the chamber containing the compressed spring. When the air pressure holding the spring compressed is lowered or released, the spring expands and applies the shoes against the drum. Air spring brakes will not provide any vehicle braking when the brakes are out of adjustment (Chapter 6). Infrequently wedge spring brakes are used in connection with hydraulic drum brakes for trucks.

## 2.4.2 Disc Parking Brakes

Rear disc brakes use either the *drum-in-hat* parking brake or the *integral* parking brake caliper design. A drum-in-hat design for a floating caliper design is illustrated in Figure 2-21. A small duo-servo brake installed inside the hub of the brake rotor is applied by cable through foot or hand application by the driver. The high brake torque output of the duo-servo brake allows a cost-effective parking brake design for modern passenger cars. Any potential variations in brake torque left-to-right are of no safety significance for the drum-in-hat parking brake.

Integral parking calipers combine the service and parking brake function into one unit as illustrated in Figure 2-22. Most systems are cable-to-lever actuated and adjust pad clearance automatically through normal service brake application. Most designs are complicated to ensure finely tuned application,

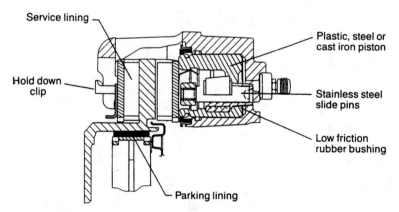

*Figure 2-21. Floating caliper with drum-in-hat parking brake (ITT-Teves).*

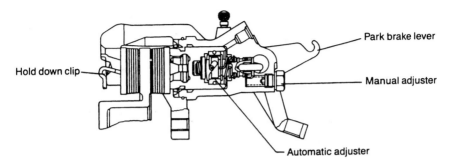

*Figure 2-22. Integral parking brake.*

particularly when the parking brake is applied while the brake pedal is applied. Due to its involved design, and when considering corrosion and maintenance problems, not all designs work well under all foreseeable circumstances.

## 2.5 Disc Brake Installation

Most calipers are installed so that the mounting bolts are located vertically above each other, with the caliper either located in front (nine o'clock) or aft (three o'clock) of the rotor. This installation ensures that the air bleeding valve is located at the highest point of the caliper, an important detail for proper brake maintenance.

The front or aft location has a significant effect on the hub bearing forces sustained during severe braking.

With the aft location, the bearing force will increase to more than twice its normal value for a deceleration of 1 g. The front location may increase bearing forces to a value equal to four times the normal level. For front- and four-wheel-driven vehicles using press fit hub and shaft connections, the bearing force may be too high, possibly resulting in hub failure and wheel separation during braking. In addition, significant shaft bending during severe braking may cause undesirable vibrations and noise.

Two opposing calipers will increase the bearing force only by approximately 170% of its normal value.

The maximum outer diameter of the brake rotor is a direct function of the rim size. Other factors are negative scrub radius, and the degree to which the contour of the caliper is optimized to the rim shape. The following table presents guidelines for choosing a rotor and drum diameter for brake system design purposes.

| Rim size (in.) | 13 | 14 | 15 | 16 | 17 |
|---|---|---|---|---|---|
| Inside Flange Diam. (mm) | 329.4 | 354.8 | 380.2 | 405.6 | 436.5 |
| Outer Rotor Diam. (mm) | 225/256 | 245/278 | 270/308 | 295/330 | 320/360 |
| Inner Drum Diam. (mm) | 230 | 250 | 280 | 300 | 325 |

## 2.6 Disc and Drum Brake Comparison

The major advantage of the disc brake is its ability to operate with little fade at high temperatures of up to 1073 to 1173 K (1500 to 1600°F). Heating of the brake rotor increases its thickness thereby causing no loss in brake fluid volume, i.e., no increased pedal travel or soft pedal feel. In the case of air disc brakes, the thicker rotor prevents running out of pushrod travel, a significant safety problem for air brakes.

An additional important benefit of disc brakes is their linear relationship between brake torque and pad/rotor friction coefficient. For example, a 10% increase in pad friction coefficient increases the brake torque by 10%. For a typical duo-servo brake, a similar friction rise increases brake torque by as much as 30 to 35%.

Drum brakes are highly temperature sensitive. A maximum temperature of 673 to 700 K (750 to 800°F) should not be exceeded. Not only are the friction coefficients affected, but the drum diameter increases with increasing temperatures. At 648 K (700°F) typical passenger drum brake diameters may increase by 1 to 1.5 mm (0.05 to 0.06 in.), with a correspondingly longer wheel cylinder piston travel sufficient to increase pedal travel by 30 to 40% of its normal value. In addition, the larger drum diameter causes improper contact between lining and drum, which results in lining/drum pressure peaks and thus higher local lining temperatures, and a variation in brake torque output. Brake drums for S-cam brakes have increased pushrod travels of 12 mm (0.5 in.) for a temperature increase of 590 K (600°F) over the cold pushrod travel value.

The specific brake torque or brake factor, defined as the ratio of drum drag to application force of one shoe (Eq. [2-10]), is a general indicator of the ability of a brake to produce torque for different lining/drum friction coefficients. More details are presented in Section 2.8. A brake factor comparison of different drum brake designs with the disc brake is shown in Figure 2-23. Inspection of this figure reveals that the duo-servo brake has the highest brake factor for any given friction coefficient. Although this characteristic is desirable for parking brake design, it may prove unsafe when used as a rear brake for the service brake. Duo-servo brakes should not be used on front brakes due to the severe left-to-right brake unbalance potential. The straight brake factor line for the disc brake reveals its linear relationship to the friction coefficient. The two-trailing shoe drum brake has a brake factor curve similar to that of the disc brake. Although no applications are envisioned, future designs may make use of this design by providing a stable service rear brake, and a higher brake factor parking brake.

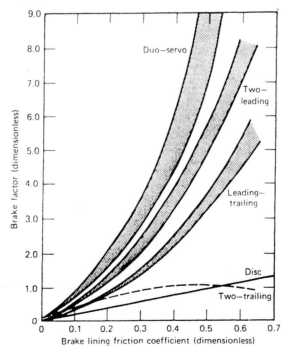

*Figure 2-23. Brake factor as a function of lining friction coefficient.*

## 2.7  Brake Lining Materials

Brake lining materials are an important component for the safe and consistent performance of the braking system of a vehicle. Since many details are trade secrets, the following paragraphs present a basic overview only.

### 2.7.1  Lining Makeup

Today's automotive linings are either asbestos (organic), semi-metallic, or asbestos-free materials. The following definitions are generally accepted for classifying different lining compositions:

*Asbestos* linings contain any amount of asbestos materials while the matrix is an organic binder.

*Non-asbestos* linings consist of three basic components: the mass or matrix holding the lining together (phenolic materials); the fibers to provide strength (steel or other metal fibers); the modifiers to control heat conduction (copper or similar metals). Due to their metal content, non-asbestos linings are commonly called semi-metallic linings.

The basic makeup of lining compositions falls into the following four basic groups:

*Fiber* provides the necessary rigidity and strength for the lining. In the case of dry mix types, the fiber holds the preform together. For high-temperature friction materials the fiber also provides thermal stability. Fiber materials include asbestos, steel wool, or aramid fibers.

*Fillers* are inexpensive minerals to extend lining life, fill space, and minimize cost. Filler materials are barytes, clay, calcium carbonate, or, in the case of metallic linings, finely ground metal powders. The fillers may cause scratching and scoring of the friction surface of the rotor.

*Binders* are the glues holding the lining materials together. Binder materials most commonly used are phenolformaldehydes. Curing of the binder occurs at temperatures up to 230°C (450°F) for several hours. When no complete curing is obtained, the lining may have reduced heat resistance.

*Friction modifiers* generally are elastomers that improve mechanical and wear properties, curing agents, and others to affect friction levels. Brass, zinc, or other metals are added to control abrasive properties and to "clean" swept surfaces of the rotor.

Asbestos—derived from the Greek word meaning non-diminishing or eternal—occurs naturally in the form of fibers. When ground to small particle size and inhaled, asbestos is a cancer-causing agent. The human cells coming in contact with the fine particle dust are agitated and develop cancer cells, mostly involving the human lung. Notwithstanding that some research suggests that the asbestos brake dust is too small to cause cancer, asbestos used in brakes becomes more and more limited with federal regulations prohibiting its use by the mid-'90s.

## 2.7.2 Lining Friction and Classification

All linings begin to disintegrate at the friction surface due to high temperatures developed from the heat-generation process. Due to nonuniform pressure distributions between lining and drum or pad and rotor, and other surface irregularities, pad friction surface temperatures will not be uniform over the pad contact area. Areas of higher temperature will have lower friction levels than those with lower temperatures. An exact analytical prediction of the pad/rotor friction coefficient is not possible at the present time. However, close estimates based on test data can be made.

The basic brake system design is based on the brake torque performance achieved with "cold" brakes. A brake is considered cold when its temperature is less than 366 K (200°F). Most lining friction coefficients will increase as brake temperatures rise to approximately 423 to 473 K (300 to 400°F). At elevated temperatures near 523 to 588 K (500 to 600°F) and above, linings tend to exhibit fade, i.e., their friction coefficient decreases below its cold value. Good linings will recover to their intended design levels after cooling. At extremely low temperatures, friction coefficients tend to decrease below the cold value. Since brake systems have to perform safely under all foreseeable operating conditions, the proper selection of a lining material can be a challenge, particularly for drum brakes, and even more so for duo-servo brake designs.

Test procedures have been developed to measure lining friction at different temperatures and to classify ranges. SAE J661 procedure is used to determine the cold (366 K or 200°F) and hot (588 K or 600°F) friction levels of lining material samples, one inch square in size, when used with a drum made of a particular material to a particular set of dimensions. Two letters are used to mark cold and hot friction coefficients. The first letter represents an average value of the normal (cold) friction, the second hot friction. The higher the letter, the higher the coefficient of friction, so that:

C    refers to friction coefficients less than 0.15

D    0.15 to 0.25

E    0.25 to 0.35

F    0.35 to 0.45

G    0.45 to 0.55

H    over 0.55

Z    unclassified

For example, a lining edge code FE indicates that the normal coefficient of friction is between 0.35 and 0.45, say, 0.38, and when heated to 588 K (600°F) between 0.25 and 0.35, say, 0.34.

It is important to recognize that the SAE J661 lining classifications using fairly broad ranges of friction coefficients may cause errors when used in the design analysis of an existing braking system. Calculations determining brake lockup sequence require reasonably accurate brake factor computations and, hence, lining friction coefficients. Simply using any value within the specified letter range will not be acceptable. As a minimum, the actually measured average friction coefficients used to establish the friction range should be considered in the design analysis. Since only a lining sample area of 25.4 by 25.4 mm (one inch by one inch) is actually tested, additional differences may exist between the classification friction coefficient and the effective lining friction coefficient actually experienced by the drum brake.

Albin Burkman of General Motors used a laboratory test method to determine that moisture may have a significant effect on lining friction coefficient (Ref. 9). He concluded that the friction coefficients are higher during periods of high humidity than under dry conditions, and lower when the brakes are flooded by water.

Brake torque and brake factor can be measured directly with special torque hubs, or indirectly from vehicle deceleration tests. If such data are available, they should be used as a basis for the brake system design analysis.

### 2.7.3 Performance Requirements of Linings

Lining wear should be kept at a minimum, but will vary from driver to driver. Lining wear for disc and drum brakes will be different. Under normally expected driving conditions, disc brake pads are expected to last between 30,000 and 50,000 km (20,000 and 30,000 miles), drum brakes between 50,000 and 80,000 km (30,000 and 50,000 miles). Disc brake rotors should last for two to three sets of pad changes. Similar performance is required for drum brakes.

A certain amount of rotor and drum wear is desirable so that possible corrosion residues and material deposits caused by severe braking are removed during normal braking.

Brake linings and, in particular, disc brake pads, should have a certain amount of porosity to minimize the effect of water on the friction coefficient. These porous openings should not store contaminants such as salt or wear particles that affect friction. The metal components of lining materials in connection with water will corrode the swept surface of the rotor. Although a vehicle that is used every day produces brake temperatures that evaporate any water present, vehicles parked for a long time in a moisture-rich environment may experience severe brake pad/rotor corrosion problems. The electrochemical reactions penetrate the rotor sufficiently deep to change the basic rotor surface. The results are brake noise and vibrations or brake pedal pulsations, which can only be eliminated by installing new rotors. In most cases, turning or grinding of the rotor will not cure the problem permanently.

Friction modifiers, added to the basic lining material to improve performance, may, under certain thermal localized conditions, accumulate as small deposits in the friction surface of the lining. They are up to 6 mm (0.25 in.) long and approximately 1.5 mm (1/16 in.) wide. Since they are hard, the rotor surface will show signs of severe abrasion, making an early replacement necessary. The problem can be mitigated by using less aggressive modifiers and improved mixing and production methods for lining materials.

Mechanical strength of linings has twofold importance, namely, strength to resist exterior loads, and structural integrity.

The mechanical strength of a lining, frequently measured by its shear strength, is mainly determined by the fiber content, and more specifically by the surface area per unit weight. Asbestos has a surface area up to 30 m$^2$/g (1270 ft$^2$/lb). Replacement fibers have been developed that reach half the value of asbestos; however, at a fairly high cost. The structural integrity of a lining is achieved by minimizing residual stresses and thermal expansion, and increasing heat resistance. Asbestos, although currently replaced more and more by asbestos-free materials, meets most of the requirements stated.

Front-end vibrations during braking are caused by brake torque fluctuations resulting from brake rotors with nonuniform thickness or out-of-round drums. Rotor thickness variations are caused by a number of factors, including manufacturing or maintenance defects, lining material deposits on the rotor surface coming in contact with the lining, and localized metallurgical changes resulting in high spots caused by extreme thermal conditions. Since the entire front end is potentially involved, brake torque fluctuation near or at the natural frequency may cause loss of vehicle control. Generally, braking at low deceleration over a long time, i.e., from high speeds, may produce more vibrations than high effectiveness stops for a short time.

Lining material compression should be as small as possible to minimize pedal travel requirements for braking. In isolated cases and related to the author by brake engineers, disc brake replacement pads of low quality could be compressed sufficiently far so that the brake pedal touched the floor during severe braking.

Lining material damping is its ability to suppress vibration, the source of brake squeal and noise. Even for a lining material with optimum damping characteristics, brake squeal may occur, particularly under dry conditions. To prevent brake squeal in disc brakes, the back of backing plates is covered by special paint or grease which firms up after exposure to air, or thin metal plates are inserted, all intended to increase damping to reduce high oscillation peaks of the brake pads.

## 2.8 Brake Torque Analysis

The mathematical prediction of brake torque for drum brakes involves lengthy algebraic computations. Comparison between theoretical and test data shows good correlation, especially when accurate lining friction coefficient data are available. The prediction of brake torque for a typical disc brake is a simple matter since the relationship between brake torque and pad friction coefficient is linear.

### 2.8.1 Brake Torque Analysis of Disc Brakes

Brake torque is measured in Nm (lbft), i.e., it is a function of the diameter of the brake. It is convenient to express the brake torque effectiveness of a single brake by a dimensionless measure, called brake factor.

The brake factor BF is defined as the ratio of total drum or rotor drag $F_d$ to the application force $F_a$ against one shoe, or

$$BF = F_d \; / \; F_a \tag{2-10}$$

For a standard (non-self-energizing) caliper disc brake, the brake factor BF is equal to

$$BF = 2\mu_L \tag{2-11}$$

where $m_L$ = lining coefficient of friction

The sensitivity S of the disc brake is a measure of how much the brake factor changes for a given lining friction coefficient change, or, in other words, how steep the brake factor curve is. Expressed mathematically, the slope of the curve is obtained by taking the derivative of the brake factor (Eq. [2-11]) with respect to the friction coefficient. For a disc brake we have

$$S = d(BF) \; / \; d(\mu_L) = 2 \tag{2-12}$$

Self-energizing disc brakes are not used in typical automotive applications. Their basic operational principles involve a wedge effect provided by a ball-and-ramp type design as illustrated in Figure 2-24 for a fully covered disc brake design. The actuating force is the force directly pressing against the disc. This force is increased by the friction force, which causes an additional relative rotation. This leads to pushing apart of the circular brake pads and increased normal force by means of the ball-and-ramp mechanism, thus introducing self-energizing.

With the notation shown in Fig. 2-24, the friction force of one circular brake pad is given by the relationship (Ref. 10):

$$F_d = \mu_L \left[ F_a + F_d (r_m / r_k) \right] \cot \delta \quad , N \text{ (lb)}$$

or solved for the plate brake factor as

$$\frac{F_d}{F_a} = \frac{\mu_L (r_k / r_m)}{(\tan \delta)(r_k / r_m) - \mu_L} \tag{2-13}$$

where  $r_k$ = disc brake dimension, mm (in.)

$r_m$ = disc brake dimension, mm (in.)

$\delta$ = disc brake ramp angle, deg

$\mu_L$ = pad friction coefficient

Since two friction surfaces are involved, the total brake factor is

$$BF = 2(r_k / r_m) \frac{\mu_L / \mu_{L\infty}}{1 - \mu_L / \mu_{L\infty}} \tag{2-14}$$

where $\mu_{L\infty}$ = self-locking pad friction coefficient

and the self-locking limits (see Section 2.8.2.a for self-locking limit details) for the pad friction coefficient are given by

$$\mu_{L\infty} = (\tan \delta)(r_k / r_m)$$

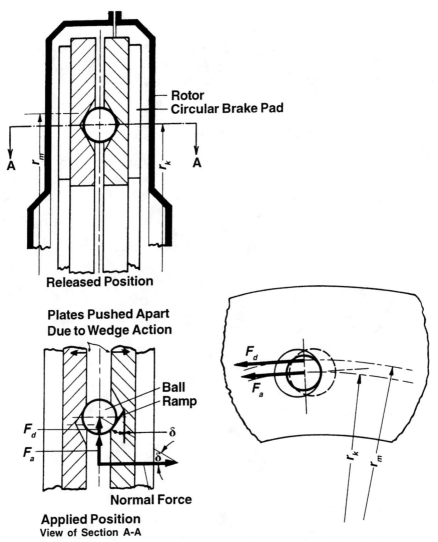

*Figure 2-24. Schematic of self-energizing fully covered disc brake.*

The sensitivity of the brake $S_B$ is expressed by

$$S_B = \frac{2 \cot \delta}{(1 - \mu_L / \mu_{L\infty})^2}$$

(2-15)

## 2.8.2 Brake Torque Analysis of Drum Brakes

One distinguishing characteristic of drum brakes is their higher brake factor when compared with disc brakes, as revealed in Fig. 2-23. Higher brake factors result from self-energizing within the brake.

### 2.8.2.a Self-Energizing and Self-Locking

A brake shoe with a single brake block is illustrated in Figure 2-25. Only the leading shoe is shown. The application force $F_a$ against the tip of the shoe pushes the brake block against the drum. The counterclockwise rotation of the drum produces a drag force $F_d$ as shown.

Moment balance around the shoe pivot point (A) yields

$$-F_a h - F_d c + F_d b / \mu_L = 0$$

where    $b$ = brake dimension, mm (in.)

$\quad\quad c$ = brake dimension, mm (in.)

$\quad\quad h$ = brake dimension, mm (in.)

$\quad\quad \mu_L$ = friction coefficient block/drum

Solving for the ratio of drum drag $F_d$ to application force $F_a$ yields the brake factor BF of the leading shoe as

$$BF_l = F_d / F_a = \mu_L h / (b - \mu_L c) \tag{2-16}$$

Inspection of Fig. 2-25 reveals that the drum drag rotates the brake shoe such that it will increase the normal force of the block pushing against the drum. This increased normal force causing an additional increase in drum drag is the self-energizing effect of the brake. The self-energizing shoe is called leading shoe.

The ratio of drum drag to application force as expressed by Eq. (2-16) will increase for smaller denominators, and will be infinite when the denominator is zero, or $b - \mu_L c = 0$.

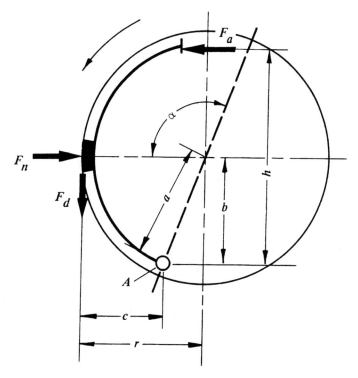

*Figure 2-25. Self-energizing in a drum brake.*

The lining friction coefficient at which the denominator will be zero for the brake geometry given is b/c, designated as $\mu_{L\infty}$. If the actual lining friction coefficient were equal to $\mu_{L\infty}$, then a brake application would cause ever-increasing self-energizing until the brake locked. Even releasing the application force would not disengage the brake block from the drum. Although self-locking generally is not a problem since b > c, brake engineers must guard against it by ensuring that neither friction levels nor brake geometries are such that self-locking may occur. If it does occur, then most likely it is only in high brake factor duo-servo brakes where adverse conditions such as moisture have significantly increased lining/drum friction coefficients.

For reversed or clockwise rotation of the drum, the leading shoe shown in Fig. 2-25 turns into a trailing shoe. The drum drag force would be directed upward attempting to "lift" the brake block off the drum, thus reducing the effect of the application force. The brake factor of the trailing shoe is given

77

by Eq. (2-16) except the minus sign in the denominator is replaced by a plus sign. The plus sign indicates the decrease in brake factor with increasing lining friction coefficients, i.e., non-self-energizing of the trailing shoe. The brake factor curve of a two-trailing shoe brake is illustrated in Fig. 2-23.

The total brake factor of the leading-trailing-type block brake is given by adding the brake factors of each shoe, resulting in

$$BF = \frac{2\mu_L h / b}{1 - (\mu_L c / b)^2} \tag{2-17}$$

The self-locking coefficient of friction at which the brake factor of the entire brake, i.e., the leading and the trailing shoe, becomes infinite is the same as for the leading shoe alone, namely $\mu_{L\infty} = b/c$. This is expected since the trailing shoe does not contribute to self-energization.

The sensitivity S of the block brake is given by the derivative of the brake factor relative to the friction coefficient in Eq. (2-17), or

$$S = \frac{d(BF)}{d(\mu_L)} = \frac{2h / b\left[1 + (\mu_L c / b)^2\right]}{\left[1 - (\mu_L c / b)^2\right]^2} \tag{2-18}$$

For h = 200 mm (8 in.), b = 100 mm (4 in.), and c = 75 mm (3 in.), the brake factor curve for the block brake is illustrated in Figure 2-26. Self-energizing of the leading shoe is clearly evident as the increasing steepness of the curve.

The brake factor analyses for actual brakes use lining pressure distributions as discussed earlier. No elastic shoe or drum deformations are included in the basic brake factor analyses that follow (Ref. 10).

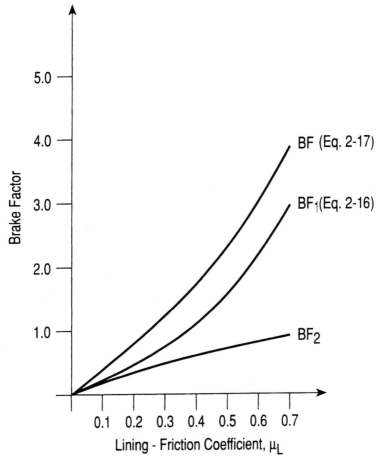

*Figure 2-26. Brake factor for block brake.*

### 2.8.2.b  Brake Factor of a Leading-Trailing Shoe Brake with Pivot on Each Shoe

The schematic of one shoe is illustrated in Figure 2-27. The total brake factor is the summation of the individual brake factors of the leading shoe $BF_1$ and of the trailing shoe $BF_2$,

$$BF = BF_1 + BF_2 = F_{d1} / F_a + F_{d2} / F_a \qquad (2-19)$$

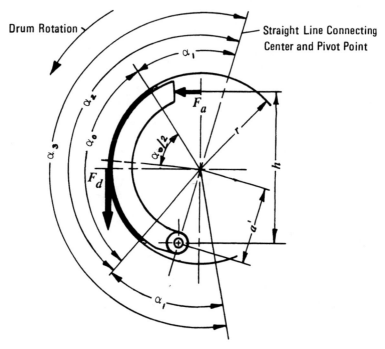

*Figure 2-27. Leading shoe with pivot.*

where $F_{d1}$ = drag force on leading shoe, N (lb)

$\quad\ F_{d2}$ = drag force on trailing shoe, N (lb)

The brake factor of the leading shoe is given by the following expression using the minus sign in the denominator:

$$BF_1 = F_{d1} / F_a$$
$$= \frac{\mu_L h / r}{(a' / r)\left[\dfrac{(\widehat{\alpha_0}) - \sin \alpha_0 \cos \alpha_3}{4 \sin(\alpha_0 / 2) \sin(\alpha_3 / 2)}\right] \pm \mu_L\left(1 + (a' / r)\cos(\alpha_0 / 2)\cos(\alpha_3 / 2)\right)}$$

$$(2\text{-}20)$$

where   $a' =$ brake dimension, mm (in.)

$\widehat{\alpha_0} =$ arc of the angle $a_0$, rad

$\alpha_1 =$ angle between beginning of lining and straight line connecting center and pivot point, deg

$\alpha_2 = \alpha_1 +$ arc angle, deg

$\alpha_3 = \alpha_1 + \alpha_2$, deg (as defined in Fig. 2-27)

The brake factor of the trailing shoe is determined by using the plus sign in the denominator of Eq. (2-19).

### 2.8.2.c  Brake Factor of a Two-Leading Shoe Brake with Pivot at Each Shoe

For this case, the brake factor can simply be determined from

$$BF = 2(BF_1) = 2(F_{d1} \, / \, F_a) \tag{2-21}$$

with $F_{d1}/F_a$ determined from Eq. (2-20) using the minus sign in the denominator.

### 2.8.2.d  Brake Factor of a Leading-Trailing Shoe Brake with Parallel Sliding Abutment

The schematic of one shoe is illustrated in Figure 2-28.  The brake factor BF is determined by Eq. (2-20).  The individual brake factors are:

For the leading shoe:

$$BF_1 = F_{d1} \, / \, F_a = \left[ \left( \mu_L D_B + \mu_L^2 E_B \right) / \left( F_B - \mu_L G_B + \mu_L^2 H_B \right) \right]_1 \tag{2-22}$$

For the trailing shoe:

$$BF_2 = F_{d2} \, / \, F_a = \left[ \left( \mu_L D_B - \mu_L^2 E_B \right) / \left( F_B + \mu_L G_B + \mu_L^2 H_B \right) \right]_2 \tag{2-23}$$

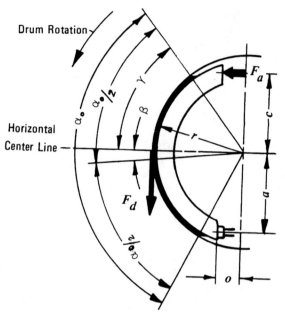

*Figure 2-28. Leading shoe with parallel sliding abutment.*

where $D_B = \left[c / r + a / r + \mu_S(o / r)\right] \cos \beta + \mu_S(c / r) \sin \beta$

$E_B = \mu_S(c / r) \cos \beta - \left[c / r + a / r + \mu_S(o / r)\right] \sin \beta$

$F_B = \dfrac{\widehat{\alpha}_0 + \sin \alpha_0}{4 \sin(\alpha_0 / 2)}\left[a / r + \mu_S(o / r)\right]$

$G_B = \cos \beta + \mu_S \sin \beta$

$H_B = F_B - (\mu_S \cos \beta - \sin \beta))$

$o$ = brake dimension, mm (in.)

$r$ = drum radius, mm (in.)

$\alpha_0$ = arc angle of lining, deg

$\beta$ = angle between center of arc angle and horizontal center line, deg

$\gamma$ = angle between beginning of lining and horizontal center line, deg

$\mu_s$ = friction coefficient at shoe tip and abutment

The value of $\mu_s$ is associated with the sliding friction between the tip of the shoe and the abutment. For steel on steel, $\mu_s \approx 0.2$ to $0.3$. The angle $\beta$ is positive when $\gamma > \alpha_0/2$, and negative when $\gamma < \alpha_0/2$.

### 2.8.2.e  Brake Factor of a Two-Leading Shoe Brake with Parallel Sliding Abutment

The brake factor can be determined from the general expression for two-leading shoe brakes, Eq. (2-19), with the brake factor of one shoe determined by Eq. (2-22).

### 2.8.2.f  Brake Factor of a Leading-Trailing Shoe Brake with Inclined Abutment

A schematic of a typical shoe is illustrated in Figure 2-29. The total brake factor may be determined from Eqs. (2-19), (2-22), and (2-23) with the abutment friction coefficient $\mu_s$ replaced by $(\mu_s + \tan\Psi)$, where $\Psi$ is the inclination angle of the abutment in deg.

### 2.8.2.g  Brake Factor of a Two-Leading Shoe Brake with Inclined Abutment

The total brake factor may be determined from Eqs. (2-19), (2-21), and (2-22) with $\mu_s$ replaced by $(\mu_s + \tan\Psi)$ where $\Psi$ is the inclination angle of the abutment.

### 2.8.2.h  Brake Factor of a Duo-Servo Brake with Sliding Abutment

The schematic is illustrated in Figure 2-30. The relationships shown earlier can be used to determine the brake factor. In this case, however, the internal application force $F_{ax}$ of the primary shoe, designated by 1, becomes the actuation force of the secondary shoe, designated by 2.

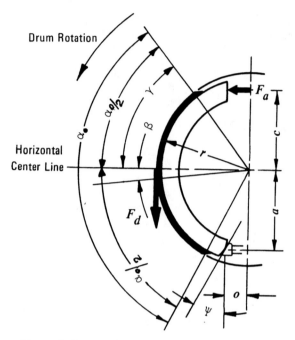

*Figure 2-29. Leading shoe with inclined abutment.*

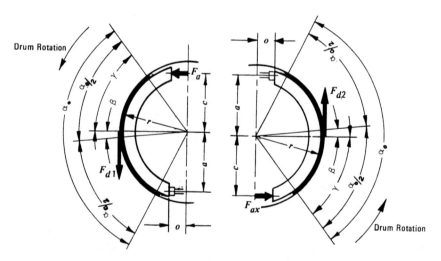

*Figure 2-30. Duo-servo brake with sliding abutment.*

The total brake factor BF is determined by

$$BF = BF_1 + BF_2 = F_{d1} / F_a + F_{d2} / F_a$$

$$= F_{d1} / F_a + (F_{d2} / F_{ax})(F_{ax} / F_a) \qquad (2\text{-}24)$$

where $\quad F_{d1} / F_a = \left(\mu_L D_B + \mu_L^2 E_B\right) / \left(F_B - \mu_L G_B + \mu_L^2 H_B\right)$

$$F_{d2} / F_{ax} = \left(\mu_L D_B + \mu_L^2 E_B\right) / \left(F_B - \mu_L G_B + \mu_L^2 E_B\right)$$

The relative support force $F_{ax}/F_a$ is determined from a moment balance about the center of the brake, and can be expressed as

$$F_{ax} / F_a = c / a + \left(F_{d1} / F_a\right)(r / a) \qquad (2\text{-}25)$$

### 2.8.2.i  Brake Factor of a Duo-Servo Brake with Pivot Support

A schematic is shown in Figure 2-31. The total brake factor can be determined from Eqs. (2-24) and (2-25) with the brake factor $BF_1$ of the primary shoe given by Eq. (2-22) and the brake factor $BF_2$ of the secondary shoe given by Eq. (2-20); the minus sign is used in the denominator.

### 2.8.2.j  Brake Factor of Air S-Cam Brake

The basic S-cam brake configuration is a leading-trailing shoe design as discussed in Section 2.2.1. Resulting from the fixed actuation of the cams, the reaction forces between cam and rollers are oriented such that the effectiveness of the leading shoe is decreased, while that of the trailing shoe is increased. The average brake factor of the S-cam brake can be expressed as (Ref. 11)

$$BF = \frac{4(BF_1)(BF_2)}{BF_1 + BF_2} \qquad (2\text{-}26)$$

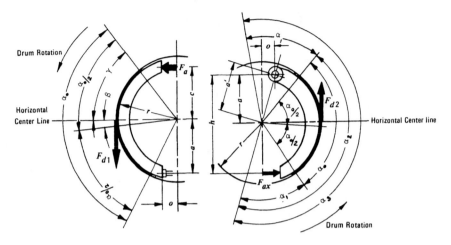

*Figure 2-31. Duo-servo brake with pivot*

where $BF_1$ = brake factor of leading shoe determined by Eq. (2-20), using the minus sign in the denominator

$BF_2$ = brake factor of the trailing shoe determined by Eq. (2-20), using the plus sign in the denominator

For example, for $BF_1 = 1.7$ and $BF_2 = 0.6$, the S-cam brake factor is 1.77, and not 2.3 as it would be for a hydraulic wheel cylinder or floating actuation against the shoe.

### 2.8.2.k  Brake Factor of Wedge Brake

Brake factors of single-chamber wedge brakes are computed by the regular leading-trailing shoe brake equation. For parallel abutment see Section 2.8.2.d. The brake factor of a dual-chamber or two-leading shoe brake is computed by the relationships presented in Section 2.8.2.e.

Example 2-1: Compute the brake factor and brake sensitivity of a commonly used duo-servo brake with sliding abutment on the primary shoe and pivot anchor at the secondary shoe. The schematic and geometrical information are illustrated in Fig. 2-31. Use the brake data that follow.

| Primary Shoe | Secondary Shoe |
|---|---|
| $a = 100$ mm (4 in.) | $a = 100$ mm (4 in.) |
| $c = 100$ mm (4 in.) | $a = 100$ mm (4 in.) |
| $o = 38$ mm (1.5 in.) | $h = 200$ mm (8 in.) |
| $r = 127$ mm (5 in.) | $o = 0$ mm |
| $\alpha_0 = 126$ deg | $r = 127$ mm (5 in.) |
| $\widehat{\alpha_0} = 2.2$ rad | $\alpha_0 = 126$ deg |
| $\beta = 3$ deg | $\widehat{\alpha_0} = 2.2$ rad |
| $\mu_s = 0.2$ (steel on steel) | $\alpha_1 = 24$ deg |
| | $\alpha_2 = 150$ deg |

The total brake factor BF may be computed by Eqs. (2-20), (2-22), and (2-24) with the brake factor $BF_1$ of the primary shoe given by Eq. (2-22) and the shoe factor $BF_2$ of the secondary shoe given by Eq. (2-20).

Substitution of the appropriate data of the primary shoe into Eq. (2-22) yields:

$$BF_1 = F_{d1} / F_a = \frac{\mu_L(1.67) + \mu_L^2(0.073)}{0.726 - \mu_L(1.01) + \mu_L^2(0.579)} \tag{2-27}$$

where $F_a$ = brake shoe application force, N (lb)

$\quad\quad F_{d1}$ = drag force due to primary shoe, N (lb)

$\quad\quad \mu_L$ = friction coefficient between lining and drum

Eq. (2-27) presents the variation of the brake factor of the primary shoe with lining friction coefficient $\mu_L$. Eq. (2-22) is used to derive Eq. (2-27) since the primary shoe of the brake to be analyzed is supported by a parallel sliding abutment.

Eq. (2-27) may be evaluated for different values of $\mu_L$ giving the values listed in Table 2-1.

**TABLE 2-1**

**$BF_1$ vs. $\mu_L$**

| $\mu_L$ | 0.1 | 0.2 | 0.3 | 0.4 | 0.5 | 0.6 |
|---|---|---|---|---|---|---|
| $BF_1 = F_{d1}/F_a$ | 0.266 | 0.616 | 1.068 | 1.639 | 2.332 | 3.131 |

The secondary shoe is actuated by the support force between the primary and secondary shoe. Since the brake factor is defined by the ratio of drum drag to application force produced by the wheel cylinder, the brake factor of the secondary shoe must be computed in two steps. First the shoe factor is determined by Eq. (2-20) with the support force of the primary shoe used as actuation force of the secondary shoe. Then the shoe factor is modified by means of Eq. (2-24) to yield the brake factor of the secondary shoe. Substitution of the appropriate data into Eq. (2-20) with $\alpha_3 = \alpha_1 + \alpha_2 = 174$ deg yields

$$F_{d2} / F_{ax} = \mu_L (1.6) / \left[ 0.67535 - \mu_L (1.019) \right] \qquad (2\text{-}28)$$

where $F_{ax}$ = application force of secondary shoe, N (lb)

$\quad\quad F_{d2}$ = drag force due to secondary shoe, N (lb)

The minus sign is used to determine the shoe factor of the secondary (or leading) shoe. Eq. (2-28) may be evaluated for different values of $\mu_L$, yielding the values given in Table 2-2.

**TABLE 2-2**

**$F_{d2}/F_{ax}$ vs. $\mu_L$**

| $\mu_L$ | 0.1 | 0.2 | 0.3 | 0.4 | 0.5 | 0.6 |
|---|---|---|---|---|---|---|
| $F_{d2}/F_{ax}$ | 0.279 | 0.679 | 1.299 | 2.390 | 4.824 | 15.012 |

Since the brake factor is defined as the ratio of total drum drag to the application force $F_a$ at the wheel cylinder, the shoe factor of the secondary shoe must be modified to yield the brake factor of the secondary shoe (Eq. [2-24])

$$BF_2 = (F_{d2} / F_{ax})(F_{ax} / F_a) \tag{2-29}$$

The ratio $F_{ax}/F_a$ is determined by Eq. (2-24)

$$F_{ax} / F_a = (c / a) + (F_{d1} / F_a)(r / a)$$
$$= 1.0 + (F_{d1} / F_a)(1.25) \tag{2-30}$$

where   a = brake dimension, mm (in.)

   c = brake dimension, mm (in.)

The ratio $F_{ax}/F_a$ assumes different values for various values of $\mu_L$. Using the values $F_{d1}/F_a$ from Table 2-1 in Eq. (2-30) gives the values of $F_{ax}/F_a$ listed in Table 2-3.

### TABLE 2-3

#### $F_{ax}/F_a$ vs. $\mu_L$

| $\mu_L$ | 0.1 | 0.2 | 0.3 | 0.4 | 0.5 | 0.6 |
|---|---|---|---|---|---|---|
| $F_{ax}/F_a$ | 1.333 | 1.770 | 2.335 | 3.049 | 3.915 | 4.914 |

The brake factor $BF_2$ of the secondary shoe can now be determined by Eq. (2-29). Values of $BF_2$ for various values of $\mu_L$ are given in Table 2-4.

### TABLE 2-4

#### $BF_2$ vs. $\mu_L$

| $\mu_L$ | 0.1 | 0.2 | 0.3 | 0.4 | 0.5 | 0.6 |
|---|---|---|---|---|---|---|
| $BF_2 = (F_{d2}/F_{ax})(F_{ax}/F_a)$ | 0.372 | 1.202 | 3.036 | 7.287 | 18.870 | 73.579 |

The total brake factor BF is obtained by adding the individual shoe brake factors, yielding the data in Table 2-5.

**TABLE 2-5**

**BF vs. $\mu_L$**

| $\mu_L$ | 0.1 | 0.2 | 0.3 | 0.4 | 0.5 | 0.6 |
|---|---|---|---|---|---|---|
| BF | 0.638 | 1.818 | 4.103 | 8.926 | 21.202 | 76.890 |

The brake factor is illustrated in Figure 2-32. Inspection of the brake factor curves of the individual shoes reveals that both shoes are self-energizing, but that the secondary shoe contributes the most to the total brake factor.

Brake sensitivity $S_B$ is defined as the ratio of change in brake factor to the associated change in lining friction coefficient. In some simple cases, the brake sensitivity may be expressed by mathematical equations (Eqs. [2-15] and [2-18]). Most drum brakes in use today require complicated relationships for the computation of brake sensitivity. For these cases, an approximate value of brake sensitivity may be obtained graphically from the brake factor curve.

For the sample problem, the approximate slope of the brake factor curve at various values of lining friction coefficient can be determined from Fig. 2-32. For example, for $\mu_L = 0.15$

$$S_B = \Delta BF / \Delta \mu_L = (1.818 - 0.637) / (0.2 - 0.1) = 11.8$$

where $\Delta BF$ = brake factor change

$\Delta \mu_L$ = lining friction coefficient change

The brake sensitivities of Table 2-6 may be obtained by the same procedure. The $S_B$ values for $\mu_L = 0.5$ and 0.6 are too large to be determined from the brake factor curve shown in Fig. 2-32.

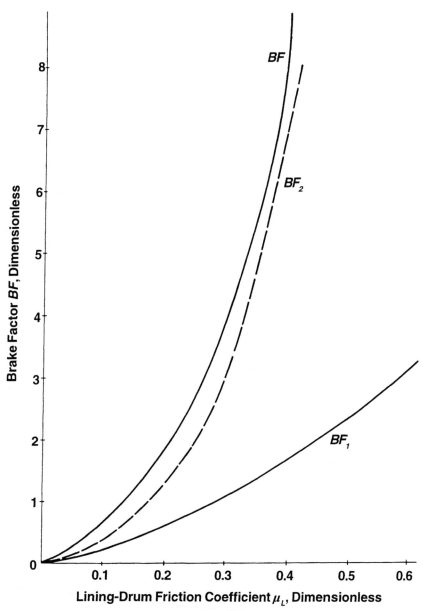

*Figure 2-32. Brake factor characteristic of a duo-servo drum brake.*

**TABLE 2-6**

$S_B$ vs. $\mu_L$

| $\mu_L$ | 0.1 | 0.2 | 0.3 | 0.4 | 0.5 | 0.6 |
|---------|-----|-----|-----|-----|-----|-----|
| $S_B$ | 8 | 4 | 32 | 80 | - | - |

A graphical representation of the data from Table 2-6 is shown in Figure 2-33. Brake sensitivity values normally should not exceed 30. Higher values could lead to severe side-to-side or front-to-rear brake imbalance. Lower values of brake sensitivity are obtained by lowering the lining friction coefficient $\mu_L$ with a corresponding decrease in brake factor. The brake factor decrease reduces the gain of the brake system. The gain of the brake system can be increased again by increasing drum or rotor radii, wheel cylinder size, or by altering the proportional valve characteristics.

## 2.9 Effect of Shoe and Drum Stiffness on Brake Torque

The derivation of the brake factor in the previous paragraphs was based on a rigid shoe and drum. All elastic deformation and wear was assumed to occur in the lining material. Test results show a significant effect of brake shoe elasticity on brake torque. Experimental data obtained for the "rigid" and "elastic" brake shoe geometries shown in Figure 2-34 are presented in Figure 2-35 (Ref. 12). Although both shoes have identical dimensions as far as brake factor calculations are concerned, their actual brake force production is different. Reasons for this difference are found in the change in pressure distribution between lining and drum in the case of the elastic brake shoe. As indicated by Eq. (2-2), in the case of a rigid shoe the pressure distribution is approximated by $p = E a \varphi \sin\alpha / d_{Lo}$. An elastic shoe produces a pressure distribution that has higher pressure concentrations at or near the ends of the linings. The pressure distribution may be approximated by

$$p = (a\varphi E / d_{Lo})(2 \sin \alpha + \cos 2\alpha) \quad , \quad N/m^2 \text{ (psi)} \qquad (2-31)$$

Application of this pressure distribution to the brake factor analysis under consideration of an elastic shoe yields fairly complicated equations for predicting

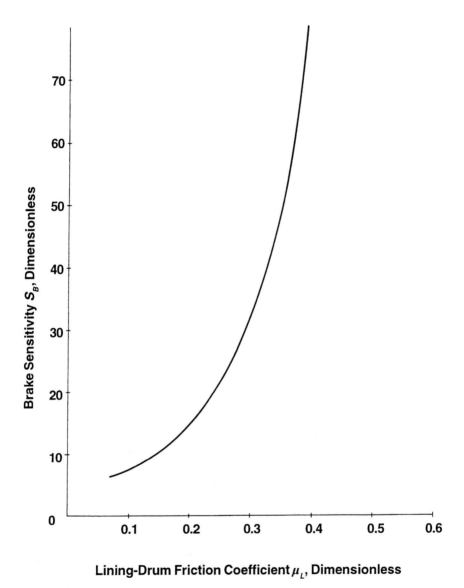

*Figure 2-33. Brake sensitivity*

brake torque. The analysis is made difficult by the complicated designs found in many brake shoes which prevent the establishment of a simple equation for the elastic deformation.

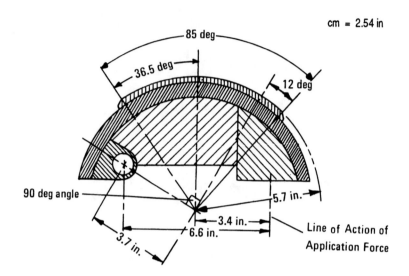

**Rigid Brake Shoe**

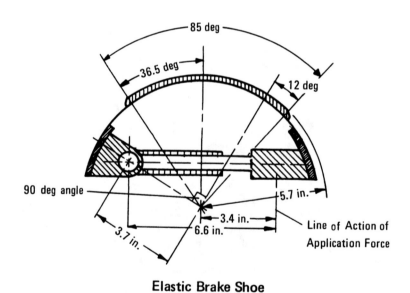

**Elastic Brake Shoe**

*Figure 2-34. Brake shoes of different stiffness.*

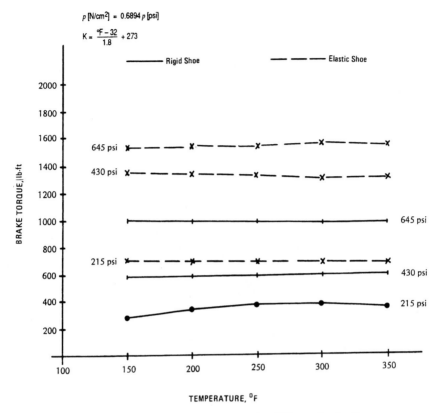

*Figure 2-35. Brake torque vs. temperature for different brake line pressures (215 psi, 430 psi, 645 psi).*

The effect of the difference in pressure distribution may be analyzed by increasing angle β (Fig. 2-28) from a typical value of 3 deg to 30 or 40 deg. This change would effectively alter the pressure distribution so as to concentrate pressure near the end of the lining. Application of this change to the brake factor equations yields significantly higher brake factors at moderate values of lining friction coefficients. The undesirable side effect is increased lining wear.

## 2.10 Analysis of External Band Brakes

In the past, medium- to heavy-weight tractors and trucks were equipped with emergency or parking brakes mounted directly on the driveshaft. These driveshaft-mounted brakes were either external band brakes or, more frequently, duo-servo type drum brakes. Only external band brakes are discussed in this section.

The major disadvantages of driveshaft-mounted parking brakes are the potential for excessive oil contamination because they are located directly behind the engine and transmission, and extremely low thermal capacity. In addition, external band brakes have high bearing forces.

For the band brake shown in Figure 2-36 the following equilibrium conditions apply (Ref. 10):

$$K\ell - S_1 a_1 - S_2 a_2 = 0 \quad , \quad \text{Nmm (lbin.)}$$

$$S_1 - S_2 = F_d \quad , \quad \text{N (lb)}$$

$$S_1 / S_2 = e^{\mu_L \alpha_B}$$

where  $a_1$ = brake dimension, mm (in.)

$a_2$ = brake dimension, mm (in.)

$F_d$ = drum drag force, N (lb)

$K$ = application force, N (lb)

$\ell$ = brake dimension, mm (in.)

$S_1$ = band force, N (lb)

$S_2$ = band force, N (lb)

$\alpha_B$ = band angle, deg

$\mu_L$ = lining friction coefficient

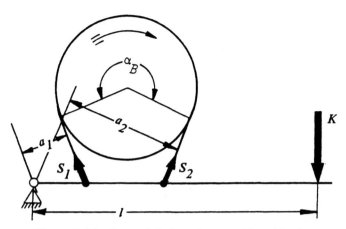

*Figure 2-36. General design of external band brake.*

In terms of the brake factor as defined earlier, the ratio of $F_d$ to $F_a$ represents the gain of the brake. The application force $F_a$ for a simple band brake ($a_1 = 0$, $a_2 = 0$) is given by

$$F_a = K\ell / a \quad , \quad N \text{ (lb)} \tag{3-32}$$

the brake factor and brake sensitivity are given in the following paragraphs for most common band brakes illustrated in Figures 2-37 through 2-39 (Ref. 9).

The band brake shown in Fig. 2-37 yields a brake factor for clockwise rotation of

$$BF = F_d / F_a = e^{\mu L \alpha_B} - 1 \tag{2-33}$$

and brake sensitivity of

$$S_B = \alpha_B e^{\mu L \alpha_B} \tag{2-34}$$

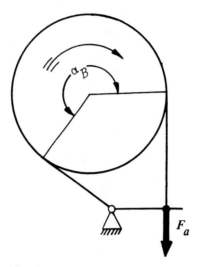

Figure 2-37.  *Single application external band brake.*

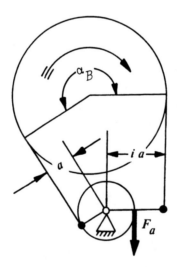

Figure 2-38.  *Opposing application external band brake.*

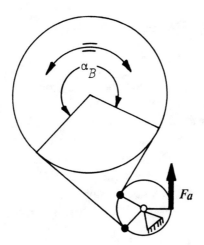

*Figure 2-39. In-line application external band brake.*

For counterclockwise rotation of the same band brake, the results are

$$BF = F_d / F_a = (e^{\mu_L \alpha_B} - 1) / (e^{\mu_L \alpha_B}) \qquad (2\text{-}35)$$

$$S_B = \alpha_B / e^{\mu_L \alpha_B} \qquad (2\text{-}36)$$

For the band brake shown in Fig. 2-38, the results are

$$BF = F_d / F_a = (e^{\mu_L \alpha_B} - 1) / (i - e^{\mu_L \alpha_B}) \qquad (2\text{-}37)$$

$$S_B = \alpha_B e^{\mu_L \alpha_B} (i - 1) / \left(i - e^{\mu_L \alpha_B}\right)^2 \qquad (2\text{-}38)$$

where   i = ratio of application arm as defined in Fig. 2-38

For the brake shown in Fig. 2-39, the results for brake factor and sensitivity are

$$BF = F_d / F_a = (e^{\mu_L \alpha_B} - 1) / (e^{\mu_L \alpha_B} + 1) \qquad (2\text{-}39)$$

$$S_B = 2\alpha_B e^{\mu_L \alpha_B} / \left(e^{\mu_L \alpha_B} + 1\right)^2 \qquad (2\text{-}40)$$

## 2.11  Auxiliary Brakes

### 2.11.1 Basic Concept

An auxiliary brake is a continuous brake in which the retarding torque is not generated by friction between two sliding surfaces such as linings and a drum. Auxiliary brakes may be divided into two classes, namely engine brakes and propeller shaft retarders. In the case of the engine brake, the retarding torque transmitted to the driven wheels of the truck or tractor can be interrupted by disengaging the clutch or placing the transmission in neutral. The propeller shaft brake, once applied, can be disconnected from the retarded wheels only through release of the control lever.

### 2.11.2 Exhaust Brake

The engine of a vehicle in motion will, if the throttle is closed, exert a retarding force on the vehicle as a portion of the kinetic energy is absorbed by the frictional, compressive, and other mechanical losses in the engine. This retarding force is, however, very limited, and various methods have been devised for increasing the effectiveness of the engine as a brake. One such improvement consists of increasing the compressor action of the engine by closing off the exhaust, hence, the name exhaust brake. This type of brake consists of a throttle in the exhaust system which can be closed by mechanical, electrical, or pneumatic means. The brake torque generated at the driven wheels depends on the gearing and engine speed. In general, at moderate and high velocities, the primary braking system also must be applied because the generated brake torque is limited to about 70% of the motor drive torque. The major limiting design factor of an exhaust brake is associated with the exit valve springs. Increased pressure in the exhaust system tends to overcome the valve springs, forcing the valves to stay open and consequently limiting the compressor action of the engine. Depending on size, gasoline engines have brake horsepowers of 96 to 184 kW (130 to 250 hp).

The retarding torque of a four-cycle combustion engine, $M_e$, either gasoline or diesel, without any special exhaust devices may be computed from the approximate expression (Ref. 13)

$$M_e = 8 \times 10^{-5} p_m V_e \quad , \quad Nm \qquad (2\text{-}41)$$

$$\left[ M_e = 0.0065 p_m V_e \quad , \quad lbft \right]$$

where  $p_m$ = average retarding pressure in combustion chamber, $N/m^2$ (psi)

$V_e$ = engine displacement, liter (in.$^3$)

The average retarding pressure associated with engine braking ranges from approximately 4 to $6.5 \times 10^5$ $N/m^2$ (45 to 75 psi) for gasoline engines and 3 to $5 \times 10^5$ $N/m^2$ (60 to 95 psi) for diesel engines. The upper values are associated with high levels of revolutions per minute of the engine crankshaft speed, the lower values with lower levels.

The retarding force $F_{ret}$ at the driven wheels of the vehicle is

$$F_{ret} = M_e \rho / \eta_T R \quad , \quad N \text{ (lb)} \qquad (2\text{-}42)$$

where   R = effective tire radius of driven wheels, m (ft)

$\eta_T$ = efficiency of transmission

$\rho$ = transmission ratio between engine and wheels

## 2.11.3 Engine Brake

Further improvement in engine brake torque can be achieved by altering the camshaft timing such that the compressor action of the engine is increased. Domestically, Jacobs is a major manufacturer of engine brakes, hence the trade name "Jake Brake."

During the normal cycle, the diesel engine stores energy in the compressed air during the compression stroke, and returns most of that energy to the crankshaft as an air spring during the expansion stroke. Engine brakes open the exhaust valves at or near top dead center of the compression stroke. The compressed air is blown into the exhaust manifold and exhaust stack. The total energy absorption is a function of the base horsepower of the engine, the engine speed, the air mass in the cylinder at the beginning of the compression stroke, and the valve timing. The turbocharger still turns at 30 to 40% of its normal speed during engine braking. Depending on the basic engine size and horsepower, brake horsepowers range from 145 to 242 kW (200 to 330 hp) or more. A rule of thumb is that a vehicle will travel downhill at the same speed and gear it traveled uphill with only 5 to 10% of the braking provided by the foundation brakes.

The operation of an engine brake is controlled by dash-mounted switches, one for "on" and "off," and, if used, as many as three for varying degrees of engine braking by activating only two, four, or six cylinders. Microswitches are activated by the movement of the fuel and clutch pedals. When the dash switch is in the "on" position, any time the gas pedal is released the engine brake is activated. Conversely, any time the clutch pedal is depressed while the engine brake is on, the engine brake is deactivated.

## 2.11.4 Hydrodynamic Retarder

The hydrodynamic retarder is a device that uses viscous damping as the mechanism for producing a retarding torque. The viscous damping or internal fluid friction is transformed into thermal energy and dissipated by a heat exchanger. In its design, the hydrodynamic retarder is similar to a hydraulic clutch; however, its turbine or drive rotor is stationary. The retarding torque is produced by the rotor which pumps the fluid against a stator. The stator reflects the fluid back against the rotor, and a continuous internal pumping cycle is developed. The reaction forces and, hence, the retarding torque are absorbed by the rotor, which is connected to the drive wheels of the vehicle. The magnitude of the retarding torque depends on the amount of fluid in the retarder and the pressure level at which it is introduced into the retarder.

The application of the retarder may result from a hand lever movement or a combined service brake/retarder control, such as the foot pedal as shown in Figure 2-40. Depending on the level of applied control force, compressed air travels over the relay valve to the charge tank and control valve. The compressed air in the charge tank forces the retarder fluid into the hydrodynamic brake, simultaneously disconnecting the line between the control valve and the retarder. For a given control input force, the control valve allows a constant retarding torque to develop. The degree of fluid application to the retarder determines the amount of fluid and fluid pressure and, consequently, retarding torque.

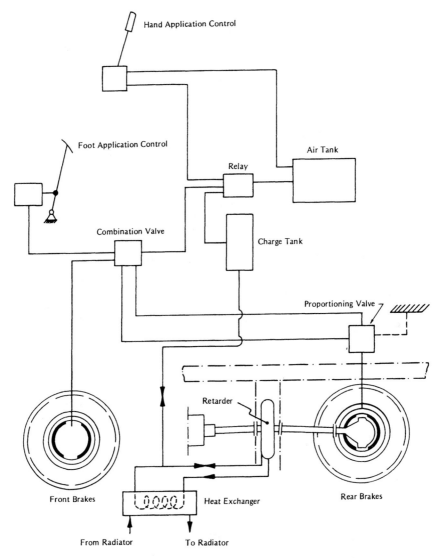

*Figure 2-40. Integrated foundation brake/retarder control system.*

One important advantage of this type of retarder is that the retarding force is greater at higher vehicle speeds. Hydrodynamic retarders operate independently of the engine, clutch, transmission, or electrical power supply. They are connected to the drive axle and represent an almost indestructible no-wear braking element when designed properly. Skidding at the wheels is impossible because the retarding torque approaches zero with decreasing retarder driveshaft speed. Depending on the downhill operating conditions, the retarder may absorb all or a portion of the vehicle braking energy. For economic reasons, the retarding capacity has to be a function of intended vehicle use. Unlike U.S. regulations, European safety standards require retarders in certain classes of commercial vehicles.

Besides the mid-mount hydraulic retarder produced by Thompson, several manufacturers offer hydraulic retarders with their automatic transmission, including Allison, Mack (Dynatard), Voith, and Caterpillar (Brakesaver). When installed in the automatic transmission, the retarder is mounted between the transmission torque converter and the range gear assembly.

The advantages of hydraulic retarders include faster trip times, increased lining life, reduced expenses for running gear, less noise than engine brakes, and reduced driver fatigue.

## 2.11.5 Electric Retarders

The principle of the electric retarder is based on the production of eddy currents within a metal disc rotating between two electromagnets which develop a retarding torque on the rotating disc. When the electromagnets are partially energized, the retarding torque is reduced. When the electromagnets are not energized, the retarding torque is zero. The eddy currents heat the disc. The cooling of the disc is accomplished by convection heat transfer with ventilated rotors. Initially, all retarding energy is absorbed by and stored in the rotor material. Only at elevated temperatures does cooling occur. The major problem of the eddy current retarder is associated with the necessity of high brake temperatures for efficient convective cooling capacity—similar to that experienced with friction-type wheel brakes. The high temperatures cause a decrease in retarding effectiveness due to the demagnetizing of the rotor. Depending on the particular material composition involved, this limiting temperature lies near 1000 K (1350°F). To limit the demagnetizing effects, the operating temperatures should not exceed values of 643 to 753 K (700 to 900°F). At these levels, a reduction in retarding effectiveness of approximately 20 to 30% exists.

Electric retarders using permanent magnets have been developed with significantly reduced weight.

### 2.11.6 Comparison of Retarders

Basic engine braking without any special provisions to increase the retarding effect will produce approximately 45 to 50% of the base power of the engine; exhaust brakes approximately 60 to 70%; Jake brakes approximately 90 to 95%; and hydraulic retarders approximately 175 to 200%.

For example, a vehicle combination weighing 355,918 N or 36,281 kg (80,000 lb) traveling at 48 km/h (30 mph) on a downgrade of 7% would require a continuous retarding effect of 327 kW (447 hp; see Eq. [3-12]). With an engine horsepower rating of 257 kW (350 hp), only the hydraulic retarder would be able to absorb the brake power developed. The Jake brake would only absorb approximately 238 kW (324 hp), making application of the service brakes necessary to absorb the remaining 90 kW (123 hp). If the driver traveled at a lower speed, the Jake brake may be sufficiently strong to absorb the total brake power.

## 2.12  Analysis of Sealed Brakes

The sealed brake is designed to accomplish two basic functions:

1. Absorb and dissipate the kinetic energy of the vehicle at its maximum speed.

2. Protect the brake against damage from adverse environment such as ice, snow, water, mud, dirt, and dust.

Sealing the brakes allows the vehicle to be operated in an off-road environment by keeping the internal brake parts free from outside contamination. The exclusion of abrasive elements results in greater safety, high reliability, and increased life for lining and metal surfaces. This reduces vehicle down time required for replacement and servicing of brake parts due to normal wear. An important consideration is the rubbing speed of the seals. Sealed brakes are often designed so that the seal is resting on the sliding surface only at low speeds, while at higher speeds centrifugal force or air pressure forces the seal off the sliding surface. Since high vehicle speeds in off-road terrain are unlikely, the life of the seals is increased significantly with speed-sensitive sealing designs.

Dissipation of heat generated during braking is accomplished by circulating cooling fluid around the metal surfaces of the brake. A study of the energy absorption capacity of the various fluids shows that oil has an outstanding capacity, and that air is reasonably close to oil. Water is also good, but its freezing characteristics make it a poor candidate. A eutectic mixture of water and ethylene glycol eliminates the freezing problem, but has only about one-third the energy absorption capacity of oil and one-half that of air, as shown in Table 2-7.

An example of an oil-cooled disc brake system is represented by a sealed-oil-cooled, multi-disc, self-adjusting service brake. The brake assembly is made up of a set of multiple disc plates, an annular piston or pistons, a coolant pump, a face type seal assembly, a labyrinth seal assembly, an automatic adjuster sleeve, and a piston return spring or springs assembled within the brake housing and cover. The brake is actuated by hydraulic pressure from a hydraulic supply system acting on the annular piston(s).

## TABLE 2-7

### Energy Absorption Capacity of Various Fluids

| Fluid | State | Energy Absorption Capacity | |
|-------|-------|------|------|
| | | Nm/kg | ft·lb/lbm |
| Air | Gas | 221,126 | 74,000 |
| Water | Liquid | 185,978 | 62,200 |
| Oil | Liquid | 279,505 | 99,500 |
| Mixture - water and ethylene glycol | Liquid | 98,670 | ~33,000 |

As braking action takes place between stationary and rotating discs, cooling oil is pumped through the assembly to absorb the heat generated. The coolant is pumped either by an integral turbine type pump which is driven by a floating gear between two driven discs, or by an external pump. The cooling oil flows from the brake assembly through piping to an oil-to-water heat exchanger mounted within the truck radiator. Piping returns the cooled oil to the brake disc pack cavity, completing the cooling oil hydraulic loop. A "zero" line also is routed from the oil reservoir to the pump inlet to eliminate the possibility of cavitation.

During braking, the cooling oil pump delivers maximum oil flow under low pressure to the heat exchanger. However, when the brake actuating force is removed, releasing the brake disc, the pump drive gear settles into a neutral position between its two adjacent driven discs. The increased operating clearances between the pump and the discs reduces the output effectiveness of the coolant pump. The resulting reduced cooling flow provides sufficient after cooling with negligible spin losses when the brakes are not applied.

Assembly sealing is divided into two elements: (1) sealing the hydraulic supply system from the cooling oil system and (2) sealing internal components from the external environment.

To provide adequate cooling at the surface of the disc plates, the cooling oil must be distributed uniformly to all the discs and directed through the discs to obtain maximum heat transfer from the discs to the oil. Uniform distribution of oil to all discs in the stack is obtained by design of plate hubs and housing for optimum coolant flow and by proper location of oil inlet and outlet connections to the brake housing. The direction of oil flow must be from the outside periphery to the inside diameter of the discs to counter the natural pumping action of the plates and ensure even distribution of cooling oil across the face of the discs. Experience has shown that oil flowing from inside to outside diameter tends to channel in a few grooves in the lined discs, resulting in local hot spots. Oil flow through the discs is provided by grooves cut in the friction material. The engine radiator is used frequently as a heat exchanger for cooling the liquid.

In specialty vehicles such as earth moving equipment, sealed oil-cooled brakes use a large reservoir in the wheel rather than a pump circulating the coolant.

The temperature analysis of liquid-cooled brakes is similar to that of an engine heat exchange process. The heat generated at the wheel brakes is dissipated to the ambient air by means of a heat exchanger or radiator. The heat generation at the friction brake is determined from Eq. (3-13) or (3-15).

For example, for the downhill braking mode the energy $q_{0,RB}$ absorbed by one brake of the rear axle of a two-axle vehicle is

$$q_{0,RB} = \frac{WV(G - R_r)\phi \times 3600}{(778)(2)} \quad , \quad \text{BTU/h} \qquad (2\text{-}43)$$

where  $G$ = road gradient

$R_r$ = tire rolling resistance

$V$ = speed, m/s (ft/s)

$W$ = weight, N (lb)

$\phi$ = rear axle brake force divided by total brake force, d'less

For continued braking, the capacity of the liquid cooled brakes is not limited, provided the radiator heat transfer is

$$h_{rad}A_{rad}\Delta T_{rad} = \overset{\overset{\text{no. of}}{\text{brakes}}}{\sum} q_{0,B} \quad , \quad \text{NM/h (BTU/h)} \qquad (2\text{-}44)$$

where  $A_{rad}$ = cooling area of radiator, m$^2$ (ft$^2$)

$h_{rad}$ = convective heat transfer coefficient of the radiator, Nm/hkm$^2$ (BTU/h·°F·ft$^2$)

$q_{0,B}$ = energy absorbed per single brake, Nm/h (BTU/h)

$\Delta T_{rad}$ = mean temperature difference of cooling liquid and air in radiator, K (°F)

A rough estimate indicates that the cooling capacity of the engine radiator is approximately 90 to 100% of the engine horsepower. Additional cooling of about 10 to 20% is provided by external heat transfer from the wheel brake surfaces and connecting lines being exposed to convective air flow.

The heat transfer coefficient of the engine radiator is dependent on the speed and assumes values between 408,880 and 613,320 Nm/hkm$^2$ (20 and 30 BTU/h·°F·ft$^2$) for vehicle speeds of 97 km/h (60 mph).

For a sealed brake of another design, air was selected as the coolant for the brakes since its heat absorption capacity based on mass compares favorably with other suitable fluids. A ventilated or fan type disc was selected to aid in circulating the cooling air and to increase the heat exchange surface by the area of the radial fan blades. This cooling action is needed to disperse the heat that has been stored during the stop from the disc. When the vehicle is regaining speed, the ventilated disc has about twice the dissipation rate of a solid disc. The greater kinetic energy conversion capacity of the ventilated disc makes it most appropriate for the system.

After air was selected as a coolant the following design problems were considered:

1. Since air is a gas, the volume of fluid to be circulated in the brake system is much greater than for a liquid. Therefore, the air ducting is larger than hydraulic tubing.

2. The sealing problem was not considered to be serious because it is necessary to circulate air through the brake enclosure. The air being expelled past a simple labyrinth at a slight pressure prevents contaminants from entering the brake assembly.

3. Cooling is accomplished by circulating the air from the blower around the disc to dissipate the heat generated by each stop within the time the vehicle can be accelerated again to full speed.

4. Dust generated by the friction material is blown from the enclosed brake.

5. The ducts bringing the cooling air into the sealed brake assembly are flexible, to allow front wheel steering movement and the relative motion between the axles and the vehicle frame.

The thermal design of the brake rotor is based on Eq. (3-36) for a continued downhill brake operation. The convective heat transfer coefficient required for sufficient cooling may be determined from Section 3.1.6. The heat transfer coefficient necessitates a minimum level of air convected over the rotor surfaces. Consequently, the blower must satisfy the requirements for sufficient air flow as well as for sufficient air pressure to keep the brakes free from contamination. The minimum air pressure required to push water out of the brake depends on the water depth through which the vehicle may travel and the specific weight of the water or mud.

# CHAPTER 3

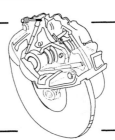

# Thermal Analysis of Friction Brakes

*In this chapter the basic relationships for predicting brake temperature as a result of single, repeated, or continued brake application are presented. Equations for the computation of the convective heat transfer coefficient of drum and disc brakes are given. Finite difference techniques are discussed for the case of a one-dimensional analysis.*

*Relationships for predicting thermal surface stress of solid rotors are developed. Solution outlines for the ventilated rotor are discussed. Fundamentals of thermal rotor failure are reviewed. Thermal design measures helpful for proper brake sizing are presented.*

## 3.1 Temperature Analysis

### 3.1.1 Braking Energy and Braking Power

During braking, the kinetic and potential energies of a moving vehicle are converted into thermal energy through friction in the brakes.

For a vehicle decelerating on a level surface from a higher velocity $V_1$ to a lower velocity $V_2$ the braking energy $E_b$ is

$$E_b = (m / 2)(V_1^2 - V_2^2) + (I / 2)(\omega_1^2 - \omega_2^2) \quad , \quad \text{Nm (lbft)} \qquad (3-1)$$

where  I = mass moment of inertia of rotating parts, kgm$^2$ (lbfts$^2$)

  m = vehicle mass, kg (lbs$^2$/ft)

  $V_1$ = velocity at begin of braking, m/s (ft/s)

$V_2$ = velocity at end of braking, m/s (ft/s)

$\omega_1$ = angular velocity of rotating parts at begin of braking, 1/s

$\omega_2$ = angular velocity of rotating parts at end of braking, 1/s

If the vehicle comes to a complete stop, then $V_2 = \omega_2 = 0$ and Eq. (3-1) becomes

$$E_b = mV_1^2 / 2 + I\omega_1^2 / 2 \quad , \quad Nm \ (lbft) \tag{3-2}$$

When all rotating parts are expressed relative to the revolutions of the wheel, then with V = Rw, Eq. (3-2) becomes

$$E_b = \frac{m}{2}(1 + \frac{I}{R^2m})V_1^2 \approx \frac{kmV_1^2}{2} \quad , \quad Nm \ (lbft) \tag{3-3}$$

where   k = correction factor for rotating masses ($k \approx 1 + I/R^2m$)

R = tire radius, m (ft)

Typical values of k for passenger cars range from 1.05 to 1.15 in high gear to 1.3 to 1.5 in low gear. Corresponding values for trucks are 1.03 to 1.06 for high gear and 1.25 to 1.6 for low gear.

Braking power $P_b$ is equal to braking energy divided by the time t during which braking occurs, or

$$P_b = d(E_b) / dt \quad , \quad Nm/s \ (lbft/s) \tag{3-4}$$

If the deceleration a is constant, then the velocity V(t) is given by

$$V(t) = V_1 - at \quad , \quad m/s \ (ft/s) \tag{3-5}$$

where   a = deceleration, m/s² (ft/s²)

t = time, s

Eqs. (3-3) through (3-5) yield the brake power as

$$P_b = kma(V_1 - at) \quad , \quad Nm/s \; (lbft/s) \quad\quad (3-6)$$

Inspection of Eq. (3-6) reveals that braking power is not constant during the braking process. At the beginning of braking $(t = 0)$, brake power is a maximum, decreasing to zero when the vehicle stops.

The time $t_s$ for the vehicle to come to a stop is

$$t_s = V_1 / a \quad , \quad s \quad\quad (3-7)$$

The average braking power $P_{bav}$ excluding tire slip over the braking time $t_s$ for a vehicle coming to a stop is

$$P_{bav} = kmaV_1 / 2 \quad , \quad Nm/s \; (lbft/s) \quad\quad (3-8)$$

Example 3-1: A vehicle weighing 22,240 N (5000 lb) decelerates to a stop from a speed of 30.5 m/s (100 ft/s) at 7.6 m/s² (25 ft/s²). Use k=1.

With Eq. (3-3), the braking energy is

$$E_b = \frac{(1)(22,240)(30.5)^2}{(2)(9.81)} = 1,054,473 \; Nm$$

$$\left[ E_b = \frac{(1)(5000)(100)^2}{(2)(32.2)} = 776,398 \; lbft \right]$$

The average braking power is

$$P_{bav} = \frac{(1)(22,240)(7.6)(30.5)}{(2)(9.81)} = 262,754 \; Nm / s = 262.7 \; kW$$

$$\left[ P_{bav} = \frac{(1)(5000)(25)(100)}{(2)(32.2)} = 194,099 \text{ lb-ft / s} = 353 \text{ hp} \right]$$

The braking time is $t_s = 30.5/7.6 = 4$ s. Consequently, the average brake power of 262.7 kW (353 hp) is produced during the relatively short time period of 4 s. The maximum power at the onset of braking is 525.4 kW (706 hp).

For a vehicle traveling downhill while decelerating, the brakes have to absorb kinetic and potential energy as illustrated in Figure 3-1. Using energy balance, the braking energy is

$$E_b = Wh + (km / 2)(V_1^2 - V_2^2) \quad , \quad \text{Nm (lbft)} \qquad (3\text{-}9)$$

where   h  =  vertical drop of vehicle mass, m (ft)

W  =  vehicle weight, N (lb)

For continued braking at constant speed, Eq. (3-9) becomes with $V_1 = V_2$

$$E_b = Wh \quad , \quad \text{Nm (lbft)} \qquad (3\text{-}10)$$

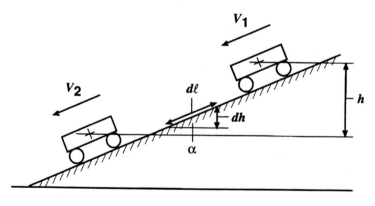

*Figure 3-1. Kinetic and potential energy on grade.*

Braking power during continued braking is obtained by differentiating energy with respect to time, or

$$P_b = d(E_b) / dt = [d(E_b) / dh](dh / dt) \quad , \quad \text{Nm/s (lbft/s)} \quad (3\text{-}11)$$

With the grade expressed by angle $\alpha$ and the actual distance traveled on the highway expressed by $\ell$ (Fig. 3-1), the change in height and road distance are related to the slope by

$$\sin \alpha = dh / d\ell$$

and Eq. (3-11) may be rewritten as

$$P_b = WV \sin \alpha \quad , \quad \text{Nm/s (lbft/s)} \quad (3\text{-}12)$$

When using imperial units in the temperature analyses it becomes convenient to express average braking power for a vehicle coming to a complete stop (Eq. [3-8]) in thermal units rather foot-pounds per second as follows:

$$q_0 = \frac{k(1 - s)V_1 aW(3600)}{2(778)} \quad , \quad \text{BTU/h} \quad (3\text{-}13)$$

where  $a$ = deceleration in g-units

$k$ = correction factor for rotating masses

$q_0$ = average braking power, BTU/h

$s$ = tire slip, defined by the ratio of the difference between vehicle forward speed and circumferential speed to vehicle forward speed

$V_1$ = initial velocity of vehicle, m/s (ft/s)

$W$ = vehicle weight, N (lb)

The tire slip accounts for the energy absorbed by the tire/roadway due to partial slipping of the tire. In the extreme, when the brake is locked, no energy will be absorbed by the brake, i.e., $s = 1.00$.

For the general case of decelerating on a downgrade, Eq. (3-13) may be modified by adding $\sin(\alpha)$ to deceleration a. For example, for a vehicle decelerating at 0.5 g on a downgrade of $\alpha = 10$ deg, the brakes "feel" an effective deceleration of $0.5 + \sin(10) = 0.674$. For uphill braking, the slope effect is subtracted from the actual deceleration because the brakes do not have to absorb all of the kinetic energy, namely that portion transformed into potential energy.

The maximum brake power $P_{b(0)}$ produced at the onset of braking is equal to (see Figure 3-2)

$$P_{b(0)} = 2P_{bavg} \quad , \quad Nm/h \tag{3-14}$$

$$[q_{(0)} = 2q_0 \quad , \quad BTU \,/\, h]$$

where $q_{(0)}$ = braking power at begin of braking (t = 0), BTU/h

During continued downhill brake application at constant speed, the power absorbed by the vehicle brakes, when expressed in imperial thermal units, is obtained from Eq. (3-12) as

$$q_0 = \frac{WV(G - R_r)3600}{778} \quad , \quad BTU/h \tag{3-15}$$

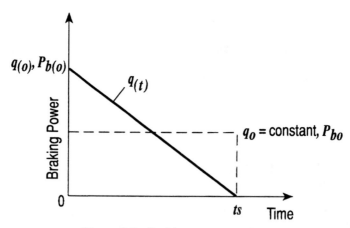

Figure 3-2. Braking power vs. time.

where  G = road gradient (percentage)

      $R_r$ = tire rolling resistance coefficient

      V = vehicle velocity, ft/s

      W = vehicle weight, lb

## 3.1.2 Braking Power Absorbed by Lining and Drum

The analysis of brake temperatures requires an accurate determination of both the total energy absorbed by the brakes and how this energy is distributed between lining and drum or disc.

The distribution of braking energy between lining and drum cannot be predicted readily. The braking or thermal energy distribution is related directly to the thermal resistance associated with both sides of the interface where the heat is generated. It is assumed that the heat transfer into the drum or rotor and lining or pad may be determined from the equivalent resistance network. For the steady-state conditions this may be expressed as

$$q_R'' \, / \, q_P'' = \Sigma R_P \, / \, \Sigma R_R \qquad\qquad (3\text{-}16)$$

where  $q_P''$ = heat flux into pad, Nm/hm$^2$ (BTU/h-ft$^2$)

      $q_R''$ = heat flux into rotor, Nm/hm$^2$ (BTU/h-ft$^2$)

      $R_P$ = thermal resistance to conductive heat flow in pad, hK/Nm (h°F/BTU)

      $R_R$ = thermal resistance to conductive heat flow into rotor, hK/Nm, (h°F/BTU)

For short brake application times, the lining and drum may be considered as semi-infinite solids. Under these conditions, no cooling of the brakes occurs because the temperature at the cooling surfaces has not increased. The requirement of identical temperatures at the interface and the fact that the total heat generation equals the heat absorbed by the lining and drum yields with Eq. (3-16)

$$\frac{q_R''}{q_P''} = \left(\frac{\rho_R c_R k_R}{\rho_P c_P k_P}\right)^{1/2} \tag{3-17}$$

where $c_P$ = pad specific heat, Nm/kg K (BTU/lb$_m$°F)

$c_R$ = rotor specific heat, Nm/kg K (BTU/lb$_m$°F)

$k_P$ = pad thermal conductivity, Nm/mh K (BTU/hft°F)

$k_R$ = rotor thermal conductivity, Nm/mh K (BTU/hft°F)

$\rho_P$ = pad density, kg/m$^3$ (lb$_m$/ft$^3$)

$\rho_R$ = rotor density, kg/m$^3$ (lb$_m$/ft$^3$)

It becomes convenient to express the portion of the total heat generation absorbed by the drum or rotor in terms of the material properties. The requirement that the total heat generated equals $q_R'' + q_P''$ and Eq. (3-17) yield for the relative braking energy $\gamma$ absorbed by the drum or rotor

$$\gamma = \frac{q_R''}{q_R'' + q_P''} = \frac{1}{1 + \left(\dfrac{\rho_P c_P k_P}{\rho_R c_R k_R}\right)^{1/2}} \tag{3-18}$$

For continued braking or repeated braking, Eq. (3-18) assumes a more complicated form due to the convective heat transfer occurring as a result of higher brake temperatures. The schematic is illustrated in Figure 3-3 for a disc brake. For steady-state conditions, no additional energy will be stored in the rotor. Consequently, the thermal resistance associated with the rotor is given by

$$\sum R_R = 1 / (h_R A_R) \quad , \quad \text{hK/Nm (h°F/BTU)} \tag{3-19}$$

where $A_R$ = area of cooling surface, m$^2$ (ft$^2$)

$h_R$ = convective heat transfer coefficient, Nm/hm$^2$K (BTU/h°Fft$^2$)

The thermal resistance $R_P$ associated with the pad is

$$\Sigma R_P = 1 / (h_P A_P) + \delta_P / (k_P A_P)$$

$$+ \delta_S / (k_S A_P) \quad , \quad hK/Nm \ (h°F/BTU) \tag{3-20}$$

where $A_P$ = pad surface, m$^2$ (ft$^2$)

$\quad$ $h_P$ = convective heat transfer coefficient of the pad, Nm/hm$^2$K (BTU/h°F ft$^2$)

$\quad$ $k_P$ = thermal conductivity of pad material, Nm/hmK (BTU/h°F ft)

$\quad$ $k_S$ = thermal conductivity of pad support, Nm/hmK (BTU/h°F ft)

$\quad$ $\delta_P$ = pad thickness, m (ft)

$\quad$ $\delta_S$ = pad support thickness m (ft)

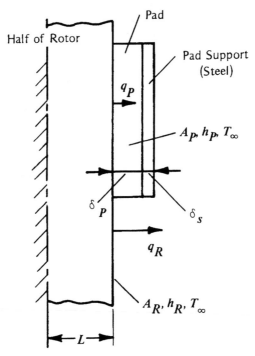

*Figure 3-3. Heat distribution for continued braking.*

With the heat distribution factor defined in Eq. (3-18), Eqs. (3-19) and (3-20) yield

$$\gamma = \frac{q_R''}{q_R'' + q_P''} = \frac{1}{1 + \dfrac{\Sigma R_R}{\Sigma R_P}}$$

and the heat distribution to the rotor is given by

$$\gamma = \left(1 + \frac{h_P k_P k_S A_P}{h_R A_R (k_P k_S + \delta_P h_P k_S + \delta_S h_P k_P)}\right)^{-1} \tag{3-21}$$

For brakes attaining high temperatures, thermal radiation may contribute substantially to cooling the brakes. As is shown in Section 3.1.7, radiation heat transfer may be included by increasing the effective convective heat transfer coefficient.

## 3.1.3 Simplified Temperature Analysis in a Single Stop

In a single stop with high heat generation, i.e., high deceleration levels, the braking time may be less than the time required for the heat to penetrate through the drum or rotor material. Under these conditions no convective brake cooling occurs and all braking energy is assumed to be absorbed by the brake and lining.

For drum brakes the heat penetration time $t_b$ to reach the outer drum surface is given by (Ref. 14)

$$t_b = L^2 / 5a \quad , \quad h \tag{3-22}$$

where   a = thermal diffusivity = $k/(\rho c)$, m²/h (ft²/h)

   c = specific heat of drum or rotor material, Nm/kgK (BTU/lb$_m$°F)

   k = thermal conductivity of drum material, Nm/hmK (BTU/hft°F)

   L = drum thickness, m (ft)

$$t_b = \text{time for heat to penetrate, h}$$

$$\rho = \text{drum density, kg/m}^3 \text{ (lb}_m/\text{ft}^3)$$

The heat flux penetration time expressed by Eq. (3-22) can also be used to determine the time until the heat flux has reached the midpoint in a solid disc brake. For that case L would be half the rotor thickness.

Eq. (3-22) may be rewritten in terms of penetration time measured in seconds and typical drum material properties as

$$t_b = 0.0127(L_d)^2 \quad , \quad s \tag{3-23}$$

$$\left[ t_b = 8.19(L_d)^2 \quad , \quad s \right]$$

where $L_d$ = drum thickness expressed in mm (in).

For example, a truck drum brake having a thickness between the friction and cooling surface of 25 mm (1 in.) would only experience an increase in cooling surface temperature after 8.19 s of braking. In a typical maximum effectiveness stop involving heavy commercial vehicles with a deceleration of 0.5 g, the heat penetration time $t_b$ to the outer surface would only be exceeded for braking speeds of 144 km/h (90 mph) or higher.

For smaller drum brakes or ventilated disc brakes with smaller wall thicknesses, the heat penetration time will be shorter, thus raising the cooling surface temperature above its initial level. Notwithstanding, the convective cooling will be significantly lower than the heat stored in the rotor during the short duration of braking.

If a linearly decreasing braking power similar to Eq. (3-6) is assumed, the surface temperature as a function of time may be expressed as

$$T(L, t) - T_i = (5 / 4)^{1/2}(q''_{(0)} / k)(at)^{1/2}(1 - 2t / 3t_s) \quad , \quad K \,(°F) \tag{3-24}$$

where   a = thermal diffusivity, m²/h (ft²/h)

       k = thermal conductivity, Nm/hm K (BTU/hft°F)

     $q''_{(0)}$ = heat flux into drum or rotor surface existing immediately after begin of braking, Nm/h m² (BTU/hft²)

      t = braking time, h

     $T_i$ = initial temperature, K (°F)

     $t_s$ = time to stop vehicle, h

It should be noted that $q''_{(0)}$ is the braking power per unit area absorbed by the drum or rotor, i.e., only that portion conducted into the drum material and not the total amount of brake power generated by the brake.

Differentiation of Eq. (3-24) with respect to time indicates a maximum of the surface temperature at $t = t_s/2$. Thus, the maximum surface temperature $T_{max,L}$ in a single stop without ambient cooling may be expressed as

$$T_{max,L} - T_i = (5 / 18)^{1/2} \frac{q''_{(0)}(t_s)^{1/2}}{(\rho c k)^{1/2}} \quad , \quad K \ (°F) \qquad (3\text{-}25)$$

where   c = specific heat of drum or rotor material, Nm/kgK (BTU/lb$_m$°F)

and $q''_{(0)}$ is determined by Eq. (3-14), however, divided by the swept area of the brake rotor.

Inspection of Eq. (3-25) reveals that for a specified heat flux $q''_{(0)}$ and braking time $t_s$, the maximum brake drum or rotor temperature will decrease for increased values of density, specific heat, and thermal conductivity. Lowering the heat flux by increasing the swept area of the brake will also decrease the maximum surface temperature.

Typical material properties for drum or disc, and asbestos-based lining and pad material are listed in Table 3-1 below. Semi-metallic pad materials will have increased values for ρ and k.

## TABLE 3-1A

### Brake Design Values, SI Units

|   | Lining | Pad | Drum or Disc | Units |
|---|--------|-----|--------------|-------|
| $\rho$ | 2034 | 2595 | 7228 | kg/m³ |
| c | 1256 | 1465 | 419 | Nm/kgK |
| k | 4174 | 4362 | 174,465 | Nm/hKm |
| a | 0.00163 | 0.0011 | 0.0576 | m²/h |

## TABLE 3-1B

### Brake Design Values, Imperial Units

|   | Lining | Pad | Drum or Disc | Units |
|---|--------|-----|--------------|-------|
| $\rho$ | 127 | 162 | 455 | lbm/ft³ |
| c | 0.30 | 0.35 | 0.10 | BTU/lbm°F |
| k | 0.67 | 0.7 | 28 | BTU/h°F ft |
| a | 0.0176 | 0.0124 | 0.615 | ft²/h |

Example 3-1: Compute the maximum front disc brake temperature of a passenger car decelerating at 0.80 g from a speed of 128 km/h (80 mph) without brake lockup. Use the data that follow: W = 20,003 N (4500 lb), percent braking on front brakes 72%, heat distribution onto rotor 0.90, tire slip 8%, swept area of one rotor side 323 cm² (50 in.²), initial brake temperature 311 K (100°F).

Solution: The average braking power of the entire vehicle is computed from Eq. (3-8) (or Eq. [3-13] for Imperial units):

$$P_{bavg} = \frac{1(1 - 0.08)(20003)(0.8)(9.81)(35.6)}{2(9.81)} = 262,055 \text{ Nm / s}$$

$$\left[ q_0 = \frac{1(1 - 0.08)(4500)(0.8)(117.3)(3600)}{2(778)} = 898,838 \text{ BTU / h} \right]$$

The stopping time is computed from Eq. (3-7):

$$t_b = 35.6 / [0.8(9.81)] = 4.54 \text{ s} = 0.001261 \text{ h}$$

$$[t_b = 117.3 / [0.8(32.2)] = 4.54 \text{ s}]$$

The average braking power absorbed per hour by one half or one side of one front brake is

$$P_{bav} = (262,055)(3600)(0.72)(0.5)(0.5)(0.90) = 1.528 \times 10^8 \text{ Nm / h}$$

$$\left[ q_0 = (898,838)(0.72)(0.5)(0.5)(0.90) = 145,612 \text{ BTU / h} \right]$$

The braking power at the onset of braking is computed from Eq. (3-14) as

$$P_{b(0)} = 2(1.528 \times 10^8) = 3.056 \times 10^8 \text{ Nm / h}$$

$$\left[ q_{(0)} = 2(145,612) = 291,224 \text{ BTU / h} \right]$$

The temperature calculation in Eq. (3-25) requires the heat flux into the swept surface area, i.e., the number of Nm (BTU) per hour and per unit area. Hence, the heat flux designated by $p''_{(0)}$ is

$$p''_{(0)} = (3.056 \times 10^8) / (323 \times 10^{-4}) = 9.461 \times 10^9 \text{ Nm / hm}^2$$

$$\left[ q''_{(0)} = [291,224(144)] / 50 = 838,725 \text{ BTU / h ft}^2 \right]$$

Substitution of the appropriate data into Eq. (3-25) yields the maximum swept surface temperature increase above the initial temperature as

$$T_{max,L} - T_i = \frac{(5/18)^{1/2}(9.461 \times 10^9)(0.00126)^{1/2}}{[7288(419)(174,465)]^{1/2}} = 242 \text{ K}$$

$$\left[ T_{max,L} - T_i = \frac{(5/18)^{1/2}(838,725)(0.00126)^{1/2}}{[455(0.1)(28)]^{1/2}} = 439°\text{ F} \right]$$

The maximum temperature is 242 + 311 = 553 K (439 + 100 = 539°F). The maximum temperature at the friction surface is reached after 4.55/2 = 2.3 s from begin of braking.

Application of Eq. (3-25) to a fully laden commercial tractor-semitrailer vehicle yields a maximum brake surface temperature of approximately 300 to 400°C or 573 to 673 K (600 to 700°F), sufficiently high to produce in-stop fade.

## 3.1.4 Complete Temperature Analysis in a Single Stop

In the previous section, brake temperature was computed for one location on the rotor only, namely at the friction surface. Convective cooling was ignored. In this section brake temperatures will be computed for any location beneath the friction surface, and as a function of time. Convective cooling is included.

### 3.1.4.a Disc Temperature in a Single Stop

The derivation of the temperature equation is relatively complicated. It is accomplished by first deriving the temperature response due to a constant heat flux as observed during constant-speed downhill braking. The final temperature expression is obtained from the constant heat flux temperature and the application of Duhamel's theorem using a time-varying heat flux.

The basic physical parameters are illustrated in Figure 3-4 for a solid rotor disc brake. Both sides of the rotor are heated by the heat flux $q''_{(0)}$, and are cooled by convection $h_R$. For the solid rotor, the conditions permit an analytical solution for a constant heat flux (Ref. 15).

$$\theta_0(z, t) = \frac{q_0''}{h_R} \left[ 2\left( \frac{\theta_i h_R}{q_0''} - 1 \right) \sum_{n=1}^{\infty} \frac{\sin(\lambda_n L)}{\lambda_n L + \sin(\lambda_n L) \cos(\lambda_n L)} \right.$$

$$\left. \times e^{-a_t \lambda_n^2 t} \cos(\lambda_n z) + 1 \right] \quad , \quad K \ (°F) \qquad (3\text{-}26)$$

where $a_t$ = $k_R/(\rho_R c_R)$ = thermal diffusivity, m²/h (ft²/h)

$h_R$ = convective heat transfer coefficient, Nm/h K m² (BTU/h°F ft²)

$L$ = one-half rotor thickness, m (ft)

$n$ = numerals 1, 2, 3, …

$q_0''$ = average heat flux into rotor, Nm/h m² (BTU/h ft²)

$t$ = time, h

$T_0(z, t)$ = transient temperature distribution in rotor due to a constant heat flux, K (°F)

$T_i$ = initial temperature, K (°F)

$T_\infty$ = ambient temperature, K (°F)

$z$ = horizontal distance measured from midplane of rotor, m (ft)

$\theta_0(z, t)$ = $T_0(z, t) - T_\infty$, relative temperature of brake resulting from constant heat flux, K (°F)

$\theta_i$ = $T_i - T_a$ = initial temperature difference between brake and ambient, K (°F)

$\lambda_n$ = $n\pi/L$, 1/m (1/ft)

The value of $\lambda_n L$ is determined from the transcendental equation

$$(\lambda_n L) \tan(\lambda_n L) - h_R L / k = 0$$

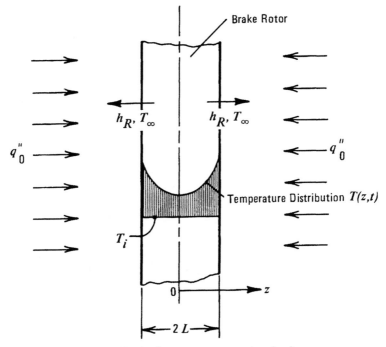

*Figure 3-4. Physical system representing brake rotor.*

Inherent in the derivation of Eq. (3-26) are the following assumptions:

1. The temperature is only a function of the coordinate normal to the friction surface and time t.

2. The heat transfer coefficient $h_R$ is constant.

3. The heat flux is in the direction normal to the friction surface.

4. The thermal properties of both friction partners are constant and evaluated at some mean temperature.

5. The ambient temperature $T_\infty$ is constant.

6. Radiative heat transfer is included in terms of an equivalent radiative heat transfer coefficient (see Section 3.1.7).

127

A few solutions to the transcendental equation of practical importance for typical brakes are presented in Table 3-2. For most temperature analyses only three terms in the summation of Eq. (3-26) are required.

### TABLE 3-2

### Coefficients for Transcendental Equation

| hL/k | $\lambda_1 L$ | $\lambda_2 L$ | $\lambda_3 L$ | $\lambda_4 L$ | $\lambda_5 L$ | $\lambda_6 L$ |
|------|--------|--------|--------|--------|---------|---------|
| 0.01 | 0.0998 | 3.1448 | 6.2848 | 9.4258 | 12.5672 | 15.7086 |
| 0.02 | 0.1410 | 3.1479 | 6.3864 | 9.4269 | 12.5680 | 15.7092 |
| 0.04 | 0.1987 | 3.1543 | 6.2895 | 9.4290 | 12.5696 | 15.7105 |

Eq. (3-26) computes the temperature response resulting from a constant heat flux at the rotor surface. When the vehicle decelerates, the heat flux varies with time. In most cases a linearly decreasing heat flux is assumed. The temperature response of the brake rotor may be obtained directly from the temperature solution shown by Eq. (3-26) associated with the time-independent heat flux $q_0''$ by application of Duhamel's theorem or superposition integral. The temperature response from a time-varying heat flux is (Refs. 14,15):

$$\theta(z, t) = \frac{q_{(0)}''}{q_0''} \theta_0(z, t) + \frac{1}{q_0''} \int_0^t \frac{dq''(\tau)}{d(\tau)} \theta_0(z, t - \tau) d\tau \quad , \quad K\ (°F) \quad (3\text{-}27)$$

where   d = differential operator

   $q_{(0)}''$ = time-varying heat flux into rotor at time t = 0, Nm/hm$^2$
        (BTU/h ft$^2$)

   $q''(\tau)$ = time-varying heat flux, Nm/hm$^2$ (BTU/h ft$^2$)

   t = time, h

   $\theta(z, t)$ = T(z, t) – $T_\infty$ = relative temperature response resulting from
        time-varying heat flux, K (°F)

   $\theta_0(z, t)$ = $T_0(z, t)$ – $T_\infty$ = relative temperature of brake resulting from
        constant heat flux, K (°F)

$T_0(z, t)$ = transient temperature distribution in rotor due to a constant heat flux, K (°F)

$\tau$ = time, h

If a time-varying heat flux

$$q''(t) = q''_{(0)}(1 - t / t_s) \quad \text{Nm/hm}^2 \text{ (BTU/h ft}^2) \quad (3\text{-}28)$$

is assumed, where t = time, h and $t_s$ = braking time to a stop, h, then integration of Eq. (3-27) with Eq. (3-26) and $\theta_i = 0$ yields the temperature response in a solid disc brake resulting from a time-varying heat flux:

$$\theta(z, t) = \frac{q''_{(0)}}{q''_0} \theta_0(z, t)$$

$$- \frac{q''_{(0)}}{t_s h_R}\left[ t - 2 \sum_{n=1}^{\infty} \frac{\sin(\lambda_n L)}{\lambda_n L + \sin(\lambda_n L)\cos(\lambda_n L)}\right.$$

$$\left. \times \left(\frac{1 - e^{-a_t \lambda_n^2 t}}{a_t \lambda_n^2}\right)\cos(\lambda_n z)\right] \quad , \quad \text{K (°F)} \quad (3\text{-}29)$$

where $q''_{(0)}$ = time-varying heat flux into the rotor at time t = 0, Nm/hm$^2$ (BTU/h ft$^2$)

$q''_0$ = average heat flux into rotor = $q''_{(0)}/2$, Nm/hm$^2$ (BTU/h ft$^2$)

$t_s$ = braking time to a stop, h

$\theta_0(z, t)$ = relative temperature of brake resulting from constant heat flux, K (°F), obtained from Eq. (3-26).

Eq. (3-29) was evaluated for a solid disc brake having an outer diameter of 317.5 mm (12.5 in.) and a rotor thickness of 12.7 mm (0.5 in.). The heat flux into one rotor side is $5.56 \times 10^9$ Nm/hm$^2$ (489,500 BTU/h ft$^2$). The convective heat transfer coefficient is 255,553 Nm/hKm$^2$ (12.5 BTU/h°F ft$^2$).

The surface-temperature response computed from Eq. (3-29) is shown in Figure 3-5, using a vehicle deceleration of 0.46 g and speeds of 80 to 97 km/h (50 to 60 mph). The braking times are approximately 5 and 6 s, respectively.

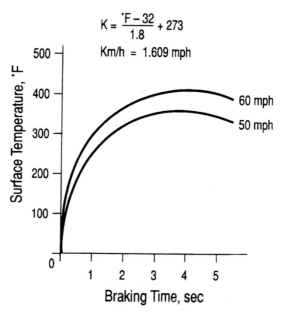

Figure 3-5. Theoretical surface temperature computed from Eq. (3-29).

With the same input parameters, the temperature distribution in the rotor computed for a stop from 97 km/h (60 mph) is illustrated in Figure 3-6. Inspection of Fig. 3-6 reveals that the temperature is nearly uniformly distributed across the width of the rotor after 5 seconds of braking. The temperature gradient existing at the surface after 1.0 s is approximately 45 K/mm (1580°F/in.).

### 3.1.4.b  Drum Temperature in a Single Stop

The temperature response of a brake drum is derived similarly to that of a disc. The major difference is that convective cooling occurs at the outer surface and not at the heat-generation surface as is the case for a solid disc brake.

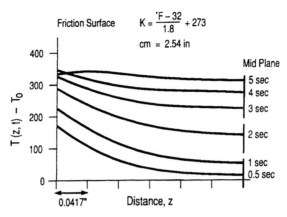

*Figure 3-6. Temperature distribution in the rotor for a stop from 97 km/h (60 mph).*

The temperature attained by a brake drum when subjected to a constant heat flux is given by (Ref. 14)

$$\theta_0(z, t) = \frac{q_0'' L}{k} \left\{ 1 - z / L + k / h_R L - 2 \right.$$

$$\left. \times \sum_{n=1}^{\infty} \frac{e^{-a_t \lambda_n^2 t} \cos(\lambda_n z)}{\lambda_n L [\lambda_n L + \sin(\lambda_n L) \cos(\lambda_n L)]} \right\} , \quad K \text{ (°F)} \quad (3\text{-}30)$$

where    k = thermal conductivity of drum, Nm/hKm (BTU/h°F ft)

       L = drum thickness, m (ft)

       z = distance measured from friction surface, m (ft)

Eq. (3-30) may also be used for computing the brake temperature of ventilated rotors when the convective heat transfer at the friction surface is negligible as in the case of a shielded rotor.

The temperature response of the brake drum resulting from a time-varying heat flux as expressed by Eq. (3-28) is given by

$$\theta(z, t) = \frac{q_{(0)}''}{q_0''} \theta_0(z, t) - \frac{q_{(0)}'' L}{kt_s}$$

$$\times \left[ t\left(1 + \frac{k}{h_R L} - \frac{z}{L}\right) - 2 \sum_{n=1}^{\infty} \frac{\sin(\lambda_n L)}{\lambda_n L + \sin(\lambda_n L)\cos(\lambda_n L)} \right]$$

$$\times \left( \frac{1 - e^{-a_t \lambda_n^2 t}}{a_t \lambda_n^2} \right) \cos(\lambda_n z) \quad , \quad K \ (°F) \qquad (3\text{-}31)$$

where $\theta_0(z, t)$ is obtained from Eq. (3-30).

## 3.1.5 Temperature Analysis for Repeated Braking

During repeated brake applications, the vehicle is decelerated at a given deceleration from, e.g., 97 km/h (60 mph) to a lower or zero speed, after which the vehicle is accelerated again to test speed and the next braking cycle is carried out. Brake pumping involves repeated brake application from one single speed until the vehicle stops. Brake temperatures attained during brake pumping will be less than those achieved during repeated braking because the braking power is lower.

The brake temperatures attained during repeated braking may be computed from simple analytical solutions, provided the braking power, cooling intervals, and braking times remain unchanged during the braking test. Under these conditions, the equations for computing the temperature increase during repeated brake applications may be expressed in a simple form. Assumptions are that the drum or disc can be treated as a lumped system, and that the heat transfer coefficient and thermal properties are constant. In the lumped analysis, the temperature is assumed to be uniform throughout the drum or rotor, making it a function of time only and not of space.

If the braking time is considerably less than the cooling time, then the cooling during braking may be neglected. In this case the drum or disc temperature will increase uniformly by (Ref. 10)

$$\Delta T = \frac{q_0 t_s}{\rho_R c_R v_R} \quad , \quad K \; (^\circ F) \tag{3-32}$$

where $c_R$ = specific heat, Nm/kg K (BTU/lb$_m$ $^\circ$F)

$\quad\quad q_0$ = braking power absorbed by the rotor, Nm/h (BTU/h)

$\quad\quad t_s$ = braking time to a stop, h

$\quad\quad v_R$ = rotor volume, m$^3$ (ft$^3$)

$\quad\quad \rho_R$ = rotor density, kg/m$^3$ (lb$_m$/ft$^3$)

The lumped formulation results in a differential equation describing the cooling of the brake after a brake application

$$\rho_R c_R v_R dT \,/\, dt = -h_R A_R (T - T_\infty) \quad , \quad Nm/h \; (BTU/h) \tag{3-33}$$

where $A_R$ = rotor surface, m$^2$ (ft$^2$)

$\quad\quad h_R$ = heat transfer coefficient Nm/hKm$^2$ (BTU/h$^\circ$F ft$^2$)

$\quad\quad T$ = temperature at time t , K ($^\circ$F)

$\quad\quad T_\infty$ = ambient temperature, K ($^\circ$F)

With an initial temperature of $T_i$, integration of Eq. (3-33) yields a cooling temperature response

$$\frac{T(t) - T_i}{T_i - T_\infty} = e^{(-h_R A_R t)/(\rho_R c_R v_R)} \tag{3-34}$$

An analysis combining heating by means of Eq. (3-32) and cooling by means of Eq. (3-34) may be developed to derive the temperatures of a brake after the first, second, third, or nth brake application. The relative brake temperature before the nth brake application is

$$\left[T(t) - T_\infty\right]_b = \frac{\left\{1 - e^{[-(n_a-1)h_R A_R t_c]/[\rho_R c_R v_R]}\right\}\left\{e^{(-h_R A_R t_c)/(\rho_R c_R v_R)}\right\}\{\Delta T\}}{1 - e^{-(h_R A_R t_c)/(\rho_R c_R v_R)}} \, , \, K(°F)$$

(3-35)

where $n_a$ = number of brake applications

$t_c$ = cooling time cycle time, h

The relative brake temperature after the nth application is

$$[T(t) - T_\infty]_a = \frac{\left[1 - e^{(-n_a h_R A_R t_c)/(\rho_R c_R v_R)}\right][\Delta T]}{1 - e^{(-h_R A_R t_c)/(\rho_R c_R v_R)}} \, , \quad K \, (°F) \quad (3-36)$$

The limit values of the temperature before and after braking for a large number of cycles ($n_a \to \infty$) may be obtained from Eqs. (3-35) and (3-36) by dropping the term involving the factor $n_a$.

Example 3-2: Federal Motor Vehicle Safety Standard 105 requires burnishing of the brakes at GVW from a speed of 64 km/h (40 mph) at a deceleration of 3.66 m/s² (12 ft/s²) for 200 stops. The cycle distance is 1.61 km (1 mile). The approximate cooling cycle time is 88 s.

Compute the average rear brake temperature after the fifth, tenth, and 200th stop. Use the data that follow: 15% of total brake power absorbed by one rear brake, brake drum volume $v_R = 0.00057$ m³ (0.02 ft³), brake cooling area $A_R = 0.051$ m² (0.55 ft²), convective heat transfer coefficient h = 367,992 Nm/hKm² (18 BTU/h°F ft²), vehicle weight W = 16,458 N (3700 lb).

Solution: The average braking power per rear brake is computed by Eq. (3-8) (Eq. [3-13] is used for Imperial units):

$$P_{bav} = \frac{1(16,458)(3.66)(17.78)(3600)(0.15)}{2(9.81)} = 2.94 \times 10^7 \, Nm \, / \, h$$

$$\left[ q_0 = \frac{1(58.6)(12)(3700)(3600)(0.15)}{2(778)(32.2)} = 28,061 \text{ BTU / h} \right]$$

The brake application time is computed as $t_s = V/a = 60/(3.6 \times 3.65) = 4.9$ s or 0.00136 h (58.6/12 = 4.9 s).

The average temperature increase per stop is computed by Eq. (3-32) as

$$\Delta T = \frac{(2.94 \times 10^7)(0.00136)}{(7288)(419)(0.00057)} = 23\text{K}$$

$$\left[ \Delta T = \frac{(28,061)(0.00136)}{(455)(0.1)(0.02)} = 41.9° \text{F} \right]$$

The brake temperature after the 5th brake application is computed from Eq. (3-36) as

$$T_5 - T_\infty = (23)$$

$$\times \frac{1 - e^{-[5(367,992)(0.051)(88)]/[7288(419)(0.00057)(3600)]}}{1 - e^{-[367,992(0.051)(88)]/[7288(419)(0.00057)(3600)]}} = 72.7 \text{ K}$$

$$\left[ \begin{array}{l} T_5 - T_\infty = (41.9) \\ \\ \times \dfrac{1 - e^{-[5(18)(0.55)(88)]/[455(0.1)(0.02)(3600)]}}{1 - e^{-[18(0.55)(88)]/[455(0.61)(0.02)(3600)]}} = 131.8° \text{F} \end{array} \right]$$

Adding the ambient temperature of 299 K (80°F) yields the brake temperature of 378 K (221.8°F) after the 5th brake application.

The corresponding brake temperatures for the 10th and 200th stop are 392 K and 399 K (246.5°F and 259°F), respectively. Inspection of the results reveals that the final brake temperature level is reached within the first few brake applications.

## 3.1.6 Temperature Analysis for Continued Braking

When the brakes are applied during a long downhill descend, cooling while braking must be considered. Similar to the lumped temperature formulation of Section 3.1.5, the temperature response of a drum or disc during continued braking is computed by

$$T(t) = [T_i - T_\infty - q_0 / h_R A_R] e^{(-h_R A_R t)/(\rho_R c_R v_R)}$$
$$+T_\infty + q_0 / h_R A_R \qquad , \qquad K \ (°F) \qquad (3-36)$$

where  $q_0$ = braking power absorbed by the rotor, Nm/h (BTU/h)

$t$ = time during which brakes are applied, h

<u>Example 3-3</u>: Compute the average brake temperature of a tractor-semi-trailer descending a 7% grade at a constant speed of 32.2 km/h (20 mph). Neglect any engine retardation. Compute the brake temperatures after 1.6, 3.2, and 8 km (1, 2, and 5 miles) of operation. Use the data that follow: vehicle weight W = 355,840 N (80,000 lb), tire rolling resistance coefficient 0.01, brake drum volume 0.00793 m³ (0.28 ft³), cooling area 0.372 m² (4 ft²), convective heat transfer coefficient 265,772 Nm/hKm² (13 BTU/(h°F ft²)), relative braking power per one tractor rear brake 0.11, initial brake temperature 338 K (150°F), ambient temperature 283 K (50°F).

<u>Solution</u>: The braking power that must be absorbed continuously by one rear brake is computed by Eq. (3-12) (Eq. [3-15]) as

$$P_b = (355,840)(0.11)(0.07 - 0.01)(32.2)(1000) = 7.56 \times 10^7 \ Nm \ / \ h$$

$$\left[ \begin{array}{l} q_0 = [(80,000)(0.11)(0.07 - 0.01)(3600)(20)(1.466)] / 778 \\ = 71,648 \ BTU \ / \ h \end{array} \right]$$

The brake application time at 32.2 km/h (20 mph) for one mile is 0.05 h. Substitution of the appropriate data into Eq. (3-36) yields the brake temperature after 1.61 km (1 mile) of braking as

$$T_t = [338 - 283 - 7.56 \times 10^7 / (0.372)(265,772)]$$

$$\times e^{-[265,772(0.372)(0.05)]/[7288(419)(0.00793)]}$$

$$+ 283 + 7.56 \times 10^{-7} / (0.372)(265,772) = 469K$$

$$\begin{bmatrix} T_t = [150 - 50 - 71,648 / (4)(13)] \\ \times e^{-[(13)(4)(0.05)]/[(455)(0.11)(0.28)]} \\ + 50 + 71648 / (4)(13) = 386° F \end{bmatrix}$$

For the 3.22-km (2-mile) and 8-km (5-mile) brake temperature calculations only the brake application times change to 0.1 h and 0.25 h, respectively. When those changes are made in Eq. (3-36), the brake temperature after 3.2 km (2 miles) is 576 K (578°F); after 8 km (5 miles), 792 K (967°F). Brake temperatures exceeding 588 K (600°F) generally involve significant brake fade.

### 3.1.7 Convective Cooling of Brakes

#### 3.1.7.a Fundamentals of Convective Cooling

The computation of brake temperature requires information on the convective heat transfer coefficient, which varies with vehicle speed. In many cases it is sufficient to evaluate the heat transfer coefficient at some mean speed.

Textbooks on heat transfer provide a large number of empirical equations for predicting the convective heat transfer coefficient for a variety of test conditions and geometries. These equations generally apply to discs or drums not obstructed by tire and rim or disc caliper.

At the outset it should be stated that any relationship expressing the convective heat transfer coefficient will yield only approximate results. A difference between predicted and measured temperature levels of 10 to 30% may be considered normal. Often "excellent" correlation is obtained by adjusting the convective heat transfer coefficient until agreement between prediction and measurement is achieved.

It has been shown that experimental results of a cooling analysis can be represented by the product of dimensionless numbers raised to some power (Ref. 16), i.e.,

$$Nu = C\, Re^m\, Pr^n \qquad\qquad (3\text{-}37)$$

where   C = heat transfer constant

   $c_a$ = specific heat of air, Nm/kg K (BTU/$lb_m$°F)

   $h_R$ = convective heat transfer coefficient, Nm/hK$m^2$
          (BTU/h°F $ft^2$)

   $k_a$ = thermal conductivity of air, Nm/hKm (BTU/h°F ft)

   $L_c$ = characteristic length, m (ft)

   $m_a$ = mass flow rate of air, $m^3$/s ($ft^3$/s)

   m = heat transfer parameter

   n = heat transfer parameter

   Nu = $h_R L_c/k_a$ = Nusselt number

   Pr = $3600\, c_a m_a/k_a$ = Prandtl number

   Re = $V\rho_a L_c/m_a$ = Reynolds number

   V = vehicle speed, m/s (ft/s)

   $\mu_a$ = viscosity of air, kg/ms ($lb_m$/ft s)

   $\rho_a$ = density of air, kg/$m^3$ ($lb_m/ft^3$)

The constant C in Eq. (3-37) is a function of the geometry of the brake and assumes different values for brake drums, solid rotors, and ventilated rotors. For ventilated rotors the value of C depends on the shape of the vanes used for ventilation.

The heat transfer parameter m is a function of the type of flow, i.e., turbulent, laminar, or transition flow. For most practical cases m is a function of vehicle velocity and the associated brake rotor angular velocity. The heat transfer parameter n depends on the thermal properties of the surrounding air. Because these properties are a function of temperature, the Prandtl number effect is nearly constant for most cases and is often included in the constant C of Eq. (3-37). The characteristic length $L_c$ is either a length or diameter depending on the definition of the Nusselt or Reynolds number.

### 3.1.7.b  Heat Transfer Coefficient for Drum Brakes

For a brake drum fully exposed to the air flow, the heat transfer coefficient $h_R$ is (Ref. 13)

$$h_R = 0.1(k_a / D)\, Re^{2/3} \quad , \quad Nm/hKm^2 (BTU/h°F\ ft^2) \qquad (3-38)$$

where   D  =  drum diameter, m (ft)

   $k_a$  =  thermal conductivity of air, Nm/hKm (BTU/h°F ft)

For example, a 381 mm or 0.381 m (15 in.) diameter drum moving through air at a speed of 97 km/h (60 mph) at an ambient temperature of 311 K (100°F) will have a convective heat transfer coefficient of approximately 183,996 Nm/hKm$^2$ (9 BTU/h°F ft$^2$).

Road test data obtained from testing of heavy vehicles equipped with drum brakes indicate that the convective heat transfer coefficient may be expressed by functional relationship of the form (Ref. 10)

$$h_R = 18808 + 67073\beta V e^{-0.01V} \quad , \quad Nm/hKm^2 \qquad (3-39)$$

$$\left[ h_R = 0.92 + \beta V e^{-V/328} \quad , \quad BTU / h°F\ ft^2 \right]$$

where   $h_R$  =  convective heat transfer coefficient, Nm/hKm$^2$ (BTU/h°Fft$^2$)

   V  =  vehicle speed, m/s (ft/s)

   $\beta$  =  0.70 for front drum brake, Nms/hKm$^3$ (BTU s/h°F ft$^3$)

       =  0.30 for rear drum brake Nms/hKm$^3$ (BTU s/h°F ft$^3$)

The corresponding values of β associated with the heat transfer from the brake shoes inside the brake assembly were found to be 0.15 and 0.06, respectively. When the vehicle is braked to rest, the convective cooling capacity is reduced to that of natural convection indicated by 18,808 Nm/hKm (0.92 BTU/h°F ft) in Eq. (3-39).

### 3.1.7.c  Heat Transfer Coefficient for Solid Discs

For solid, non-ventilated disc brakes the convection heat transfer coefficient associated with laminar flow may be approximated by

$$h_R = 0.70(k_a / D) \, Re^{0.55} \quad , \quad Nm/hKm^2 \, (BTU/h°F \, ft^2) \quad (3\text{-}40)$$

where  $D$ = outer diameter, m (ft)

For Re > 2.4 × 10$^5$ the flow characteristics will be turbulent and the heat transfer coefficient may be expressed as

$$h_R = 0.04(k_a / D) \, Re^{0.8} \quad , \quad Nm/hKm^2 \, (BTU/h°F \, ft^2) \quad (3\text{-}41)$$

Eqs. (3-40) and (3-41) were obtained from experimental data collected with a disc brake system of a light truck (Ref. 13).  Use the data of the previous example: a 381 mm (15 in.) outer diameter rotor at 97 km/h (60 mph) will have a convective heat transfer coefficient of approximately 408,880 Nm/hKm$^2$ (20 BTU/h°F ft$^2$).  For the example chosen, the transition from laminar to turbulent flow lies at about 38.4 km/h (24 mph). Consequently, the convective heat transfer coefficient at 32 km/h (20 mph) is computed by Eq. (3-40) to be approximately 143,108 Nm/hKm$^2$ (7 BTU/h°F ft$^2$).

A comparison of the computed heat transfer coefficients indicates clearly that a disc brake has a higher convective heat transfer coefficient than a drum brake.

It should be noted that Eqs. (3-40) and (3-41) were obtained from experiments with two calipers located horizontally 180 deg apart at the 3 and 9 o'clock positions. The particular location of the caliper relative to the air flow may have an effect on the cooling capacity of disc brakes.

### 3.1.7.d  Heat Transfer Coefficient of Ventilated Disc Brakes

Ventilated disc brakes generally exhibit convective heat transfer coefficients approximately twice as large as those associated with solid discs. The cooling effectiveness associated with the internal vanes tends to decrease somewhat for higher speeds due to the increased stagnation pressure of the air.

For estimating purposes the following relationship may be used to obtain the heat transfer coefficient inside the vanes of the brake rotor (Ref. 17)

$$h_R = 0.023 \, [1 + (d_h / \ell)^{0.67}] \text{Re}^{0.8} \, \text{Pr}^{0.33}$$

$$\times \, (k_a / d_h), \, \text{Nm/hKm}^2 \, (\text{BTU/h°F ft}^2) \tag{3-42}$$

where  $d_h$ = hydraulic diameter, m (ft)

$\ell$ = length of cooling vane, m (ft)

$\text{Re}$ = $\rho_a d_h / m_a) V_{average}$

$V_{average}$ = average velocity, m/s (ft/s)

Eq. (3-42) is valid for $\text{Re} > 10^4$, i.e., for turbulent flow. The hydraulic diameter is defined as the ratio of four times the cross-sectional flow area (wetted area) divided by the wetted perimeter, as illustrated in Figure 3-7. For vanes with varying cross-sectional size, an average hydraulic diameter is determined from the dimensions of the inlet and outlet locations on the vane.

The velocity associated with the Reynolds number is the air flow velocity existing in the vanes which is not identical to the forward speed of the vehicle.

For low values of velocity, laminar flow will exist in the vanes. For $\text{Re} < 10^4$ the convective heat transfer coefficient may be approximated by (Ref. 17)

$$h_R = 1.86(\text{Re Pr})^{1/3}(d_h / \ell)^{0.33}$$

$$\times \, (k_a / d_h) \quad , \quad \text{Nm/hKm}^2 \, (\text{BTU/h°F ft}^2) \tag{3-43}$$

141

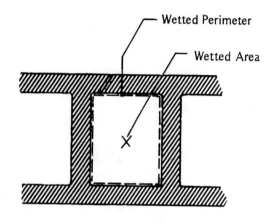

**Section A—A**

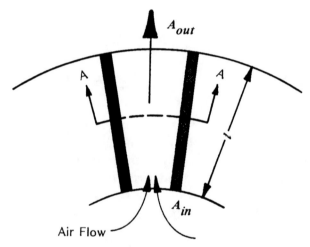

*Figure 3-7. Ventilated disc.*

The average velocity through the cooling vanes can be computed by

$$V_{average} = (V_{in} + V_{out}) / 2 \quad , \quad \text{m/s (ft/s)} \tag{3-44}$$

where $A_{in}$ = inlet area, m² (ft²)

$A_{out}$ = outlet area, m² (ft²)

$$d = \text{inner diameter, m (ft)}$$

$$D = \text{outer diameter, m (ft)}$$

$$n_T = \text{revolutions per minute, 1/min (rpm)}$$

$$V_{in} = 0.0158 n_T (D^2 - d^2)^{1/2} \quad , \quad \text{m/s}$$

$$[V_{in} = 0.052 n_T (D^2 - d^2)^{1/2} \quad , \quad \text{ft / s}]$$

$$V_{out} = V_{in}(A_{in} / A_{out}) \quad , \quad \text{m/s (ft/s)}$$

The air flow rate $m_a$ is determined by

$$m_a = 0.00147 n_T [(D^2 - d^2) A_{in}]^{1/2} \quad , \quad \text{m}^3\text{/s} \tag{3-45}$$

$$\left[ m_a = 0.052 n_T [(D^2 - d^2) A_{in}]^{1/2} \quad , \quad \text{ft}^3 / s \right]$$

If the ventilated rotor is exposed to air, i.e., the friction surfaces are not shielded, then the convective heat transfer coefficient is obtained by the summation of the heat transfer coefficients of Eqs. (3-41) and (3-42), or (3-40) and (3-43).

### 3.1.7.e Radiative Heat Transfer

At higher brake temperatures the radiative cooling capacity of the brakes has to be considered. A radiative heat transfer coefficient $h_{R,rad}$ may be defined by (Ref. 17)

$$h_{R,rad} = \frac{\sigma \varepsilon_R (T_R^4 - T_\infty^4)}{T_R - T_\infty} \quad , \quad \text{Nm/hKm}^2 \text{ (BTU/h°F ft}^2) \tag{3-46}$$

where $T_R$ = rotor surface temperature, K (°R)

$\quad\quad\, T_\infty$ = ambient temperature, K (°R)

$\varepsilon_R$ = rotor surface emissivity

$\sigma$ = Stefan-Boltzmann constant = $3.56 \times 10^{-5}$ Nm/m²K

$$= 0.1714 \times 10^{-8} \text{ BTU/h ft}^2\text{°R}$$

Evaluation of Eq. (3-46) using $\varepsilon_R = 0.55$, a value typical of machined cast iron surfaces of brake rotors, yields the radiative heat transfer characteristics illustrated in Figure 3-8. It is apparent that significant radiation cooling does not occur until high brake temperatures are attained. However, for hot brakes with the vehicle traveling at low speed, radiation cooling may be the predominant cooling mechanism.

## 3.1.8 Computer-Based Brake Temperature Analysis

Only the basic elements of finite difference temperature computations are shown. The reader interested in transient two- or three-dimensional systems is referred to a standard text on heat transfer (Ref. 16). Furthermore, the numerical methods for solving steady-state temperature problems are not discussed here, because the steady-state temperatures are of less importance in most braking analyses than time-dependent temperature distributions.

In the unsteady-state system the initial temperature distribution is known; however, the variation of temperature with time must be determined.

The system, i.e., the drum or disc thickness is divided into a number of discrete nodal points as illustrated in Figure 3-9 for a one-dimensional temperature analysis. In Fig. 3-9 the temperature is analyzed only as a function of distance x and time t. Application of the first law of thermodynamics, or energy balance, to each individual node results in a set of algebraic equations whose solution will yield individual nodal temperatures for each finite time interval. It is therefore necessary to deduce the temperature distribution at some future time from a given distribution at an earlier time, the earliest time being associated with the known initial temperature distribution. The relationship expressing heat conduction between two nodes is known as Fourier's Conduction Law and may be expressed in the form of an exact integral

$$q_{ij} = \int_{\Delta y} -k(\partial T / \partial x)bdy \quad \text{Nm/h (BTU/h)}$$

$$\approx -k(dT / dx)_{average}b\Delta y \approx -k(\Delta T / \Delta x)_{av}b\Delta y \quad (3\text{-}47)$$

144

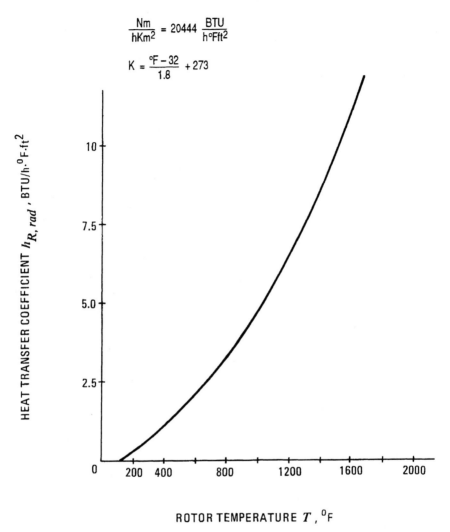

$$\frac{Nm}{hKm^2} = 20444 \; \frac{BTU}{h\,^\circ F ft^2}$$

$$K = \frac{^\circ F - 32}{1.8} + 273$$

*Figure 3-8. Radiative heat transfer coefficient as function of temperature.*

where   b = width of plate, m (ft)

   $q_{ij}$ = heat flow between nodal points i and j, Nm/h (BTU/h)

   x = horizontal distance between two adjacent nodal points, m (ft)

   y = vertical distance between two adjacent nodal points, m (ft)

$\partial T/\partial x$ = temperature gradient, K/m (°F/ft)

145

The distances $\Delta x$, $\Delta y$, and b designate control volume size, and k the thermal conductivity of the material. Eq. (3-47) may be rewritten in the form of the temperature of the two nodal points

$$q_{ij} = -\frac{k(T_j - T_i)b\Delta y}{\Delta x} \quad , \quad \text{Nm/h (BTU/h)} \tag{3-48}$$

where $T_i$ = temperature of node i, K (°F)

$T_j$ = temperature of node j, K (°F)

For two-dimensional temperature problems, where $T = f(x, y, t)$ for example, and a square grid size with $\Delta x = \Delta y$, the basic heat conduction between two nodal points becomes

$$q_{ij} = k(T_i - T_j)b \quad , \quad \text{Nm/h (BTU/h)} \tag{3-49}$$

For one-dimensional systems such as a solid disc brake, the basic heat conduction equation with $\Delta y$ equal to unity becomes

$$q_{ij} = \frac{k_R(T_i - T_j)b}{\Delta x} \quad , \quad \text{Nm/h (BTU/h)} \tag{3-50}$$

With the mass contained in the control volume of thickness $\Delta x$, $d_m = r_R \Delta x b(1)$, and the change in enthalpy, $\Delta h = c_R \Delta T$, the first law of thermodynamics applied, e.g., to the interior node 2, results in the expression (Figure 3-9)

$$\rho_R \Delta x b c_R[(T_2' - T_2) / \Delta t]$$
$$= k_R[(T_1 - T_2) / \Delta x]b + k_R[(T_3 - T_2) / \Delta x]b \quad , \quad \text{Nm/h (BTU/h)}$$

where $c_R$ = specific heat of rotor, Nm/kgK (BTU/lb$_m$°F)

$k_R$ = thermal conductivity of rotor, Nm/hmK (BTU/hft°F)

$T_1$ = temperature at node 1, K (°F)

$T_2$ = temperature at node 2, K (°F)

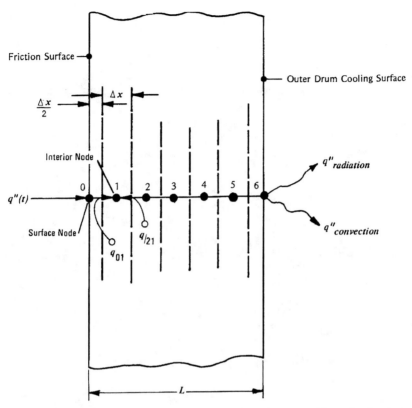

*Figure 3-9. Thermal model for finite difference computation (drum brake shown).*

$T_3$ = temperature at node 3, K (°F)

$\rho_R$ = rotor density, kg/m³ (lb_m/ft³)

Here $T_2'$ represents the temperature attained by node 2 after the time interval $\Delta t$ has elapsed. Solving for $T_2'$ yields

$$T_2' = (1 / M)(T_3 + T_1) + [1 - (2 / M)]T_2 \quad , \quad K \ (°F) \qquad (3\text{-}51)$$

where $a_t$ = thermal diffusivity, m²/h (ft²/h)

$M = (\Delta x)^2/(a_t \Delta t)$

$t$ = time interval, h

Eq. (3-51) may be expressed for any arbitrary interior point n in the form

$$T_n' = (1 / M)(T_{n+1} + T_{n-1}) + [1 - (2 / M)]T_n \quad , \quad \text{K (°F)} \quad (3\text{-}52)$$

Application of the first law of thermodynamics to a surface point yields (Fig. 3-8)

$$T_0' = \left(1 - \frac{2N + 2}{M}\right)T_0 + \frac{2NT_\infty}{M} + \frac{2T_1}{M}$$

$$+ \frac{2\Delta x q_R''}{k_R M} - \frac{2\Delta x q_{rad}''}{kM} \quad , \quad \text{K (°F)} \quad (3\text{-}53)$$

where $h_r$ = convective heat transfer coefficient, $Nm/hm^2K$ ($BTU/hft^2°F$)

$k_R$ = thermal conductivity of rotor, $Nm/hmK$ ($BTU/hft°F$)

$M = (\Delta x)^2/(a_t \Delta t)$

$N = h_R \Delta x/k_R$

$q_{rad}''$ = radiation heat flux away from surface

$= \varepsilon\sigma(T_0^4 - T_\infty^4) \quad , \quad Nm/hm^2$ ($BTU/h\ ft^2$)

$q_R''$ = heat flux absorbed by the rotor computed from Eq. (3-55), $Nm/hm^2$ ($BTU/hft^2$)

$T_\infty$ = ambient temperature, K (°R)

$\varepsilon_R$ = emissivity $\approx 0.8$ for black drum surface, 0.55 for metallic disc surface

$\sigma$ = Stefan-Boltzmann constant, $3.56 \times 10^{-5}\ Nm/m^2K$ ($0.174 \times 10^{-8}\ BTU/hft^2°R$)

Mathematical stability conditions require that M be chosen

$$M \geq 2N + 2 \quad (3\text{-}54)$$

Otherwise the coefficient of T assumes negative values resulting in an unstable temperature solution. The time step $\Delta t$ and grid size may be chosen arbitrarily, provided the conditions in Eq. (3-54) are satisfied.

148

The instantaneous heat flux $q_R''$ per rotor friction surface may be determined from

$$q_R'' = \mu_L p_\ell A_{wc} \eta (1 - s) V(r_i / R)[(3600\gamma) / A_j] \quad , \quad Nm/hm^2 \quad (3\text{-}55)$$

$$\left[ q_R'' = \mu_L p_\ell A_{wc} \eta (1 - s) V(r_i / R)[(3600\gamma) / (778 A_j)] \quad , \quad BTU / h\,ft^2 \right]$$

where   $A_j$ = friction area of one rotor side, $m^2$ ($ft^2$)

$\quad A_{wc}$ = wheel cylinder area, $cm^2$ ($in.^2$)

$\quad p_\ell$ = brake line pressure, $N/cm^2$ (psi)

$\quad r_i$ = distance of nodal point i from center of rotor, m (ft)

$\quad R$ = tire radius, m (ft)

$\quad s$ = tire slip

$\quad V$ = vehicle speed, m/s (ft/s)

$\quad \gamma$ = heat distribution to the rotor computed from Eq. (3-18) or (3-21)

$\quad \eta$ = wheel cylinder efficiency

$\quad \mu_L$ = pad/rotor friction coefficient

The velocity in Eq. (3-55) may be determined by Eq. (3-5) with the deceleration either specified or computed from vehicle brake system parameters as discussed in Chapter 5 or 6.

For the two-dimensional solid brake rotor illustrated in Figure 3-10 and an initial temperature of 297 K (75°F), a vehicle speed of 97 km/h (60 mph) and a 324 mm (12.75 in.) outer diameter solid disc, the temperature response as a function of rotor radius r and 0.5 and 5 seconds after brake application is shown in Figure 3-11. Note that the inner radius of the swept surface is approximately 102 mm (4 in.) from the center of the rotor. Inspection of the temperature curves reveals that, initially, the brake temperature is not, near the end of braking, only a slight function of the rotor radius. Underlying in the temperature calculations for this two-dimensional problem was a non-uniform

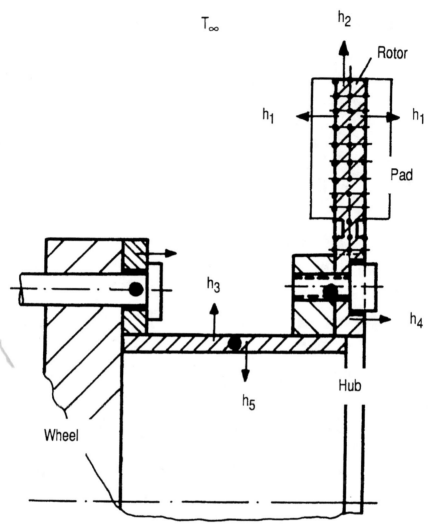

*Figure 3-10. Physical system for two-dimensional finite-difference analysis.*

pad pressure distribution with higher pressures at the inner and lower at the outer radius. This non-uniform pressure distribution between pad and rotor frequently results after burnishing or use of brakes. However, as discussed in Chapter 2, the particular design of the caliper may have a specific effect on pad pressure distribution and, hence, temperature distribution.

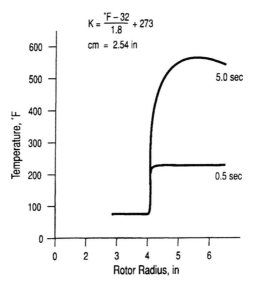

*Figure 3-11. Theoretical radial temperature distribution attained in a stop from 60 mph ($T_i = 75$, $P = non\text{-}uniform$).*

## 3.2 Thermal Stress Analysis

Thermal stresses result from non-uniform temperature distributions. In addition, mechanical stresses may arise from body deformations or body forces. In most practical thermal stress problems, it is permissible to separate the temperature problem from the stress problem and to solve both consecutively.

### 3.2.1 Thermal Stress in Disc Brake Rotors

The approximate compressive stress $\sigma$ developed in the surface layer of an infinite flat plate as a result of a sudden temperature increase is (Ref. 15)

$$\sigma = -\left(\frac{E}{1-\nu}\right)\alpha_T\Delta T \quad , \quad N/m^2 \text{ (psi)} \tag{3-56}$$

where   E = elastic modulus, $N/m^2$ (psi)

   $\Delta T$ = temperature difference, K (°F)

151

$\alpha_T$ = thermal expansion coefficient, m/K m (in./°F in.)

$v$ = Poisson's ratio

The solid rotor of a disc brake can be treated as a thin flat plate, as illustrated in Figure 3-12. The thermal stress in the rotor can be analyzed based on the following limitations:

1. Surface traction is negligible.

2. Body forces are negligible.

3. Temperature is a function of thickness z and time t only.

4. Temperature distribution is symmetrical.

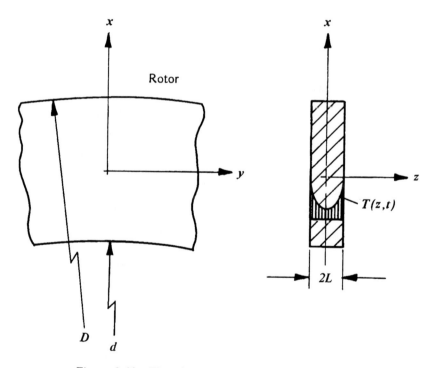

*Figure 3-12. Flat plate representation of brake rotor.*

The stress analysis of a free plate yields the expression for computing the thermal stresses in the rotor

$$\sigma_x = \sigma_y = \frac{\alpha_T E}{1 - v}\left[-T(z) + \frac{1}{L}\int_0^L T(z)dz\right] \quad , \quad \text{N/m}^2 \text{ (psi)} \quad (3\text{-}57)$$

where   $L$ = one-half rotor thickness, m (ft)

$\quad$ $T(z)$ = temperature distribution over z, K (°F)

$\quad$ $\sigma_x$ = stress in x-direction, N/m$^2$ (psi)

$\quad$ $\sigma_y$ = stress in y-direction, N/m$^2$ (psi)

The thermal stresses $\sigma(z, t)$ produced by a linearly decreasing heat flux are determined from the temperature response given by Eq. (3-29) for a solid rotor and Eq. (3-57) as

$$\sigma(z, t) = \frac{q''_{(0)}}{q''_0}\sigma_0(z, t) + \frac{2q''_{(0)}\alpha_T E}{t_s(1 - v)h_R}$$

$$\times \sum_{n=1}^{\infty}\left\{\frac{\sin(\lambda_n L)}{\lambda_n L + \sin(\lambda_n L)\cos(\lambda_n L)} \times \frac{1 - e^{-a_t\lambda_n^2 t}}{a_t\lambda_n^2}\right.$$

$$\times \left.\left[\frac{\sin(\lambda_n L)}{(\lambda_n L)} - \cos(\lambda_n z)\right]\right\} \quad , \quad \text{N/m}^2 \text{ (psi)} \quad (3\text{-}58)$$

where   $L$ = one-half rotor thickness, m (ft)

$\quad$ $t_s$ = stopping time, h

$\quad$ $\sigma_0(z, t)$ = stress produced by a constant heat flux, N/m$^2$ (psi)

The stress $\sigma_0(z, t)$ produced by a constant heat flux, required in Eq. (3-58), is computed from the constant heat flux temperature response and Eq. (3-57) as

$$\sigma_0(z,t) = \frac{2\alpha_T E q_0''}{(1-\nu)h_R} \sum_{n=1}^{\infty} \left\{ \frac{\sin(\lambda_n L)e^{-a_t \lambda_n^2 t}}{\lambda_n L + \sin(\lambda_n L)\cos(\lambda_n L)} \right.$$

$$\left. \times \left[ \frac{\sin(\lambda_n L)}{(\lambda_n L)} - \cos(\lambda_n z) \right] \right\} \quad , \quad \text{N/m}^2 \text{ (psi)} \qquad (3\text{-}59)$$

The theoretical thermal stresses at the surface of a solid rotor computed by Eq. (3-58) for stops from 97 and 128 km/h (60 and 80 mph) are illustrated in Figure 3-13. Inspection of the curves reveals a maximum near a braking time of 1 s.

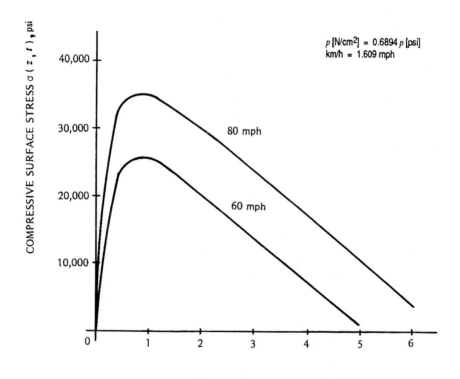

Figure 3-13. Thermal stresses at the surface attained in stops from 97 to 128 km/h (60 to 80 mph).

154

Since thermal shock and subsequent surface cracking are functions of the initial temperature gradient at the swept rotor surface, Eq. (3-58) also may be used to approximate the thermal stresses produced in a ventilated rotor. In this case L equals the flange thickness.

## 3.2.2 Thermal Stresses in Brake Drums

The detailed equations for predicting thermal stresses in brake drums are very complicated. Approximate equations may be given for drums with the ratio of drum width to drum radius much less than unity. However, this approach would exclude a large number of brakes, especially those for heavy vehicles.

A rough estimate of the thermal stresses produced in a single stop may be obtained from Eq. (3-56).

## 3.2.3 Thermal Rotor Failure

Normal braking operation will cause fine, hairline cracks laterally across the swept surface of brake drums. This is a normal condition caused by constant low-energy heating and cooling of the braking surface.

Excessive heating and cooling may cause the drum to crack all the way through the entire drum wall from the swept surface to the cooling surface.

Heat-spotted drums show hard, slightly raised, dark-colored spots on the braking surface with uneven wear. Heat (or hot) spots are caused by localized high-temperature-caused material changes of the structure.

Bluish, discolored swept braking surfaces are caused by excessive heat exceeding 533 to 588 K (500 to 600°F), possibly by constant low shoe drag or by overheating in severe fade.

Rotor surface cracking will occur when the stress exceeds the strength of the material. The occurrence of surface rupture is affected by thermal stress, number and frequency of braking cycles, surface condition due to machining, corrosion, and rotor geometry.

For a given rotor material having a certain strength, the tendency of the surface to rupture will be decreased if for a given heat flux the temperature gradient at the surface is small, the cooling is not extremely rapid, the thermal

155

expansion coefficient and the elastic modulus are decreased, and the rotor is designed so that thermal expansion is maximally unconstrained. Rotor deformation due to hub mouth widening should be minimized. The temperature gradient, again for a given heat flux, depends during the first few brake applications on the thermal properties of the rotor material.

For example, a high thermal conductivity results in a less marked temperature gradient. The rate of change of temperature at a given location in the rotor will be less pronounced for increased values of specific heat and material density. The rotor thickness has a twofold effect on the stress state: A thicker rotor produces higher temperature gradients and tends to be more rigid, thus producing more marked constraints on free thermal expansion. Consequently, thicker disc brake rotors have lower allowable heat flux values than thinner rotors (Eq. [3-61]).

In single brake applications wherein the rotor surface was not subjected to excessive temperature, surface cracks (if they occurred) developed generally in a radial direction. Subsequent braking cycles with the brake operating below certain maximum temperatures produced a stress pattern that was mainly a further development of the original cracks caused by thermal shock.

The other failure mode appears to be associated with severe surface temperatures, resulting in partial surface melting and dislocation of particles at the surface. Those conditions might produce hot spots, resulting in randomly oriented localized stress patterns. Stress patterns of this type may also be caused by localized interference of heat transfer, resulting from subsurface porosities. The surface porosities may be produced by material tearing or opening of subsurface porosities, and localized geometry changes due to temperature-induced microstructural changes of the base material.

Excessive heating alone generally is of no severe consequence to the rotor material. Maximum cooling rates may have significant effect on the performance of the rotor. This condition occurs when a vehicle is subjected to successive high effective stops and is then driven at high speeds to cool the brakes. With improper rotor materials martensite is formed causing surface thickness variations. Martensite occupies a larger volume than pearlite. The results are brake shudder and brake torque variations due to different friction coefficients between martensite, pearlite, and bainite. High-carbon-content cast iron up to 3 to 3.5% C generally yields favorable rotor performance.

## 3.3 Thermal Design Measures

Experience gained by engineers over the years in the design, testing, and use of automotive brakes has led to a number of design guides relative to the thermal performance of drum and disc brakes. These measures are generally based on commonly used materials for rotors and pads or linings. After brakes have been designed based on these guides, testing must always be done to ensure that the actual thermal performance falls within expected safety limits.

### 3.3.1 Allowable Heat Flux into Drum or Rotor

#### 3.3.1.a Brake Drums

In general, thermal surface cracking of the drum friction surface has not been observed when the heat flux into the swept area of the drum is kept below a certain value as defined by Eq. (3-60)

$$q_{D, \text{ allowable}} = q_0 \phi_i \ / \ 3600 A_s < 170 \quad , \quad \text{Nm/cm}^2\text{s} \quad (3\text{-}60)$$

$$[< 150 \quad , \quad \text{BTU/ft}^2\text{s}]$$

where $q''_{D,\text{allowable}}$ = allowable heat flux into drum, Nm/cm²s (BTU/ft²s)

$A_s$ = swept area of rotor or drum, cm² (ft²)

$q_0$ = braking power, Nm/h (BTU/h)

$\phi_i$ = brake force of ith brake divided by total brake force

In general, passenger-car drums showing crack widths of 0.7 mm (0.027 in.) and lengths of 50 mm (~2 in.) are still serviceable.

#### 3.3.1.b Disc Brake Rotors

In earlier discussions on thermal stress it was indicated that the thickness of the disc brake rotor has an influence on rotor failure. A rotor material will have improved thermal performance with higher thermal endurance strength, higher thermal conductivity, decreased elastic modulus, and decreased thermal expansion coefficient. Thicker rotors tend to exhibit higher potential for thermal stress and, hence, rotor failure. Disc brake rotors generally have sufficient thermal endurance when the heat flux is limited to a value computed by (Ref. 2)

$$q''_{R,\text{allowable}} = 28.8(439 - 0.46T) / L \quad , \quad Nm/cm^2s \qquad (3\text{-}61)$$

$$\left[ q''_{R,\text{allowable}} = (439 - 0.46T) / L \quad , \quad BTU / ft^2s \right]$$

where   T = brake temperature, °C (°F)

      L = rotor thickness, mm (in.)

The actual heat generation between pad and swept rotor surface for a disc brake is not uniformly distributed. It is a function of how uniformly the entire pad surface is pressing against the rotor surface. The actual swept rotor surface involved in heat generation is not identical to the geometrical swept surface. For fixed caliper disc brakes the swept surface to be used is only 70 to 75%, for floating caliper disc brakes only 50 to 65% of the geometrical swept surface.

For example, for an average brake temperature of 623 K or 350°C (662°F) and a rotor thickness of 22 mm (0.85 in.), Eq. (3-61) yields an allowable heat flux of 364 Nm/cm²s (327 BTU/ft²s). The actual heat flux produced during a stop is computed by Eq. (3-60), however, with the swept area adjusted by the correction factor indicated above. The velocity and deceleration used for computing the heat flux produced should be the maximum speed and deceleration attainable. If the heat flux produced exceeds the heat flux allowable, then the disc brake rotor will exhibit a limited endurance relative to thermal cracking. High-performance vehicles equipped with ABS should be able to withstand a minimum of 100 high effectiveness stops without rotor surface failure.

Since most vehicles in public use are not subjected to repeated severe brake applications, a deceleration level of 0.7 g may be used relative to Eq. (3-61) in connection with the maximum speed attainable. With a disc size based on Eq. (3-61), thermal surface performance should generally be unlimited for vehicles in normal use.

The theoretical temperature increase of the swept surface mass of the rotor should not exceed certain limits. The mass is computed by the product of swept surface area, rotor thickness in the case of a solid rotor, and rotor material density. For ventilated rotors three times the thickness of one rotor plate should be used as effective rotor thickness.

The braking energy is computed by Eq. (3-9), adjusted for one front (or rear) brake. The theoretical temperature increase is computed by an expression similar to Eq. (3-32), where density times rotor volume is equal to the swept surface rotor mass, and c is the specific heat of the rotor material.

For example, for one front brake on a level road, the theoretical temperature increase $T_{th,F}$ is:

$$T_{th,F} = \frac{(1 - \Phi)}{2} \left[ \frac{W\left(V_1^2 - V_2^2\right)}{2g\rho_R c_R v_R} \right] \quad , \quad K \ (\degree F) \qquad (3\text{-}62)$$

where $c_R$ = rotor specific heat, Nm/kgK (BTU/lb$_m$°F)

$\quad\quad g$ = gravitational constant, 9.81 m/s$^2$ (32.2 ft/s$^2$)

$\quad\quad V_1$ = initial velocity, m/s (ft/s)

$\quad\quad V_2$ = final velocity, m/s (ft/s)

$\quad\quad v_R$ = rotor volume, m$^3$ (ft$^3$)

$\quad\quad W$ = vehicle weight, N (lb)

$\quad\quad \rho_R$ = rotor density, kg/m$^3$ (lb$_m$/ft$^3$)

The percentage rear braking is designated by the Greek letter $\Phi$. See Section 7.4 for a detailed explanation.

$T_{th}$ is a theoretical temperature only since heat transfer to the brake pad and all other cooling mechanisms have been ignored. $T_{th}$, however, yields a good comparison measure for the evaluation of a brake system based on the temperature limits stated below (Ref. 2).

For $T_{th}$ less than 500 K (440°F) the rotor dimensions are sufficient for most passenger cars.

For $T_{th}$ between 500 and 600 K (440 and 630°F) high-performance passenger vehicles still have enough brake size.

$T_{th}$ greater than 600 K (630°F) must be avoided.

## 3.3.2 Horsepower into Lining or Pad

Brakes generally do not exhibit significant in-stop fade if the power $q_p''$ absorbed by the lining or pad is kept less than a certain value as defined by Eq. (3-63)

$$q_p'' = \lambda\phi_i P_{bav} / A_p < 400 \text{ Nm} / \text{cm}^2\text{s (drum)} \qquad (3\text{-}63)$$

$$< 2000 \text{ Nm/cm}^2\text{s (disc)}$$

$$\left[ q_p'' = \lambda\phi_i P_{bav} / 555A_p \begin{array}{l} < 500 \text{ hp} / \text{ft}^2\text{s (drum)} \\ < 2500 \text{ hp} / \text{ft}^2\text{s (disc)} \end{array} \right]$$

where $A_p$ = lining or pad rubbing area of leading or secondary shoe or brake pad, cm² (ft²)

$P_{bav}$ = braking power, Nm/s (lbft/s) from Eq. (3-8)

$\lambda$ = relative portion of braking power absorbed by an individual brake shoe

= 0.5 for two-leading shoe brake

= 0.7 for leading-trailing shoe or duo-servo brake

= 1.0 for disc brake

For a stop from maximum speed and fully laden vehicle and modern pad materials, total energy absorbed per one square inch of pad friction surface should be limited to values of 20,000 Nm/cm².

### 3.3.3 Lining or Pad Wear

Excessive wear of the linings or pads in normal automotive operation generally has not been observed if the product of the mean pressure $p_m$ between the lining and drum or disc and lining friction coefficient $m_L$ is kept below a certain value as defined by Eq. (3-64)

$$\mu_L p_m = \lambda Wa\phi_i(R \, / \, r) \, / \, A_{pp} < 65 \, N \, / \, cm^2 \; (95psi) \quad (drum) \qquad (3\text{-}64)$$

$$< 241 \, N/cm^2 \; (350 \, psi) \quad (disc)$$

where   a = deceleration, g-units

$A_{pp}$ = projected lining area of leading or secondary shoe

= $1.62ru$, $cm^2$ ($in.^2$)

r = effective drum or rotor radius, cm (in.)

R = effective tire radius, cm (in.)

u = effective width of brake drum swept area, cm (in.)

W = vehicle weight, N (lb)

$\mu_L$ = lining or pad friction coefficient

The maximum mechanical pressure for disc brake pads should be less than 1200 $N/cm^2$ (1750 psi).

### 3.3.4 Rotor Design Considerations

An important consideration in designing brake rotors is the expected thermal expansion and associated deformation of the entire rotor geometry. Minimizing thermal stresses results in increased thermal endurance strength and rotor life.

Thermal stress in the lateral or thickness direction of the rotor is mostly affected and controlled by material properties. Stress in the circumferential direction is a function of the deformation of the entire rotor including hub. Minimum circumferential stresses are obtained when the actual brake rotor ring is bolted to the hat section. In this design the ring can expand via the bolts without forcing its geometry change onto the hat or hub. The design is relatively expensive and not used for normal passenger cars.

For conventional ventilated rotors it is advantageous to make the rotor plates of different thickness. The outboard plate should be thicker than the inboard facing plate. This design reduces the cone-shape deformation or opening of the hub and rotor structure as well as minimizes surface cracking sensitivity.

If ventilation holes are used they must be located such that cooling air can circulate from the inboard side of the brake underneath the vehicle through the brake. In this regard wheel rim, hub design, and hub covers (if any) must be optimized for maximum brake cooling.

Thermal rotor stresses are minimized when stress raisers are eliminated as much as possible. For example, the number of cooling vanes of a ventilated rotor should always be arranged symmetrically with respect to the attachment bolts.

## 3.4  Test Results

In a brake comparison test conducted by a German auto magazine (Ref. 18), the braking performance of eight different passenger vehicles was evaluated.

The tests consisted of three maximum effectiveness stops with the vehicle loaded to GVW with cold brakes, one maximum effectiveness stop with hot brakes, and repeated braking during severe downhill travel.

The speeds in the effectiveness stops were 100 km/h (62 mph), 160 km/h (99.4 mph), and 80% of the maximum speed attainable. The downhill involved a vertical drop of 800 m (2625 feet), 26 brake applications from 90 km/h (55.9 mph) to 40 km/h (24.9 mph) with a cooling period of 10 to 20 seconds between braking.

Braking speeds, pedal force, deceleration and brake temperatures were measured in each run. The transmission was placed in neutral during each brake application.

The brakes were heated by making five repeated maximum braking effectiveness stops from 80% of the maximum attainable speed of the vehicle.

The table shows the results for five of the vehicles tested. The Porsche 968 demonstrates the best braking performance with very little decrease in deceleration between the cold 100 km/h stop and the "hot" effectiveness stop from 200 km/h (124 mph). The Lexus LS 400 required 244.5 m (802 feet) from 195 km/h (121 mph) or 87 m (285 feet) more than the Porsche. Expressed differently, when the Porsche is stopped, the Lexus still travels at a speed of approximately 116 km/h (72 mph). The new Opel Astra GT shows similar disappointing results at elevated temperatures.

The test report also noted that the pedal travel of the Lexus was near the floor after the third brake heating stop. The same German magazine also states that in connection with a long-time driving evaluation, the brake pads of the Lexus LS 400 had to be replaced on the front at 31,473 km (19,560 miles), at the front and rear at 46,645 km (28,990 miles), and again at 52,517 km (32,639 miles) with the front rotors turned.

The results clearly demonstrate that the design of a braking system for average conditions generally does not present a difficult engineering challenge. Only the use of brake components that are best suited for high thermal loads, such as fixed caliper disc brake designs which provide for a more uniform pressure distribution over the brake pad surface, will result in little or no fade during severe braking.

| | Porsche 968 | BMW 3181 | Mercedes 400 SE | Lexus LS 400 | Opel Astra GT |
|---|---|---|---|---|---|
| Max. Effectiveness Stop/ 100 km/h cold brakes | | | | | |
| Distance (m) | 37.5 | 39.0 | 39.4 | 42.9 | 44.5 |
| Decel. (m/s$^2$) | 9.6 | 9.9 | 9.8 | 9.0 | 8.7 |
| Max. Effectiveness Stop/ 160 km/h cold brakes | | | | | |
| Distance (m) | 97.8 | 107.4 | 107.4 | 109.7 | 135.3 |
| Decel. (m/s$^2$) | 10.1 | 9.2 | 9.2 | 9.0 | 7.3 |

*(continued)*

163

*(continued from previous page)*

| | Porsche 968 | BMW 3181 | Mercedes 400 SE | Lexus LS 400 | Opel Astra GT |
|---|---|---|---|---|---|
| **Max. Effectiveness Stop/ 80% of max. speed, cold brakes** | | | | | |
| Speed (km/h) | 200 | 160 | 200 | 195 | 145 |
| Distance (m) | 143.8 | 107.4 | 169.6 | 163.9 | 102.7 |
| Decel. (m/s²) | 10.2 | 9.2 | 9.1 | 8.5 | 7.9 |
| **Max. Effectiveness Stop/ 80% of max. speed, warm brakes** | | | | | |
| Speed (km/h) | 200 | 160 | 200 | 195 | 145 |
| Distance (m) | 157.5 | 101.8 | 175.6 | 244.5 | 105.3 |
| Decel. (m/s²) | 9.8 | 9.7 | 8.8 | 6.0 | 7.7 |
| **Downgrade, Front brake Temp., K** | 742 | 944 | 859 | 966 | 1015 |
| (°F) | (876) | (1240) | (1087) | (1279) | (1368) |

# 3.5 Design of Heavy Equipment Disc Brakes

The objective is the design check of a disc brake system for a heavy track vehicle to demonstrate the use of Eq. (3-42). The following vehicle data are specified:

1. Weight 66,000 lb

2. Maximum speed 45 mph

3. Maximum deceleration 0.6 g

4. Gear ratio between brake shaft and track drive sprocket 1:11

5. Track drive sprocket radius 16 in.

6. Tack rolling resistance coefficient 0.045 for operation on smooth dirt road, 0.075 for operation on off road surface

7. Mass moment of inertia per track 1000 lbin.s$^2$

The following disc brake data are specified:

1. Brake rotor weight 175 lb (ventilated rotor)

2. Outer brake rotor diameter 22.5 in.

3. Inner brake rotor diameter 12 in.

4. Maximum brake rotor revolutions, 4750 rpm at a vehicle speed of 45 mph

5. Brake rotor revolutions, 1825 rpm at 17 mph during continued downhill operations

6. Effective brake rotor radius 8.5 in.

7. Number of wheel cylinders per rotor 2

8. Wheel cylinder diameter 2 in.

9. Brake factor 0.50

10. Number of rotors 2

## 3.5.1 Mechanical Analysis

The maximum brake force $F_{x,total}$ may be obtained by

$$F_{x,total} = (66,000)(0.6) = 39,600 \text{ lb}$$

One track has to produce a braking force of 19,800 lb. Rolling resistance opposes vehicle motion. Under consideration of the rolling resistance, the braking force per track becomes

$$F_X = 19,800 - \frac{(0.045)(66,000)}{2} = 18,315 \text{ lb}$$

The brake torque $T_B$ at the rotor may be obtained by

$$T_B = F_x R \eta / \rho$$

where $F_x$ = braking force per track = 18,315 lb

$R$ = track drive sprocket radius = 16 in.

$\eta$ = mechanical efficiency = 0.95

$\rho$ = gear ratio = 1:11

$$T_B = \frac{(18315)(16)(0.95)}{(11)(12)} = 2109 \text{ lbft}$$

The kinetic energy $E_T$ produced in the 45 mph effectiveness stop by both rotors may be obtained by

$$E_T = \left(\frac{W}{2g}\right) V^2 + \left(\frac{I}{2 \times 12}\right) \omega^2$$

where $g$ = gravitational constant = 32.2 ft/s$^2$

$I$ = mass moment of inertia = 44.26 lbin.s$^2$

$V$ = velocity = 66 ft/s

$W$ = weight = 66,000 lb

$\omega$ = angular velocity = 4750/($\pi$30) = 497 rad/s

$$E_T = \frac{66,000}{(2)(32.2)}(66)^2 + \frac{(2)(44.26)}{(2)(12)}(497)^2$$

$$= 4,464,223 + 911,052 = 5,375,275 \text{ lbft}$$

In general, the equivalent mass moment of rotational inertia $I_{t,Rotor}$ at the brake rotor is obtained by

$$I_{TR} = I_R + \rho_t^2 I_d + \rho_t^2 \rho_d^2 I_e \ (\text{in.lbs}^2)$$

where  $I_d$ = drive shaft mass moment of inertia, in.lbs$^2$

$I_e$ = engine mass moment of inertia, in.lbs$^2$

$I_R$ = wheel and shaft mass moment of inertia, in.lbs$^2$

$\rho_d$ = final drive ratio

$\rho_t$ = transmission ratio

$$I_{t,Rotor} = I_R + \frac{1000}{(11)^2} = 36 + 8.26$$

$$= 44.26 \ \text{in.lbs}^2$$

The track mass moment of inertia of 1000 lb·in·s$^2$ and the mass moment of inertia of the brake rotor of 36 lb·in.·s$^2$, estimated from rotor weight and inner and outer diameter, are used.

The kinetic energy absorbed by the rotor is equal to the kinetic energy of the vehicle minus the work due to rolling resistance. The total rolling resistance work is equal to the product of rolling resistance and stopping distance, giving 335,610 lb ft when a stopping distance of 113 ft is used. Consequently, the kinetic energy absorbed by one brake rotor becomes 2,519,832 lb ft.

Use the value of brake torque above in Eq. (5-2) (solved for brake line pressure) to determine brake line pressure required for an effective stop without consideration of the pushout pressure as

$$p_1 - p_o = \frac{(2109)(12)}{(8.5)(0.5)(2)(3.14)(0.92)} = 1031 \ \text{psi}$$

The "2" in the denominator indicates that two separate wheel cylinders are used in the caliper, each having a wheel cylinder area of 3.14 in.$^2$ A brake factor of 0.50 and a wheel cylinder efficiency of 0.92 are used.

The pad friction area may be obtained from Eq. (3-63). Eq. (3-63) requires the use of Eq. (3-13). The tire slip is replaced by the track slip which is assumed to be zero. Eq. (3-13) yields a braking energy per rotor friction surface of 1,511,722 BTU/h. With $q_p'' = 2300$ hp / ft$^2$ and $\phi_i = 1$, the minimum pad area $A_p$ per rotor friction surface is 37 in.$^2$

The requirement of the secondary brake system is to hold the vehicle on an 80% slope on off-road surfaces. The track rolling resistance coefficient is approximately 0.075.

The braking force per track $F_x$ becomes

$$F_x = (W / 2)\sin\alpha - (W / 2)(0.075)$$

$$= (33,000)(0.625) - 2475 = 18,150\,\text{lb}$$

where  W = vehicle weight, lb

   $\alpha$ = incline angle, deg

The brake torque per rotor $T_B$ is

$$T_B = \frac{(18,150)(16)(0.95)}{(11)(12)} = 2090\,\text{lb ft}$$

The hydraulic pressure $p_l$ required for the production of this brake torque is

$$p_l = \frac{(2090)(12)}{(8.5)(0.60)(2)(3.14)(0.92)} = 851\,\text{psi}$$

For the secondary brake a slightly larger brake factor was assumed. The reason for this is the larger static pad-rotor friction coefficient as compared to the smaller sliding value. The secondary system uses the same wheel cylinder and brake pads for the brake force production. The actuation mechanisms are different from those of the service brake.

## 3.5.2 Thermal Analysis

The temperature response of the brake during a continued downhill brake operation must be determined for a vehicle speed of 17 mph, 10% slope, and travel distance of 6 miles. The thermal energy to be absorbed and dissipated by one brake rotor may be obtained by Eq. (3-15) as

$$q_o = \frac{(33,000)(24.9)(0.10 - 0.045)(3600)}{(778)}$$

$$= 209,122 \text{ BTU/h}$$

The time required for the continued braking process is 0.35 h or 1270.6 s.

The heat transfer coefficient of a ventilated rotor may be obtained by Eq. (3-42). The number of cooling vanes $n_v$ may be determined by the approximate relationship

$$n_v = \frac{4\pi D_o}{D_o - D_i}$$

where $D_i$ = inner rotor diameter, ft

$D_o$ = outer rotor diameter, ft

Substitution of the rotor data yields 27 vanes.

The hydraulic diameter $d_h$ is determined by the ratio of four times the cross-sectional flow area of one cooling passage divided by the wetted perimeter of one cooling passage (Fig. 3-7). By the use of a rotor width of 3.5 in., a flange thickness of 0.5 in., and a fin thickness of 0.5 in., $d_h$ is determined as

$$d_h = \frac{(4)(3.768)}{8.014} = 1.88 \text{ in.}$$

The hydraulic diameter is based on the vane dimensions existing at the average rotor diameter, i.e., 17.25 in. The cross-sectional area is determined from the product of vane width and vane circumferential dimension. For the example, the area is given by $(3.5 - 1.0) \times [17.25 \times \pi/(27) - 0.5] = 3.768$ in.$^2$ The wetted perimeter is determined from the sum of twice the vane width and twice the circumferential dimension, i.e., $(3.5 - 1.0) \times 2 + [17.25 \times \pi/(27) - 0.5] \times 2 = 8.014$ in.

The Reynolds number for Eq. (3-42) can be determined from the hydraulic diameter, the density and viscosity of the cooling air, and the average velocity of the cooling air through the vanes. The average velocity may be determined by Eq. (3-44). The inlet velocity $V_{in}$ is determined by the outer and inner rotor diameter, and the revolutions per minute of the rotor as (Eq. [3-44])

$$V_{in} = (0.052)(1825)[(1.875)^2 - (1)^2]^{1/2}$$

$$= 150.5 \text{ ft/s}$$

The outlet velocity $V_{out}$ is determined by the inlet velocity and the inlet and outlet areas. By the use of a ratio of inlet area to outlet area of 0.534, the outlet velocity is

$$V_{out} = (150.5)(0.534) = 80.38 \text{ ft / s}$$

The average velocity determined by Eq. (3-44) is 115.45 ft/s.

The convective heat transfer coefficient obtained by Eq. (3-42) is 24.9 BTU/h °F ft$^2$. The thermal properties of the air were evaluated at an assumed expected mean temperature of 500°F. The parameters used in Eq. (3-42) are $d_h$ = 1.88 in., $\ell$ = 5.25 in., Re = 48,069, Pr = 0.683, $k_a$ = 0.0231 BTU/h °F ft. The Reynolds number is computed for an air density of 0.0412 lbm/ft$^3$, air viscosity of $1.89 \times 10^{-5}$ lbm/ft s, hydraulic diameter of 0.191 ft, and an average velocity of 115.45 ft/s.

The rotor temperature may be obtained by Eq. (3-36). The rotor surface is 9.67 ft$^2$, the rotor volume is 0.385 ft$^3$, the rotor density is 455 lbm/ft$^3$, the rotor specific heat is 0.10 BTU/lbm °F, the ambient temperature is 50°F, the duration of the brake application is 0.35 h, and the initial temperature is 50°F. The rotor temperature T is determined as

$$T = \left[50 - \left(50 + \frac{209,122}{9.67 \times 24.9}\right)\right] \times \exp\left(-\frac{24.9 \times 9.67 \times 0.35}{455 \times 0.10 \times 0.385}\right) + 50 + \frac{209,122}{9.67 \times 24.9}$$

$$= 925°F$$

Inspection of Fig. 3-8 reveals a heat transfer coefficient due to radiation of approximately 4 BTU/h °F ft$^2$ at a rotor temperature of 925°F. By the use of a total heat transfer coefficient of 28.9 BTU/h °F ft$^2$ in Eq. (3-36), a rotor temperature of 801°F is determined at the end of the downhill brake application. In this analysis, it was assumed that the entire surface area of the rotor contributed to convective and radiative cooling. In order to accomplish the cooling of the swept areas of the rotor, cooling air must be blown against the rotor in addition to the self-ventilating effect of the rotor.

The rotor temperature attained in an effectiveness stop from 45 mph at 0.6 g deceleration may be obtained by Eq. (3-25) as

$$T = \frac{(0.52)(1,530,086)(3.42 / 3600)^{1/2}}{[(455)(0.10)(28)]^{1/2}} + 50°$$

$$= 737°F$$

A stopping time of 3.42 s is determined by dividing vehicle speed by vehicle deceleration. The thermal conductivity of the rotor = 28 BTU/h °F ft.

The maximum brake power produced at the outset of braking at one friction surface of one rotor for zero slip is obtained by Eq. (3-13) and Eq. (3-14)

$$q_{(o)} = \frac{(66,000)(66)(0.6)(3600)}{(4)(778)} = 3,023,445 \text{ BTU / h}$$

The braking power per one swept rotor area is obtained from

$$q_{(o)}'' = 3,023,445 / 1.976 = 1,530,086 \text{ BTU} / \text{hft}^2$$

where the swept area is determined by

$$A_s = (1.875^2 - 1^2)\frac{\pi}{4} = 1.976 \text{ ft}^2$$

If engine drag is considered in the braking analysis according to Eqs. (2-41) and (2-42), the braking power absorbed per rotor friction surface is decreased to approximately 2,437,233 BTU/h. An engine retarding moment $M_e = 546$ lb·ft is computed by Eq. (2-41) by the use of an engine displacement $V_e = 1000$ in.$^3$ and an average retarding pressure $p_m = 84$ psi. Eq. (2-42) yields a retarding force $F_{ret}$ at the track of 7678 lb due to engine drag. The retarding force due to the brakes and the engine for a deceleration of 0.6 g must be $(0.6) \times (66,000) = 39,600$ lb and, consequently, the brakes are required to produce only 31,922 lb. Hence, the engine retarding effect reduces the braking power absorbed by one rotor friction surface to approximately 2,437,233 BTU/h. Based on this reduced braking power a brake rotor temperature of approximately 604°F is determined by Eq. (3-25).

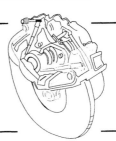

# CHAPTER 4

# Analysis of Mechanical Brake Systems

*In this chapter the basic mechanical brake system components are discussed. The physical expressions determining brake torque and vehicle deceleration are shown. Air spring brakes are discussed in Chapter 6.*

## 4.1  General Observations

Mechanical brakes use mechanical devices such as cables, rods, levers, cams, or wedges to transmit force to the wheel brakes. In current design practice, mechanical brakes are used for parking or emergency brakes. Their mechanical efficiency is low at approximately 65%. An efficiency of 65% indicates that 35% of the operator apply force is lost in terms of friction and is not available for vehicle braking. Mechanical brakes in poor condition, or designs with long, curved cable tubes, may have efficiencies below 65%. Frequent adjustment and lubrication are required for proper operation.

Time delays are relatively small, and are largely a function of the distance required to overcome shoe-to-drum clearance. The brake force build-up time, during which various load-carrying components deflect, is relatively short.

Air-actuated S-cam brakes use a number of mechanical devices between the brake chamber and the tip of the brake shoes, such as lever arm (slack adjuster), camshaft, cam, and rollers. Air-actuated wedge brakes use a wedge and rollers to apply the shoes against the drum. However, in current applications, full mechanical brakes are those systems in which the energy required for the brake shoe/drum pressing force is transmitted from the energy source to the shoes by mechanical means.

## 4.2 Wheel Brakes

A typical mechanical brake for parking brake purposes is illustrated in Figure 4-1. The cable force $F_c$ moves the lever $\ell_2$ to the left such that both shoes are spread apart to apply the brake. The individual shoe tip forces are slightly different. However, if the average actuation force is used, then the mechanical gain $r_B$ between the cable force into the brake and the average actuation force is given by

$$\rho_B = 1 / 2[(\ell_2 / \ell_1\ell_5)(\ell_3 - \ell_1) + (\ell_3 / \ell_1\ell_5)(\ell_2 - \ell_1)] \qquad (4\text{-}1)$$

where $\ell_1$ through $\ell_5$ are brake dimensions identified in Fig. 4-1.

For a typical drum parking brake the mechanical gain ranges between 2.75 and 3.

The total brake force $F_x$ at the two wheels braked by the mechanical system may be computed by

$$F_x = 2BF(F_H\rho_H\eta_H - F_s)\rho_B\eta_B(r / R) \quad , \quad N \text{ (lb)} \qquad (4\text{-}2)$$

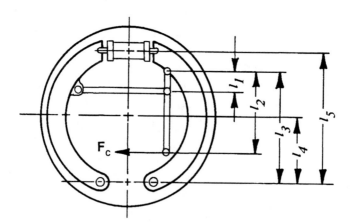

*Figure 4-1. Schematic of parking brake.*

where BF = brake factor

$F_H$ = hand or foot force, N (lb)

$F_s$ = return spring force, N (lb)

$\eta_H$ = mechanical efficiency of parking brake from hand force to cable force

$\eta_B$ = efficiency of actuation mechanism inside wheel brake

$\eta_B$ = gain of mechanical brake (Eq. [4-1])

$\eta_H$ = displacement gain between hand force and cable force

The displacement gain $\rho_H$ is equivalent to the pedal lever ratio of the foot-operated service brake. It is determined by the ratio of hand travel available (or foot travel in the case of a foot-operated parking brake) to the cable travel.

The average displacement $d_H$ of the tip of each brake shoe is determined by

$$d_H = Y_H \, / \, \rho_B \rho_H \tag{4-3}$$

where $Y_H$ = maximum available hand or foot travel for parking brake, mm (in.)

The displacement $d_H$ of each shoe tip associated with the parking brake should exceed the corresponding value obtained for the hydraulic service brake system by approximately 10 to 15%. This is to ensure adequate emergency braking in the event of excessive brake wear resulting in service brake loss due to increased pedal travel.

The deceleration a achievable with the parking brake is determined by

$$a = \frac{2BF(F_H\rho_H\eta_H - F_s)\rho_B\eta_B(r \, / \, R)}{W} \quad , \quad \text{g-units} \tag{4-4}$$

where BF = brake factor of wheel brake

r = drum radius, mm (in.)

175

$$R = \text{tire radius, mm (in.)}$$

$$W = \text{vehicle weight, N (lb)}$$

The slope angle $\alpha$ on which a vehicle can be held stationary for a given hand or foot force $F_H$ is indirectly expressed by Eq. (4-4) since $\sin\alpha = \text{deceleration } a$.

The coefficient of friction between tire and road surface must be equal to or greater than the values computed below to prevent wheel slipping.

Parking brake acting on front brakes:

$$\mu_F = \frac{S + f_{roll}}{(\ell_R / L) - (h / L)S} \tag{4-5}$$

Parking brake acting on rear brakes:

$$\mu_R = \frac{S + f_{roll}}{(\ell_F / L) + (h / L)S} \tag{4-6}$$

where $f_{roll}$ = tire rolling resistance coefficient

$h$ = center-of-gravity height above ground, mm (in.)

$L$ = wheelbase, mm (in.)

$\ell_F$ = distance from front axle to center of gravity, mm (in.)

$\ell_R$ = distance from rear axle to center of gravity, mm (in.)

$S$ = slope

## 4.3 Driveshaft-Mounted Brake

Current design practice does not use driveshaft-mounted brakes as parking brakes. In the past they were installed on medium and heavy trucks using either hydraulic or air brakes. Modern air-brake-equipped trucks use spring-actuated parking brakes. Four-wheel disc-brake-equipped medium trucks use disc integrated parking brakes.

The total braking force Fx at the wheels retarded by the driveshaft-mounted brake is determined by

$$F_x = \frac{F_H \rho_H \eta_m BF \rho_D r}{\eta_t R} \quad , \quad N \text{ (lb)} \tag{4-7}$$

where  BF = brake factor

$F_H$ = hand or foot force, N (lb)

r = drum or effective rotor radius, mm (in.)

$\eta_m$ = mechanical efficiency between hand force and cable into brake

$\eta_t$ = mechanical efficiency between vehicle transmission and driven wheels

$\rho_D$ = final drive or rear axle ratio

$\rho_H$ = displacement gain between hand force and cable force

The deceleration a achieved with the driveshaft-mounted parking brake is determined by

$$a = F_x / W \quad , \quad \text{g-units} \tag{4-8}$$

where  W = vehicle weight, N (lb)

Bureau of Motor Carrier Safety regulations require that the shaft brake stop the vehicle laden at GVW from 32.2 km/h (20 mph) in a distance of 25.9 m (85 ft) or less (average deceleration of 0.16 g). Under those conditions the theoretical brake temperature of the swept surface of the brake is approximately 477 K (400°F). If the braking speed were increased to a typical highway speed of 80.5 km/h (50 mph), then the parking brake temperature would exceed 1089 K (1500°F), clearly indicating that driveshaft-mounted parking brakes are not suitable as emergency brakes in the event the service brake has failed.

Eq. (4-7) may also be used to determine the slope on which the vehicle can be held stationary with the driveshaft-mounted parking brake for a given hand force by

$$\sin \alpha = F_x \, / \, W \qquad\qquad (4\text{-}9)$$

where  W = vehicle weight, N (lb)

$\alpha$ = slope angle, deg

Federal Motor Vehicle Safety Standard 105 requires that the vehicle when loaded at GVW remain stationary for five minutes on a 20% slope with a lever apply force of not more than 556 N (125 lb) when foot-applied, or 400 N (90 lb) when hand-applied.

The driveshaft torque $T_s$ developed by the weight of the vehicle parked on a slope is determined by

$$T_s = WR\eta_t \sin \alpha \, / \, \rho_D \quad , \quad \text{Nm (lb-in.)} \qquad\qquad (4\text{-}10)$$

For example, for W = 222,400 N (50,000 lb), a slope of 20% (11.3 deg), a tire radius of R = 0.533 m (21 in.), a rear axle ratio of 4.1 to 1, and an efficiency of $\eta_t$ = 0.90, the maximum driveshaft torque that the parking brake has to react against is obtained by Eq. (4-10) as

$$T_s = 222,400(0.533)(0.9) \sin(11.3) \, / \, 4.1 = 5098.7 \text{ Nm}$$

$$[T_s = 50,000(21)(0.9)\sin(11.3) \, / \, 4.1 = 45,163 \text{ lb - in.}]$$

The brake torque $T_B$ produced by the parking brake is determined by

$$T_B = F_H\rho_H\eta_H BFr \quad , \quad \text{Nm (lb-in.)} \qquad\qquad (4\text{-}11)$$

As long as the brake torque of the parking brake exceeds the torque developed by vehicle weight attempting to turn the driveshaft, the vehicle will remain stationary, assuming the tire-road friction coefficient is sufficiently large to prevent locked-wheel sliding (Eq. [4-6]).

178

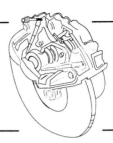

# CHAPTER 5

# Analysis of Hydraulic Brake Systems

*In this chapter the basic hydraulic brake system components are discussed. The physical expressions determining vehicle deceleration are shown. Brake booster performance is analyzed and design charts for a quick booster analysis are presented. Proportioning valves are discussed. Master cylinder sizing relationships for hydraulic brakes are shown. The use of brake fluid volume by different brake components is analyzed. The dynamic response behavior of hydraulic brake systems is discussed.*

## 5.1 Manual Hydraulic Brakes

Manual or standard brakes use only the pedal effort by the driver to press the shoes against the drum, or pads against the rotors. No additional energy source is used. Manual brakes are commonly used on small and lightweight vehicles.

Application of the pedal force displaces the foot pedal, which in turn presses the pushrod into the master cylinder. The pedal linkage is designed to produce a mechanical force advantage or gain between the pedal and the master cylinder piston, resulting in a master cylinder piston travel which is less than the foot pedal travel.

The cross-sectional area of the master cylinder and the cross-sectional areas of the wheel cylinders are chosen to produce an increase of force transmitted between the master cylinder and the wheel cylinders. (NOTE: The term wheel cylinder refers to both drum and disc brakes, except where noted otherwise.) This force increase or gain is accomplished by having the total wheel cylinder cross-sectional areas greater than the master cylinder cross-sectional area. Because the master cylinder piston travel is limited by the pedal ratio and pedal travel, the gain ratio between master cylinder and wheel cylinders is limited, too. To keep the pedal force below a certain maximum value of approximately 445 N (100 lb), brake boosters in the form of vacuum assists or pump-pressured hydro-boosts are installed.

The hydraulic brake line pressure $p_\ell$ produced by the pedal force $F_p$ is determined by

$$p_\ell = F_p \ell_p \eta_p / A_{mc} \quad , \quad \text{N/cm}^2 \text{ (psi)} \tag{5-1}$$

where $A_{mc}$ = master cylinder cross-sectional area, cm² (in.²)

$\quad F_p$ = pedal force, N (lb)

$\quad \ell_p$ = pedal lever ratio

$\quad \eta_p$ = pedal lever efficiency

A typical value for the pedal lever efficiency is 0.8, which includes the efficiency of the master cylinder including return springs.

The braking force $F_x$ per axle is obtained from the definition of the brake factor from Eq. (2-10) as

$$F_x = 2(p_\ell - p_o)A_{wc}\eta_c BF(r / R) \quad , \quad \text{N (lb)} \tag{5-2}$$

where $A_{wc}$ = wheel cylinder area, cm² (in.²)

$\quad$ BF = brake factor

$\quad p_o$ = pushout pressure, required to bring brake shoes or pads in contact with drum or disc, N/cm² (psi)

$\quad r$ = drum or effective disc radius, mm (in.)

$\quad R$ = tire radius, mm (in.)

$\quad \eta_c$ = wheel cylinder efficiency

Pushout pressures for disc brakes in good mechanical condition are small at 3.5 to 7 N/cm² (5 to 10 psi) and may be ignored in most cases. Floating caliper disc brakes with corroded slider surfaces may exhibit significantly larger pushout pressures. Pushout pressures for drum brakes are determined by the shoe return spring force and wheel cylinder area, and may assumes values as high as 70 to 172 N/cm² (100 to 250 psi). The wheel cylinder efficiency is approximately 0.96 for drum and 0.98 for disc brakes.

The wheels-unlocked deceleration a of the vehicle is determined from the summation of the braking forces of all axles or,

$$a = (2 / WR)[(A_{wc}BF\eta_c)_F(p_\ell - p_o)_F$$

$$+(A_{wc}BF\eta_c)_R(p_\ell - p_o)_R] \quad , \quad \text{g-units} \qquad (5\text{-}3)$$

where  W  =  vehicle weight, N (lb)

The subscripts F and R indicate that the wheel brake parameters $A_{wc}$, BF, and r must be evaluated for the front and rear brakes, respectively. If more than two axles are braked, then the appropriate terms are added on the right-hand side of Eq. (5-3).

For vehicles equipped with proportioning valves, the brake line pressures front and rear are not the same for pressures above the knee-point. See Eq. (5-11) for determining rear brake line pressures as a function of input or front brake line pressures.

## 5.2  Boost System Analysis

### 5.2.1  Overview and Requirements

Brake boost systems allow the driver to decelerate heavy vehicles with pedal force levels and pedal travels well within the acceptable range of the average driver. They contribute significantly to braking safety and driver comfort. The boost assist or booster factor must be optimized relative to the vehicle involved.

The following performance requirements should be observed in the design of a brake boost system:

1. The brake booster must be sensitive enough so that the operator can modulate braking effectiveness when low pedal forces are involved (low friction surfaces). Less than 13 to 22 N (3 to 5 lb) pedal force should initiate boost assist.

2. Pedal force/deceleration feedback must be provided so that the operator can gage the severity of braking through the level of . pedal force feedback.

3. The booster response time should be less than 0.1 second to reach the saturation point in the event of a rapid brake application with pedal travel rates at 1 m/s (3 ft/s).

4. A smooth transition from boost to manual braking at the saturation point should be provided so that the operator will be able to continue to increase pedal force in an emergency.

5. Reliability should be high to minimize booster failure. Booster failure may contribute to operator confusion, including abandoning of brake application in an emergency. Some drivers think that the entire brake system has failed since the pedal feels hard without boost and the associated deceleration levels are lower than expected.

6. Small size and low weight to allow optimum design location within the engine compartment.

## 5.2.2 Vacuum-Assisted Brake Booster

Vacuum-assisted hydraulic brakes, also called power brakes, use a vacuum booster as illustrated in Figure 5-1 to assist the driver effort in pressing the shoes against the drums. The common system, sometimes called mastervac, is mounted directly against the fire wall opposite the driver's foot. It is mounted between the foot pedal and the master cylinder. The assist force, acting on the pushrod which actuates the master cylinder piston, is produced by the difference in pressure across the booster piston or diaphragm with the vacuum or low pressure on the master cylinder side, and the atmospheric or high pressure on the input side. The level of assist force for a given pedal force is controlled by the reaction disc shown in Figure 5-2. The rubber-like material

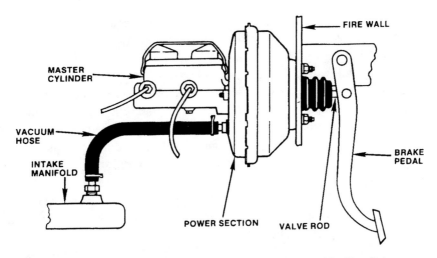

Figure 5-1. *Vacuum-booster master cylinder assembly (Bendix).*

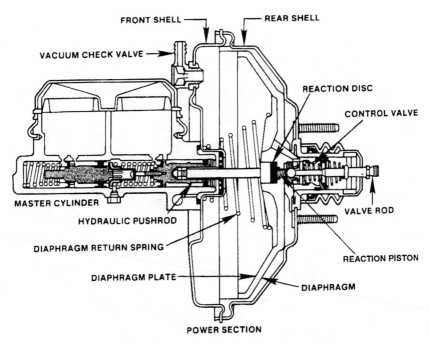

Figure 5-2. *Bendix single-diaphragm mastervac.*

of the reaction disc acts like a hydraulic fluid producing equal pressures against all surfaces it contacts. The result is a finely modulated atmospheric air pressure inlet valve with correspondingly modulated pushrod forces against the master cylinder piston. The vacuum developed in the intake manifold of gasoline engines is generally sufficient to fully actuate the booster. Diesel engines require a separate vacuum pump due to their insufficient manifold vacuum, caused by the absence of a throttle. Vacuum pumps are of the vane-, diaphragm-, or piston-type design. Vane-type vacuum pumps require engine oil lubrication to seal and produce proper vacuum.

Due to their limited amount of assist, both in terms of booster pressure and booster size, vacuum-assisted boosters generally are used for master cylinder volumes of only up to 24.6 cm³ $(1.5 \text{ in.}^3)$.

### 5.2.2.a  Mastervac Analysis

The boost ratio B is defined as the ratio of the pushrod force against the master cylinder piston to the pedal effort input into the booster,

$$B = (F_p \ell_p + F_A) / F_p \ell_p \qquad (5\text{-}4)$$

where  $F_A$  = booster force, N (lb)

   $Fp$  = pedal force, N (lb)

   $\ell_p$  = pedal lever ratio

Vacuum boosters increase the brake system gain by as much as eight or nine to one for most heavy domestic passenger cars, and approximately three to four for smaller cars. A gain of eight means that the effect of the pedal force is increased eightfold. Although this high gain permits maximum braking effectiveness with small pedal forces, in the event of a booster failure the driver will most likely not be able to produce sufficient pedal force to decelerate the vehicle at an acceptable level (Ref. 19).

The brake line pressure $p_\ell$ is determined by an expression similar to Eq. (5-1), however, modified by the boost ratio B as

$$p_\ell = F_p \ell_p \eta_p B / A_{mc} \quad , \quad N/cm^2 \text{ (psi)} \tag{5-5}$$

The boost ratio can be computed from the basic dimensions and spring forces associated with a basic mastervac as illustrated in Figure 5-2. The outer diameter of the reaction disc is $D_o$. The diameter of the reaction piston is $D_r$. The computations that follow are carried out for a single-diaphragm mastervac with 203 mm (8 in.) diameter assist piston. The diameter of the reaction disc and reaction piston are 30.7 and 18.5 mm (1.21 and 0.729 in.), respectively.

The pushrod force produced by the boost assist is computed first.

The effective booster area $A_B$ is equal to the booster area minus the pushrod area,

$$A_B = 20.3^2 \pi / 4 - 0.838^2 \pi / 4 = 323 \text{ cm}^2$$

$$[A_B = 8^2 \pi / 4 - 0.33^2 \pi / 4 = 50.18 \text{ in}^2]$$

where a pushrod diameter of 8.38 mm or 0.838 cm (0.33 in.) was assumed.

The booster force $F_B'$ for an effective vacuum of 7.928 N/cm² (11.47 psi; 80% of maximum) and a mechanical efficiency of 0.95 is

$$F_B' = 323(7.928)(0.95) = 2432.7 \text{ N}$$

$$[F_B' = 50.18(11.47)(0.95) = 546.9 \text{ lb}]$$

The effective booster force $F_B$ is smaller due to the diaphragm piston return spring force opposing the boost action. Hence,

$$F_B = 2432.7 - 155.7 = 2277 \text{ N}$$

$$[F_B = 546.9 - 35 = 511.9 \text{ lb}]$$

where a return spring force of 155.7 N (35 lb) was assumed. The computations thus far show that the booster portion produces a hydraulic pushrod force of 2277 N (511.9 lb).

The manually produced force against the hydraulic pushrod is computed next.

The rubber reaction disc acts similar to a pressurized hydraulic fluid. The pressure in the reaction disc $p_r$ is equal to the effective booster force divided by the difference in cross-sectional area of the reaction disc $A_2$ and the reaction piston $A_1$,

$$p_r = \frac{2277(4)}{(3.07^2 - 1.85^2)\pi} = 483 \text{ N / cm}^2$$

$$\left[ p_r = \frac{510.2(4)}{(1.21^2 - 0.729^2)\pi} = 696.5 \text{ psi} \right]$$

The control pressure $p_r$ is acting against any surface in contact with the reaction disc. Since the reaction piston is pushing against a portion of the reaction disc, the reaction piston force $F_r$ is equal to the reaction pressure multiplied by the reaction piston area $A_1$, hence

$$F_r = p_r A_1 = 483(1.85)^2(\pi / 4) = 1298 \text{ N}$$

$$[F_r = p_r A_1 = 696.5(0.729)^2(\pi / 4) = 290.7 \text{ lb}]$$

The reaction piston force is opposed by the reaction piston return spring force. For a 203 mm (8 in.) diameter vacuum booster the return spring force is approximately 66.7 N (15 lb). Consequently, the effort against the pushrod piston of the master cylinder produced by the foot pedal is $1298 + 66.1 = 1364$ N ($290.7 + 15 = 305.7$ lb).

The total force on the master cylinder piston and, hence, the brake line pressure producing force is equal to the sum of the effective booster force and reaction piston force, or $2277 + 1298 = 3575$ N ($510.2 + 290.7 = 800.9$ lb).

Finally, the vacuum boost ratio B is given by the ratio of pushrod force on the master cylinder piston to the reaction piston force

$$B = 3575 / 1298 = 2.75$$

$$[B = 800.9 / 290.7 = 2.75]$$

It is interesting to note that the booster factor B is also equal to the ratio of reaction disc area $A_2$ to reaction piston area $A_1$, or

$$B = A_2 / A_1 = 3.07^2 / 1.85^2 = 2.75$$

$$[B = A_2 / A_1 = 1.212 / 0.729^2 = 2.75]$$

The theoretical results may be used to construct a diagram illustrating the booster performance. In Figure 5-3, the pushrod force on the master cylinder piston versus the pedal force multiplied by the pedal lever ratio is shown. As can be seen, the booster has a maximum assist of approximately 3561 N (800.9 lb). For decelerations requiring higher brake line pressures and, hence, pushrod forces, the additional work input into the vacuum booster must come from the pedal effort, i.e., the driver. As discussed in Section 1.3.2, the booster saturation point should not be reached for decelerations less than 0.9 to 1 g.

Also shown in Fig. 5-3 are the different booster output forces as a function of different levels of vacuum.

The vacuum booster analysis presented for a given booster size may be expanded to a general analysis relating the various parameters in graphical form as shown in Figure 5-4. The use of the chart is as follows for a vehicle with the values that follow.

1. Given

   a. Pedal force $F_p$ = 289 N (65 lb)

   b. Pedal travel Y = 127 mm (5.0 in.)

   c. Brake line pressure $p_\ell$ = 896 N/cm$^2$ (1300 psi)

   d. Master cylinder volume $V_{mc}$ = 11.5 cm$^3$ (0.7 in.$^3$)

2. Find

    a. Booster work $F_A X$

    b. Boost ratio B

    c. Booster diameter

    d. Relative vacuum

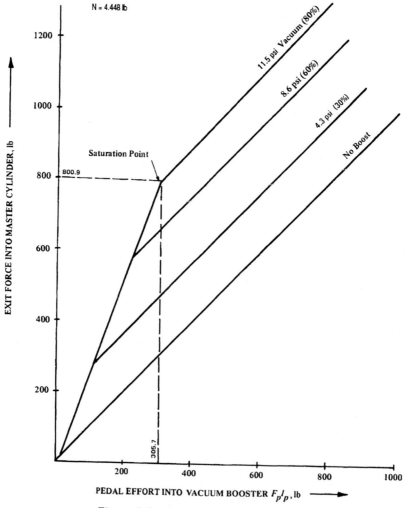

Figure 5-3. Mastervac characteristics.

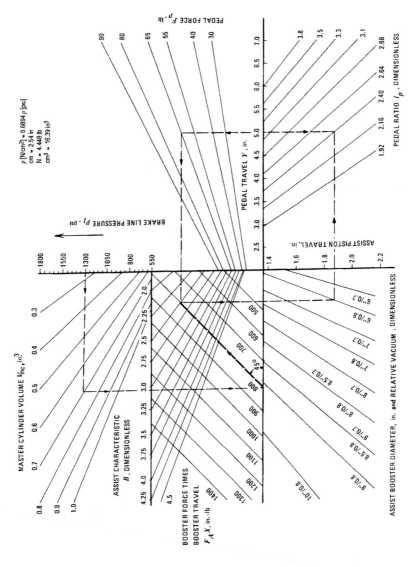

*Figure 5-4. Vacuum booster design chart.*

e. Booster piston travel

f. Pedal lever ratio $\ell_p$

The solution is illustrated by broken lines in the chart (Fig. 5-4).

3. Solution

a. Booster work $F_A X$:

(1) Draw a horizontal line from the brake line pressure $p_\ell = 896\,\text{N/cm}^2$ (1300 psi) to the line representing $V_{mc} = 11.47\,\text{cm}^3$ (0.7 in.³).

(2) From the point of intersection of the horizontal line with the line representing $V_{mc} = 11.47\,\text{cm}^3$ (0.7 in.³) drop a vertical line to the second horizontal line on the chart.

(3) The intersection of the vertical line with the second horizontal line gives the booster work $F_A X$, which, in this case, is 90.4 Nm (800 lb-in.).

b. Boost ratio B:

(1) Draw a vertical line from the pedal travel Y = 127 mm (5.0 in.) to the line representing pedal force $F_p = 289\,\text{N}$ (65 lb).

(2) From the intersection of the vertical line with $F_p = 289\,\text{N}$ (65 lb), draw a horizontal line to the left.

(3) From the point representing booster work $F_A X = 90.4\,\text{Nm}$ (800 lb-in.) draw a line extending upward at an angle of 45 deg.

(4) The intersection of the horizontal line with the one drawn at 45 deg gives a boost ratio of B = 2.5

c. Booster diameter and relative vacuum:

(1) Drop a vertical line from the point established in b(4).

(2) The intersection of this vertical line with one of the booster lines gives acceptable values of booster diameter and relative vacuum. In this case let the vertical line intersect the

line representing booster diameter and relative vacuum of 152.4 mm/0.8 (6 in./0.8).

d. Pedal ratio $\ell_p$:

  (1) Drop a vertical line from pedal travel Y = 127 mm (5.0 in.).

  (2) Draw a horizontal line through the point established in c(2).

  (3) The intersection of the vertical and horizontal lines gives the pedal ratio, in this case $\ell_p$ = 2.4.

e. Booster piston travel:

  (1) The intersection of the horizontal line established in d(2) with the vertical axis determines the booster piston travel, in this case approximately 47.5 mm (1.87 in.).

If a different booster diameter and/or relative vacuum is chosen, then the pedal travel and booster piston travel change accordingly. For example, with a booster diameter of 178 mm (7 in.) and a relative vacuum of 0.7, the pedal ratio becomes 2.88 and the booster piston travel 39 mm (1.54 in.). If desired, a pedal ratio may be selected rather than the booster diameter and/or relative vacuum. The choice of booster diameter or pedal ratio is a function of the space available for the installation of the booster or foot pedal.

### 5.2.2.b  Hydrovac Analysis

In the mastervac the amount of assist force is controlled by a rubber-like reaction disc. In the hydrovac the application is controlled by the hydraulic fluid pressure produced by the operator. The hydrovac unit can be located anywhere on the vehicle, with the frame rails near the driver's cab being the preferred location. The design of a typical hydrovac is shown in Figure 5-5, with the brakes in the applied position. The hydrovac consists of the vacuum cylinder (1) with piston (2), return spring (3), and pushrod (4). The control pipe (5) connects the left chamber of the vacuum cylinder with the lower chamber of the membrane (6) of the vacuum valve, while the right chamber of the vacuum cylinder is connected to the vacuum inlet (7) leading to the engine manifold. The right side of the vacuum cylinder is also connected to the upper side of the membrane (8). The hydraulic cylinder consists of the cylinder (9), the piston (10) equipped with a check valve (11), and the pushrod (4).

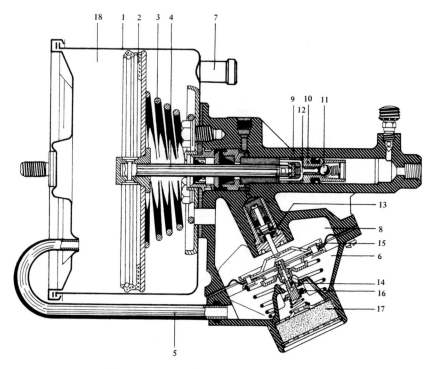

*Figure 5-5. Hydrovac in "on" position.*

In the "off" position, piston (2) is held at the left side of the vacuum cylinder by the return spring (3). In this position, the arm (12) of the piston (10) rests against the back plate, and the ball of the check valve (11) is lifted off the seat. The control piston (13) is located at its uppermost position, thus separating the control valve (14) from the seat of the membrane.

During application of the driver-operated master cylinder located at the fire wall, the line pressure is transmitted through the check valve into the brake system and to the wheel brakes. At the same time the hydraulic pressure in front of the control piston (13) begins to rise, moving the piston and membrane downward until the membrane contacts the control valve (14). At this moment the two chambers to the left and right of the vacuum piston (2) are separated. Any further motion of the membrane (15) downward will open the ambient valve (16).

The atmospheric air flows past the air filter (17) through the ambient valve (16) into the valve chamber (6) and through the control pipe (5) into the cylinder chamber (18), resulting in a rightward motion of the piston (2), pushrod (4), and piston (10). The check valve (11) will be closed as a result of the movement of the piston (10) to the right allowing the line pressure to increase and to be transmitted to the wheel brakes. The vacuum difference across piston (2) is identical to the pressure difference across the membrane (15). The position of membrane (15) is determined by the pressure in the pedal-master cylinder and the pressure differential across the membrane. Any change in pedal force will cause a corresponding change in vacuum application and, hence, pressure differential across piston (2), allowing a sensitive control of the brake application.

### 5.2.3 Hydraulic Brake Booster

In the hydraulic boost system the energy source is pressurized fluid. In most cases the steering pump is used. The brake system remains totally conventional with only the booster and plumbing added. Since two incompatible fluids are used in the two different circuits, extreme care must be taken not to contaminate one circuit with the fluid of the other. If it does occur, all seals must be replaced. Its compact size and high-pressure potential allow it to be used in virtually all applications from passenger cars to light- to medium-weight trucks. Although certain details vary between manufacturers, the hydraulic booster without accumulator is limited to master cylinder volumes of 33 to 41 cm$^3$ (2 to 2.5 in.$^3$). A schematic of a hydraulic boost system is shown in Figure 5-6. The pressure line runs from the steering pump to the brake booster, and from there to the steering gear and back to the reservoir. A spool valve in the brake booster controls the fluid flow from the steering pump. Without any brake application the fluid flow is not affected. During braking the fluid flow is restricted resulting in a corresponding pressure rise in the fluid and pressure application to the booster piston. The spool valve is designed so that the brake and steering operations do not interfere during either apply or release operation.

A reserve pressure accumulator is provided which allows two to three brake applications with the pump failed or engine stalled. In the '70s, a spring-loaded accumulator was used, either integral with the booster, or separately mounted in the engine compartment. Later a gas-charged accumulator was used for energy storage in the event of a pump failure.

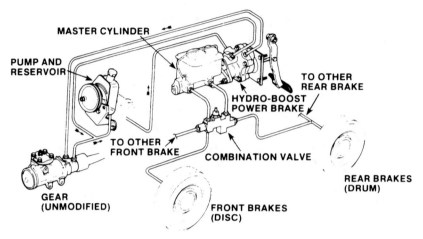

MASTER CYLINDER

PUMP AND
RESERVOIR

TO OTHER
REAR BRAKE

HYDRO-BOOST
POWER BRAKE

TO OTHER
FRONT BRAKE

COMBINATION VALVE

REAR BRAKES
(DRUM)

GEAR
(UNMODIFIED)

FRONT BRAKES
(DISC)

*Figure 5-6. Hydro-boost system (Bendix).*

For medium- to heavy-vehicle applications an electrical pump is used as a reserve energy source. In the event that the normal fluid flow from the steering pump is interrupted, the integral flow switch inside the booster closes, which energizes a power relay and provides electric power to the pump. The reserve pump then circulates the fluid throughout the system and builds up pressure as demanded. Master cylinder volumes up to 107 to 115 cm$^3$ (6.5 to 7 in.$^3$) are accommodated by the electric reserve pump design booster.

Modifications to the basic hydraulic booster system have been introduced where the steering pump charges a gas-charged accumulator, which, in turn, pressurizes brake fluid. As the driver applies pedal force, the regulated brake fluid pressure is transmitted to the wheel brakes. The advantages of this system include increased reserve capacity in the event of a pump failure, quicker response time since the brake line pressure does not have to be built up from zero, and sufficient energy source for ABS application.

The size of the accumulator is a function of the vehicle weight and the number of stops required by one accumulator charge. The schematic of a hydraulic booster is shown in Figure 5-7. The pressure $p_B$ supplied by the accumulator to the booster in addition to the pedal effort by the driver acts on the master cylinder piston, which, in turn, produces the hydraulic brake line pressure to the wheel brakes.

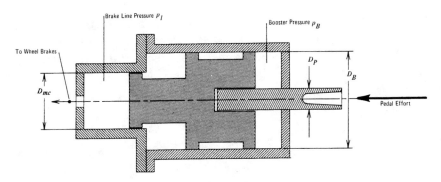

*Figure 5-7. Schematic of hydraulic booster.*

The effective input force to the booster is determined by the booster area and the pushrod cross-sectional area.

The booster input area ratio $A_R$ is given by

$$A_R = (D_B / D_p)^2 \qquad (5\text{-}6)$$

where $D_B$ = booster piston diameter, cm (in.)

$D_p$ = pushrod diameter, cm (in.)

The booster pressure ratio $p_R$ is defined by the ratio of output pressure to input pressure and may be expressed in terms of the diameters as

$$p_R = (D_B / D_{mc})^2 \qquad (5\text{-}7)$$

where $D_{mc}$ = output to master cylinder diameter, cm (in.)

The brake line pressure may be determined for a given booster pressure (or accumulator pressure) once the booster pressure ratio has been computed. With the brake line pressure determined, vehicle deceleration is computed by Eq. (5-3).

The boost circuit fluid volume, i.e., size and operating pressure range of the accumulator, are a function of the maximum accumulator pressure $p_A$ and the initial gas charge pressure $p_G$ of the gas used for energy storage by the accumulator. The volume ratio $V_R$ of the booster is defined by the ratio of the volume displaced at the booster side to the volume displaced at the output side and is determined by

$$V_R = (D_B^2 - D_P^2) / D_{mc}^2 \qquad\qquad (5\text{-}8)$$

A typical booster characteristic is shown in Figure 5-8, indicating both boosted and no-boost performance.

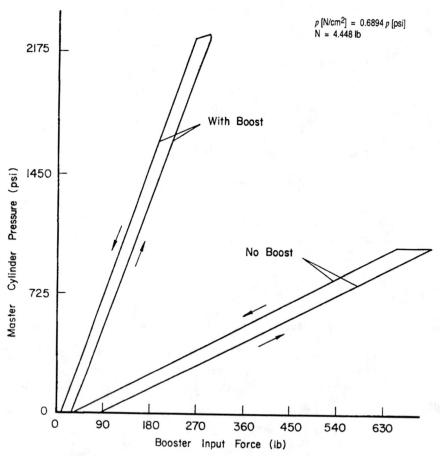

Figure 5-8. *Master cylinder pressure vs. booster input force.*

The minimum size of the accumulator required for a safe deceleration of a vehicle in successive stops may be obtained from the accumulator design chart shown in Figure 5-9.

It is assumed for the preparation of the accumulator design chart that approximately 67% of the master cylinder volume is required for an emergency stop. The example illustrated in Fig. 5-9 indicates that a vehicle having a master cylinder volume $V_{mc} = 49$ cm$^3$ (3.0 in.$^3$), five emergency stops, a volume ratio $V_R = 2.4$ computed by Eq. (5-8), and a pressure ratio $p_G/p_A = 0.35$ requires an accumulator size of approximately 623 cm$^3$ (38 in.$^3$).

If the same energy had to be stored by a vacuum-assist unit, a volume approximately 40–50 times larger than that associated with a medium-pressure accumulator, or 100–130 times larger than that associated with a high-pressure accumulator would be required.

The energy stored in the accumulator is affected by the ambient temperature. The fluid volume available for braking at high pressures decreases with decreasing temperature. For example, an accumulator having a volume of 656 cm$^3$ (40 in.$^3$) available between the pressure range of 1448 to 1792 N/cm$^2$ (2100 to 2600 psi) when operating at 353 K (176°F), provides only 246 cm$^3$ (15 in.$^3$) when the temperature is 233 K (−40°F) (Ref. 20).

For vehicles that do not have a steering pump, and to avoid the potential problems associated with the use of two different fluids, an integral accumulator/pump system has been designed. It is operated with brake fluid only. Because brake fluid does not provide adequate lubrication for long periods of operation, an intermittently operating electrical pump is used to charge the accumulator only when needed. The compact design is advantageous for installation in the crowded engine compartment. The basic design has been expanded for application to ABS braking systems.

## 5.2.4 Full-Power Hydraulic Brakes

All hydraulic brake systems discussed in earlier sections had the ability for the driver to manually apply the brakes in the event of a power or energy source failure.

In the full-power hydraulic brake system the pedal effort is only used to modulate and control the amount of assist demanded. The pedal force itself is not

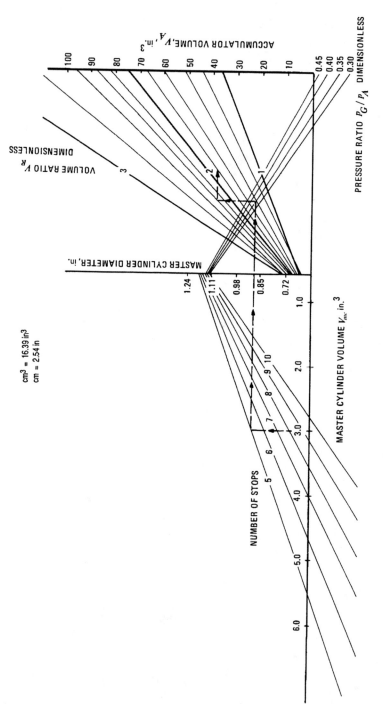

*Figure 5-9. Accumulator design chart.*

involved in pressing the brake shoes against the drum. Because no manual backup system exists except for the parking brake, the hydraulic system has many components in dual fashion. The brake system consists of one or two pumps, two accumulators, a dual circuit control valve, the hydraulic lines and hoses, and the wheel brakes. Rolls Royce uses two pumps and accumulators for each individual circuit, and mineral oil as working fluid. Extreme care must be taken not to contaminate the mineral oil with regular brake fluid because mineral oil-resistant rubber components are made of neoprene and will be damaged if they come in contact with polyglycolether-based brake fluid (see Chapter 10). The pumps used are either vane or radial piston designs. Vane pumps are generally limited to a pressure of approximately 965 N/cm$^2$ (1400 psi); extreme pressure levels may go as high as 1379 N/cm$^2$ (2000 psi). Radial piston pumps may produce pressures up to 2068 N/cm$^2$ (3000 psi).

### 5.2.5 Comparison of Brake Boost Systems

A comparison of hydraulic and full-power boost systems with vacuum-assisted brakes indicates the latter to be the most economical power source, assuming that a sufficient quantity of vacuum is available. However, exhaust emission regulations and fuel injection systems have greatly reduced the degree of vacuum available as energy source. In addition, diesel engine-powered vehicles require an additional vacuum pump. Consequently, more and more future designs will require a hydraulic energy source consisting of a pump and, if necessary, a gas-loaded accumulator. For light- to medium-weight trucks the vacuum booster capacity is generally too small for an adequate brake boost system.

## 5.3 Brake Line Pressure Control Devices

It is common knowledge that the normal or weight forces increase on the front axle and decrease on the rear axle as deceleration increases. For a complete discussion on vehicle dynamics during braking see Chapter 7.

Since the rear axle becomes lighter as deceleration increases, relatively less brake line pressure is required on the rear brakes to keep the rear brakes from locking before the front brakes. In addition, differences in load distribution for the lightly and fully laden cases require different brake torque balance front to rear for the empty and laden vehicle.

There are two basic types of brake line pressure valves, namely the *brake pressure limiter* and *brake pressure reducer*. Each of these devices can either be activated by brake line pressure or by vehicle deceleration. In addition, the point at which the brake line pressure activates the limiter or reducer valve can be load- or suspension-height-sensitive.

### 5.3.1 Brake Line Pressure Limiter Valve

A typical pressure curve of a limiter valve is shown in Figure 5-10. Up to a brake line pressure of 310 N/cm² (450 psi) chosen in this example, both front and rear brakes receive equal pressures. For brake line pressures beyond

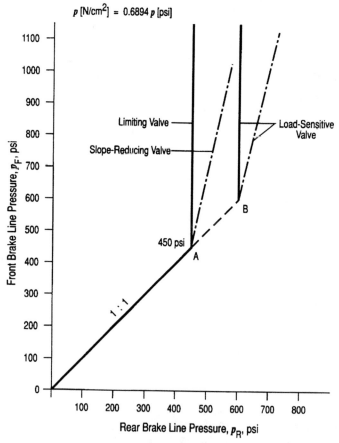

*Figure 5-10. Rear brake line pressure valves.*

$310 \ N/cm^2$ (450 psi), the rear brake line pressure remains at a constant $310 \ N/cm^2$ (450 psi), while the front brake line pressure increases. Consequently, for braking maneuvers with the front brakes locked, any further pedal force increase will not increase vehicle deceleration since the rear brake line pressure cannot be increased any further with the limiting valve.

A schematic of a limiter valve is shown in Figure 5-11. The master cylinder pressure enters the limiter valve at A. Spring force (5) holds valve (2) open and the master cylinder pressure can pass through the valve to the outlet B and to the rear brakes. As the pressure increases on top of the piston, the pressure force overcomes the spring force (5) and closes valve (2), at which point no further pressure increase occurs at outlet B, thus limiting the pressure in the rear circuit. If the brake fluid volume increases in the rear circuit due to lining wear or drum expansion, then the pressure at B decreases; valve (2) opens briefly to allow a slight pressure increase to the rear until valve (2) closes again.

When the pedal force is lowered or fully released, the pressure at B is greater than the pressure at A, and the valve seat (6) moves against the spring force (6), releasing the pressure, and valve (2) returns to its released position, permitting a free fluid flow from inlet A to outlet B.

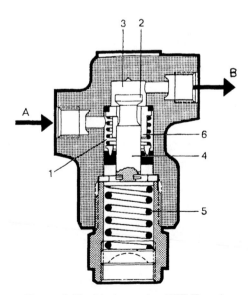

*Figure 5-11. Limiter valve (ITT-Teves).*

201

As more weight is placed in the vehicle, more braking force can be utilized by the rear brakes before locking up. This increase in rear brake line pressure is accomplished by load-sensitive or suspension-height-sensitive valves. For a given valve geometry the spring force (5) determines the switch point from the 1:1 slope to the limited condition. In load-sensitive valve designs the spring force is increased through cam and lever action as the suspension deflects as more weight is placed into the vehicle. The increased spring force moves the knee-point from point A in Fig. 5-10 to point B. For the example chosen, the rear brake line pressure is now limited to 414 N/cm² (600 psi), that is, 103 N/cm² (150 psi) more than for the lightly loaded case.

## 5.3.2 Brake Line Pressure Reducer Valves

With the reducer valve, also called the proportioning valve, the rear brake line pressure is identical to the front brake line pressure up to the knee-point pressure. For higher pressures the rear brake line pressure increases at lower amounts than that of the front brakes as illustrated in Fig. 5-10. The advantages of the reducer over the limiter valve include increasing rear brake forces after the front brakes have locked, and front and rear brake line pressures that are closer to the optimum values (Chapter 7).

A typical schematic of a reducer valve is shown in Figure 5-12. The master cylinder brake line pressure enters at A and reaches the rear brakes spaces (2, 3, and 5), and through outlet B, since the spring force (7) has pushed the differential piston (6) to the left, which opens valve (4). The difference in

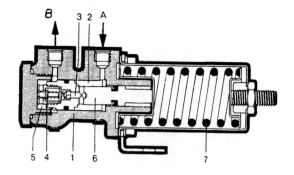

*Figure 5-12. Brake line pressure reducer valve (ITT-Teves).*

pressurized surface area of piston (6) acts against the spring force (7). When the switch point pressure is reached, the piston moves to the right until valve (4) closes. With valve (4) closed, no further pressure increase develops in space B until the increased pressure on the right side of piston (6) pushes the piston slightly to the left so that valve (4) opens again, allowing the brake line pressure to increase again at B. The piston oscillates back and forth, allowing the pressure to increase to the rear brakes in direct relationship to the ratio of the ring area to full area of piston (6). When the brakes are released, the higher pressure at B opens valve (4) against the small valve spring, and piston (6) returns to its released position.

Brake systems with front-to-rear dual hydraulic split generally have a proportioning valve bypass feature if the front circuit fails. Under these conditions the rear brakes will receive full master cylinder pressure at all levels of brake line pressure since the reducer function of the valve is locked out.

When the spring force pushing against the differential piston is made a function of the weight carried by the vehicle, the switch or knee-point can be moved to higher pressure levels. These variable knee-point valves are commonly known as load- or height-sensitive proportioning valves.

Height-sensitive proportioning valves (HSPV) were used on a number of cars, light trucks, and vans. Starting in the early '80s through the late '80s most domestic manufacturers used HSPVs on pickup trucks and vans. In many light trucks the HSPV was added to the normal combination valve system so that it would further reduce the rear brake line pressure and knee-point pressure when lightly loaded, but be ineffective when loaded. Many foreign manufacturers use HSPVs on smaller front-wheel-driven cars such as the VW Golf or Yugo. With the introduction of ABS brakes, height-sensitive brake pressure proportioning was removed from many vehicles.

The theoretical reducer knee-point pressure $p_K$ may be determined from a basic force analysis on the differential piston as

$$p_K = 4F_s / \pi D^2 (SL) \quad , \quad N/cm^2 \text{ (psi)} \tag{5-9}$$

where  D = large piston diameter, cm (in.)

$F_s$ = spring force, N (lb)

SL = reducer slope

The theoretical reducer slope SL is expressed by the area or diameter-squared ratios as

$$SL = 1 - (d / D)^2 \qquad (5\text{-}10)$$

where  d = small piston diameter, cm (in.)

D = large piston diameter, cm (in.)

It must be recognized that manufacturing tolerances and frictional factors will affect the performance of the reducer valve. Furthermore, the performance characteristic during pressure increase will differ somewhat from that achieved during pressure release.

With the reducer valve characteristic indicated in Fig. 5-10, for pressures above the knee pressure the rear brake line pressure $p_R$ is determined by

$$p_R = p_K + (p_{mc} - p_K)SL \quad , \quad N/cm^2 \text{ (psi)} \qquad (5\text{-}11)$$

where  $p_k$ = knee-point pressure, $N/cm^2$ (psi)

$p_{mc}$ = master cylinder pressure, $N/cm^2$ (psi)

SL = reducer slope

In some designs, proportioning valves involving accumulators and solenoids use the ABS signal from the rear brakes to adjust the rear brake line pressure more closely to the optimum values.

### 5.3.3 Combination Valves

In many cases, two or three different functions are combined into one valve, commonly called the combination valve.

A typical three-function combination valve is shown in Figure 5-13. The valve to the left is the metering valve, and the valve to the right the reducer or proportioning valve discussed earlier. The switch in the center is the differential pressure switch, which is activated in the event of a hydraulic leak in one of the dual brake circuits.

The metering valve is used to improve front-to-rear brake balance when braking on low friction surfaces, that is, at low brake line pressures. The metering valve prevents application of the front brakes for pressures up to approximately 51.7 to 93 N/cm² (75 to 135 psi). During this pressure increase, although the front brakes are not braking yet, the rear brake return spring forces are overcome so that rear brake shoes contact the drum at nearly the same time the front brake pads contact the rotor surface.

In some cases, a two-function combination valve is used by simply not installing either the metering or proportioning valve portion of the system.

Metering valves are used primarily for rear-wheel-driven vehicles, using the front disc/rear drum brake system. Front-wheel-driven vehicles in many cases do not use metering valves because the engine inertia turning on the front rotors requires brake line pressure in line with the rear brakes before brake lockup occurs on low friction surfaces.

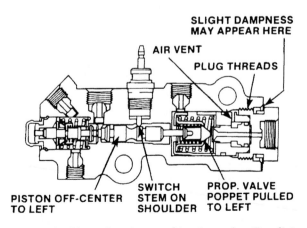

*Figure 5-13. Three-function combination valve (Bendix).*

## 5.3.4 Deceleration-Sensitive Reducer Valves

With this valve design the brake line pressure at which the rear brakes receive a reduced amount of brake line pressure is a function of vehicle deceleration. The switch point is not as clearly identified as in the case of the pressure-dependent valves discussed earlier. The pressure switch point is a function of the pedal force apply rate, the basic brake system characteristics, and the amount of any residual air present in the rear brake line circuit.

The basic designs involve a specific mass which moves due to braking inertia. In the Bendix design the mass rolls on steel balls against a preloaded spring. As soon as the spring force is overcome, the differential piston commences its pressure-reducing function. The corresponding deceleration level at the switch point is approximately 0.7 g. The inertia mass is "held" in place by a small spring/piston system to eliminate any "deceleration noise" prior to approximately 0.7 g.

In the Girling design, illustrated in the basic schematic of Figure 5-14, a steel ball travels on an incline according to the inertia force acting on the ball during braking when the switchover deceleration point has been reached. The upward movement of the ball closes off the opening to the rear circuit. Further brake line pressure increases from the master cylinder result in the reducing function of the differential piston.

A typical brake line pressure performance chart for a deceleration-sensitive valve is shown in Figure 5-15. The brake line pressures increase at equal amounts up to the switch point A. At this moment the ball has closed off the opening to the rear brakes and no further brake line pressure increase occurs at the rear brakes. As the brake line pressure from the master cylinder increases at the valve inlet and against the small area of the differential piston, the differential piston begins to move away from the ball, allowing a small amount of brake fluid to enter the rear brake circuit, resulting in a corresponding pressure increase at the rear brakes. However, the ball again closes off the rear brake opening, and the differential piston reduces the outlet brake line pressure to the rear in relationship to the ratio of the area or diameter-squared ratios, that is,

$$\text{Reducer Slope} = (d / D)^2 \qquad (5\text{-}12)$$

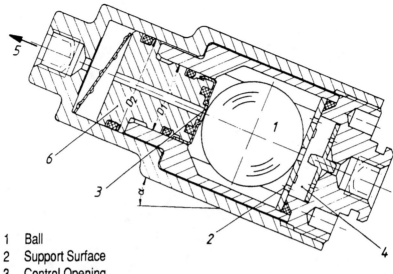

1  Ball
2  Support Surface
3  Control Opening
4  Brake Fluid
5  To Rear Brakes
6  Differential Piston

*Figure 5-14. Girling G-valve.*

where  d  = small diameter of differential piston, cm (in.)

D  = large diameter of differential piston, cm (in.)

The switch point at point A of Fig. 5-15 is reached for a deceleration a determined by

$$a = \tan \alpha \qquad (5\text{-}13)$$

where  α = installation angle relative to the horizontal baseline of the vehicle, deg

In some designs proportioning valves use the ABS lockup signal from the rear brakes to reduce the rear brake line pressure more closely to the optimum values. This is accomplished by means of solenoids and an accumulator incorporated into the proportioning valve.

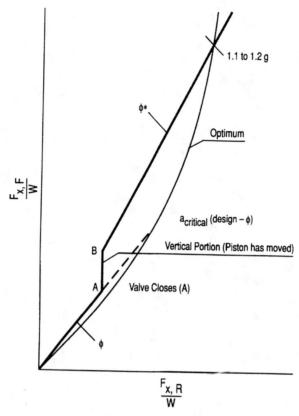

*Figure 5-15. Braking forces for G-valve.*

## 5.3.5 Step Bore Master Cylinder

The brake line pressure valves discussed earlier are installed between the master cylinder outlet and the rear brake wheel cylinders. With the standard dual or tandem master cylinder design shown in Figure 5-16 both pistons have the same diameter. Consequently, equal pressure levels are produced in either the primary (or pushrod) piston, and secondary (or floating) piston circuit. For a detailed discussion of the tandem master cylinder see Section 10.3.2.

A step bore master cylinder as shown in Figure 5-17 operates like a normal dual cylinder, but differs in bore size between the primary and secondary sections of the master cylinder. Since both pistons are pushed with equal force, the piston with the smaller bore produces larger brake line pressures.

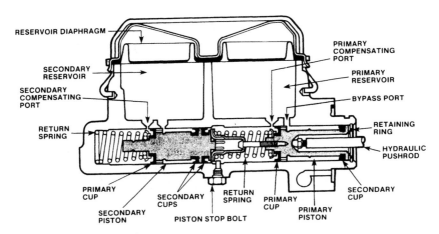

Figure 5-16. Dual-system master cylinder (Bendix).

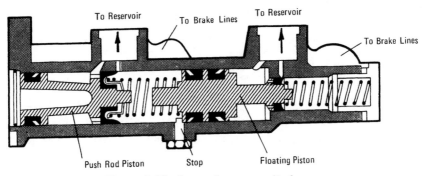

Figure 5-17. Stepped master cylinder.

The step bore section will generally be connected to the front disc brakes of a vehicle. The step bore master cylinder can be used only in the front-to-rear dual split systems (Chapter 10).

## 5.3.6 Adjustable Step Bore Master Cylinder

The adjustable master cylinder was developed from the basic step bore master cylinder. A schematic is shown in Figure 5-18. The master cylinder is a tandem or dual-circuit design with a third piston in addition to the pushrod and floating piston. A magnetic valve control is used to connect the space

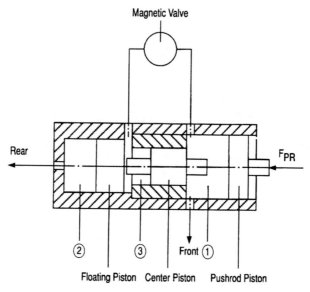

*Figure 5-18. Adjustable master cylinder.*

between the third piston and the floating piston either with the brake fluid reservoir, or directly with the pressure space between the pushrod piston and the third piston. When connected to the reservoir, the third piston transmits its force directly against the floating piston. In the other case, the third piston is functionally eliminated and the force transmission occurs hydraulically.

A typical brake line pressure diagram is shown in Figure 5-19. The two-slope brake pressure distribution provides better brake balance front to rear compared to the bilinear distribution shown in Fig. 5-10 for the reducer valves commonly used in today's passenger cars and light trucks.

When the space 3 is not connected to the reservoir, then the pressure $p_1$ in the pushrod piston space 1 is given by

$$p_1 = F_{PR} / A_{mcl} \quad , \quad \text{N/cm (psi)} \tag{5-14}$$

where $A_{mcl}$ = master cylinder cross-sectional area of space 1, cm$^2$ (in.$^2$)

$F_{PR}$ = pushrod piston force, N (lb)

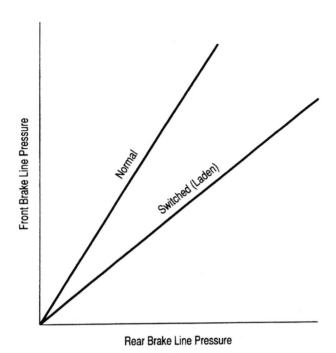

*Figure 5-19. Brake line pressure for adjustable master cylinder.*

Furthermore, the pressure in the floating piston space 2 is the same as in the pushrod piston space 1, computed by Eq. (5-14).

In the other case, the space 3 is connected to the reservoir resulting in atmospheric pressure in the space 3. The third piston mechanically transmits the force $F_3$:

$$F_3 = p_1 A_{mc3} \quad , \quad N \text{ (lb)}$$

against the floating piston, producing the pressure in the floating piston space 2 determined by

$$p_{23} = F_3 / A_{mc2} = p_1 A_{mc3} / A_{mc2} \quad , \quad N/cm^2 \text{ (psi)} \qquad (5-15)$$

where $A_{mc2}$ = cross-sectional area in space 2, cm² (in.²)

$A_{mc3}$ = cross-sectional area in space 3, cm² (in.²)

211

Just as the normal step bore master cylinder, the adjustable step bore master cylinder can be used only for the basic front-to-rear dual split systems standard on many medium- and full-size rear-wheel-driven passenger vehicles.

The adjustable master cylinder is used in connection with four-wheel ABS brakes of some modern passenger cars. When the magnetic valve is energized, the brake force distribution to the rear brakes is increased (as if the vehicle were laden) allowing for improved rear braking efficiency. Any potential rear brake lock is prevented by ABS. In the event of an ABS brake failure the master cylinder valve is deenergized, causing the master to switch to a steeper slope. The result is front brake lock before rear, providing a stable braking maneuver in the event brakes are locked.

## 5.3.7  Comparison of Brake Line Pressure Valves

As discussed in Chapter 1, an important consideration in selection of a design solution selection is system simplicity. Increased complexity will affect reliability and, to some extent, repairability. Of equal or even more importance, however, is the basic directional stability of a motor vehicle during braking with some wheels locked. For some vehicle configurations and, in particular, front-wheel-driven passenger cars, the relatively low static rear axle load may require the use of rear brake line pressure reducing valves. If possible, and if premature rear brake lockup can be prevented under all foreseeable operating conditions, a single knee-point pressure reducer or proportioning valve should be used. If that is not sufficient, a deceleration-sensitive valve should be considered. Load-sensitive valves require linkages and levers, which with time may be out of adjustment due to spring sagging, after market suspension component installation, or simple component damage underneath the vehicle.

In the design of load-sensitive devices and adjustment linkages, care must be taken that suspension deflections caused by road roughness or crossing of railroad tracks do not alter the basic knee-point setting of the valve.

Adjustable master cylinders are an important safety contribution in combination vehicles when towing a trailer. Under these conditions, the rear brake force of the tow vehicle is increased through an electrical signal from the same light circuit that activates the magnetic control valve.

# 5.4  Brake Fluid Volume Analysis

## 5.4.1 Basic Concepts

Inspection of Eq. (5-3) reveals that vehicle deceleration will increase with brake line pressure. Inspection of Eqs. (5-1) and (5-5) reveals that brake line pressure will increase with decreasing values of master cylinder cross-sectional area. Conversely, increasing the cross-sectional areas of the wheel cylinders will also increase vehicle deceleration. However, decreasing the master cylinder cross-sectional area will reduce the amount of brake fluid volume delivered, while increasing the size of the wheel cylinders will increase the amount of brake fluid required by the wheel brakes to function properly.

In the brake fluid volume analysis the minimum amount of brake fluid volume that must be delivered by the master cylinder is determined so that all fluid volume absorbing brake system components can function properly, and the brake pedal travel does not exceed an upper safe value.

Hydraulic brakes use the principle of equal pressure throughout the brake system. The schematic of this principle is illustrated in Figure 5-20. The piston to the left pressurizes the fluid with a given force. It represents the pedal effort input by the driver. The eight pistons to the right represent the wheel cylinder pistons. If they have the same cross-sectional area as the single piston, then each of the eight pistons carries a weight of 445 N (100 lb), or a total of 3558 N (800 lb). With a displacement of the single piston of, for example, 203 mm (8 in.), the eight pistons will move only 25.4 mm (1 in.). Of course, if the cross-sectional area of the single piston were decreased, a smaller force could produce the same pressure and, hence, lifting forces at the eight cylinders. With this change, the eight pistons would move a correspondingly shorter distance.

Consider the following application. A four-wheel disc brake vehicle has a master cylinder size of 19.05 mm bore diameter and 36 mm piston stroke (0.75 by 1.4 in.). The front disc brake wheel cylinder diameter is 5.71 cm (2.25 in.), that of the rear 3.81 cm (1.5 in.). The master cylinder volume is computed to be 10.1 cm$^3$ (0.618 in.$^3$). The total wheel cylinder piston area displaced by brake fluid for the four disc brakes is $4(25.6 + 11.4) = 148$ cm$^2$ $[4(3.976 + 1.767) = 22.97$ in.$^2]$. Consequently, the average wheel cylinder

N = 4.448 lb
cm = 2.54 in

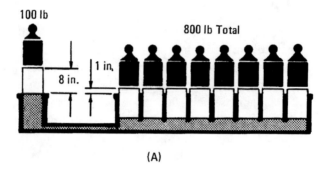

(A)

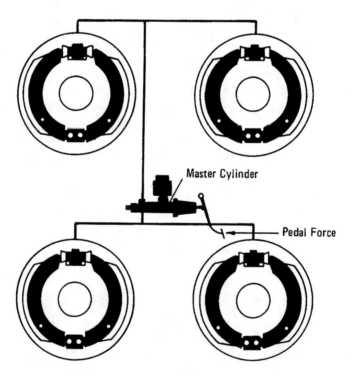

*Figure 5-20. Hydraulic brake system.*

piston travel is $10.1/148 = 0.068$ cm $= 0.68$ mm $(0.618/22.97 = 0.027$ in.$)$. If the wheel cylinder piston travel required to press the pads against the rotors exceeds the maximum of 0.68 mm (0.027 in.) available from the master cylinder, due to, for example, excessive axial rotor runout or air in the brake system, the pedal will go to the floor without any or with only very little brake line pressure buildup, resulting in total loss of braking effectiveness. An important design objective for drum brakes and disc brake calipers is to minimize the distance that the wheel cylinder pistons have to travel for safe brake operation. Automatic brake adjusters are but one design solution to accomplish this design objective.

To fully understand the importance of the minimum wheel cylinder piston travel, and how it relates to other brake components, the following brake system gain analysis is presented.

For simplicity, consider a brake system with identical brakes on each wheel. With Eq. (5-2), the total brake force $F_{x,total}$ on the vehicle can be determined by

$$F_{x,total} = n_B(p_\ell - p_o)A_{wc}\eta_c BF(r / R) \quad , \quad N \text{ (lb)} \qquad (5\text{-}16)$$

where $BF$ = brake factor

$n_B$ = number of brakes

$p_\ell$ = brake line pressure, $N/cm^2$ (psi)

$p_o$ = pushout pressure, $N/cm^2$ (psi)

$r$ = drum or rotor radius, mm (in.)

$R$ = tire radius, mm (in.)

$A_{wc}$ = wheel cylinder area, $cm^2$ (in.$^2$)

$\eta_c$ = wheel cylinder efficiency

Application of Newton's second law to Eq. (5-16) yields

$$aW = n_B(p_\ell - p_o)A_{wc}\eta_c BF(r / R) \quad , \quad N \text{ (lb)} \qquad (5\text{-}17)$$

where   a = deceleration, g-units

W = vehicle weight, N (lb)

Eq. (5-17) may be rewritten in terms of pedal force rather than brake line pressure. Pushout pressures generally are small compared to normal operating pressures and are ignored in this analysis. Eq. (5-17) becomes

$$aW = n_B(F_p \ell_p \eta_p B / A_{mc})A_{wc}\eta_c BF(r / R) \quad , \quad N \text{ (lb)} \quad (5\text{-}18)$$

where   B = boost ratio

$\ell_p$ = pedal lever ratio

$\eta_p$ = pedal lever efficiency

From a brake fluid volume analysis, it follows that the fluid displaced by the master cylinder equals the fluid absorbed by the individual wheel cylinders due to their piston travel to press the shoes against the drums. Any other brake fluid volume absorbing components such as brake hose expansion or caliper deformation, are ignored at this time. They will be considered in the detailed analysis presented in Section 5.4.3. Hence, the volume $V_{mc}$ produced by the master cylinder is

$$V_{mc} = A_{mc}X = n_s A_{wc}d \quad , \quad cm \text{ (in.)}$$

or

$$A_{wc} / A_{mc} = X / n_s d \qquad (5\text{-}19)$$

where   $A_{mc}$ = master cylinder cross-sectional area, cm$^2$ (in.$^2$)

d = wheel cylinder piston displacement, cm (in.)

$n_s$ = number of brake shoes

X = master cylinder piston travel, cm (in.)

If the ratio of wheel cylinder area to master cylinder area (Eq. [5-19]) is introduced in Eq. (5-18), the following expression results

$$aW = F_p[\ell_p B \eta_p \eta_c(X / d)BF(r / R)(n_B / n_s)] \quad , \quad N \text{ (lb)} \quad (5\text{-}20)$$

where  $a$ = vehicle deceleration, g-units

$\quad F_p$ = pedal force, N (lb)

$\quad W$ = vehicle weight, N (lb)

Inspection of Eq. (5-20) reveals that vehicle braking force $aW$ and, hence, vehicle deceleration is determined by multiplying the pedal force $F_p$ by a total systems gain consisting of six individual dimensionless component gains, and two efficiencies. The individual gains are

pedal ratio: $\ell_p = S_F/X$ ; $S_F$ = pedal travel, cm (in.)

boost ratio: $B$

displacement or hydraulic gain: $X/d$

brake factor: $BF$

radius ratio: $r/R$

shoe ratio: $n_B/n_s$

Each of the gains can be increased only to certain upper limits. Increasing the boost ratio creates problems in the case of a booster failure as well as in compliance with safety standards. Increasing the pedal ratio or master cylinder piston travel requires excessive pedal travels. Decreasing the minimum wheel cylinder piston travel available from the master cylinder volume requires extremely stiff brake system components and small clearance values between shoe and drum or pad and rotor. Increasing the brake factor results in brake torque variation and potential for left-to-right and front-to-rear brake imbalance. Increasing drum or rotor radius is limited by rim size. Decreasing tire radius is limited by the load-carrying capacity required by the maximum weight of the vehicle. Decreasing the number of shoes would be impractical with current design practice.

Typical values for a four-wheel disc brake vehicle system substituted for the individual gains into Eq. (5-20) may result in

$$aW = F_p[(4(3)(0.8)(0.96)(36 / 0.635)(0.7)(0.35)(0.5)] = 64F_p$$

$$[aW = F_p[(4(3)(0.8)(0.96)(1.42 / 0.025)(0.7)(0.35)(0.5)] = 64F_p]$$

Inspection of the numerical values reveals that the hydraulic gain equal to $36/0.635 = 56.7$ ($1.42/0.025 = 56.7$) is the most significant contribution to braking effectiveness.

Combining Eqs. (5-20) and (5-21) yields an approximate limiting empirical relationship for manual (B = 1) brakes in terms of pedal force $F_p$ and maximum pedal travel $S_p$ as:

$$F_p S_p = 0.9 \ aW \quad , \quad Ncm$$

$$[F_p S_p = 0.35 \ aW \quad , \quad lb \ in.]$$

where  a = deceleration, g-units

$S_p$ = maximum pedal travel, cm (in.)

W = vehicle weight, N (lb)

For example, for a pedal force of 580 N (130 lb) and pedal travel of 15 cm (5.9 in.) a vehicle weighing 12,251 N (2753 lb) can safely be braked at 0.8 g. If a booster were used in the braking system, then the maximum safe vehicle weight would simply be multiplied by the boost ratio B.

The approximate relationship is derived as follows.

We start with Eq. 5-3, however ignore pushout pressures, resulting in

$$aW = p_\ell\left[(A_{wc}BF)_F + (A_{wc}BF)_R\right]\frac{2\eta_c}{R}$$

The required fluid displacement $V_{mc}$ produced by the master cylinder can be expressed as

$$V_{mc} = 4\left[(A_{wc}d)_F + (A_{wc}d)_R\right](1 + v)$$

where  $v$ = relative portion of $V_{mc}$ required for hose expansion.

Substituting above $d_{min} \approx \dfrac{BF}{25}$ (Eq. [5-21]) yields:

$$V_{mc} = \frac{4}{25}\left[(A_{wc}BF)_F + (A_{wc}BF)_R\right](1 + v)$$

Combining the first and last equations yields:

$$\frac{Wa}{2(r/R)\eta_c p_\ell} = \frac{25V_{mc}}{4(1 + v)}$$

or,

$$p_\ell V_{mc} = \frac{Wa(1 + v)}{(12.5)(r/R)\eta_c}$$

Using typical values for $v = 0.3$, $r/R = 0.35$, $\eta_c = 0.85$ (conservative), yields

$$p_\ell V_{mc} = \frac{Wa(1.3)}{(12.5)(0.35)(0.85)} = (0.35)Wa$$

Since pedal work and master cylinder work are approximately equal, we have

$$F_p \cdot S_p = 0.35Wa$$

## 5.4.2 Simplified Brake Fluid Volume Analysis

Whether one uses a simplified or detailed analysis, the purpose of a brake fluid volume analysis is to determine the minimum amount of brake fluid required by the hydraulic brake system, i.e., all brake fluid absorbing components, and match that volume requirement with the volume delivered by the master cylinder. For the common dual brake systems, this means the proper determination of the individual travels of the primary and secondary piston.

Since the wheel cylinders are the major fluid users, past experience has provided design guidelines for the minimum piston travel required for safe operation of wheel brakes.

For a typical hydraulic brake system the minimum safe travel $d_{min}$ for one piston, one brake pad, or one brake shoe tip may be determined by the empirical expression

$$d_{min} \approx BF \quad , \quad mm \tag{5-21}$$

$$[d_{min} \approx BF / 25 \quad , \quad in]$$

where BF = brake factor

For example, for a front disc brake having a brake factor BF = 0.7, the minimum piston or pad travel is 0.7 mm (0.7/25 = 0.028 in.). If the master cylinder volume feeding the two front disc brakes provides a brake fluid volume sufficient to move each pad by 0.7 mm (0.028 in) as the master cylinder bottoms out, then the brake pedal travel will not be excessive under normal operating conditions.

The floating caliper disc brake design has only one piston. The single piston travel must be twice the safe amount or 1.4 mm (0.056 in.) to ensure that both pads retract properly when the brake is released.

## 5.4.3 Detailed Brake Fluid Volume Analysis

### 5.4.3.a Master Cylinder Volume Analysis

The brake fluid volume delivered by the master cylinder must be sufficiently large so that all fluid using brake components function properly, and so that the pedal travel does not exceed approximately 8.89 cm (3.5 in.) for deceleration of 0.9 to 1 g and "cold" brakes (see Section 1.3).

The sizing of a master cylinder in terms of cross-sectional area is mostly a function of braking performance requirements with the booster failed.

The master cylinder cross-sectional area required to achieve a specified deceleration with a given pedal force is computed by combining Eqs. (5-1), (5-3), and (5-11) and ignoring pushout pressures:

$$A_{mc} = \frac{2F_p \ell_p \eta_p \eta_c [(A_{wc}BFr)_F + (A_{wc}BFr)_R \, SL]}{aWR - 2(A_{wc}BFr)_R \, p_K (1 - SL)\eta_c} \quad , \quad cm^2 \ (in.^2) \quad (5-22)$$

where  $F_p$ = pedal force, N (lb)

$\ell_p$ = pedal lever ratio

$p_K$ = knee point pressure, N/cm$^2$ (psi)

$\eta_c$ = wheel cylinder efficiency

$\eta_p$ = pedal lever efficiency

The master cylinder piston travel $S_{mc}$ is given by

$$S_{mc} = S_p \, / \, \ell_p \quad , \quad cm \ (in.) \quad (5-23)$$

where  $S_p$ = maximum pedal travel, cm (in.)

The maximum master cylinder volume $V_{mc-max}$ is obtained by combining Eqs. (5-22) and (5-23):

$$V_{mc-max} = \frac{2S_p F_p}{aWR} \sum^n (A_{wc}BFr)_{F,R} \quad , \quad cm^3 \ (in.^3) \quad (5-24)$$

The brake fluid determined by Eq. (5-24), or from the actual master cylinder for a given vehicle, must cover all fluid volume "demands" of the entire brake system.

Quick take-up master cylinders as illustrated in Figure 5-21 and used in some vehicles require additional volume analysis, because the low-pressure chamber of the master cylinder provides a large amount of fluid at low pressures up to approximately 69 N/cm$^2$ (100 psi) to quickly apply the pads against the rotors. The area between the base of the master cylinder and the face of the primary piston is the low pressure chamber. When the brakes are applied, the primary piston begins to move into the bore, and the volume surrounding the piston body begins to decrease, resulting in a pressure increase of the fluid behind the primary piston. As the quick take-up valve is closed to the reservoir, the fluid is pressed past the primary seal directly in front of the primary piston into the primary circuit. At a pressure of approximately 48 to 69 N/cm$^2$ (70 to 100 psi) the quick take-up valve

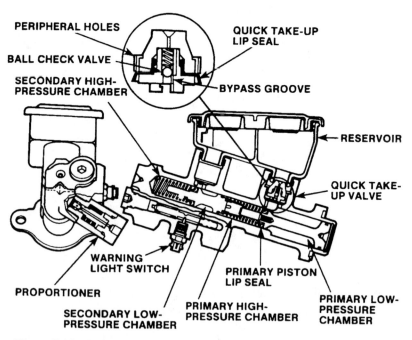

*Figure 5-21. Delco-Moraine "quick take-up" master cylinder (Bendix).*

opens, relieving the low-pressure chamber. For higher brake line pressures, the master cylinder operates as a normal dual or tandem master cylinder with each circuit producing the same pressure. The quick take-up chamber usually supplies the circuit with the larger brake fluid volume, in most cases the larger brake fluid requirements of the front low-drag disc brakes (see Section 2.2.2).

### 5.4.3.b  Individual Component Fluid Requirements

The individual fluid volumes and, hence, pedal travel losses involve the factors discussed next. The individual fluid volumes associated with the various brake components are computed for a brake line pressure that will produce a deceleration of 0.9 g for the vehicle laden at GVW with a pedal travel of not more than 8.89 cm (3.5 in.).

### 1.  *Brake Shoe or Pad Actuation*

The fluid volume required to move the pads or shoes against the rotor or drum is computed from the axial runout or gap for the disc brake, or the shoe-to-drum clearance, and the wheel cylinder area. During this portion of the pedal application only negligible brake line pressures are produced. Disc brakes and nearly all drum brakes are self-adjusting making the clearance a somewhat predictable minimum value. The inspection of an accident vehicle generally will reveal any abnormal conditions. Drum and shoe-circle diameter determine the actual lining clearance. It should be noted that many hydraulic drum brakes use a wheel cylinder piston stop, causing the piston to push against a stop in the event of excessive lining wear. When this occurs, the brake shoes are not or are only partially pressed against the drum without producing sufficient brake torque. Although the pedal may feel firm because high brake line pressures are developed, the braking effectiveness may be reduced significantly.

Excessive axial runout of disc brake rotors will cause increased pad and caliper piston pushback when the brakes are released. The total gap between the pads and the rotors may be greater than the fluid available from the master cylinder for that circuit. Proper, i.e., quick, pumping of the brakes will normally result in brake pedal rise due to fluid bypass around the primary seals from the reservoir into the brake system, as illustrated in Figure 5-22. In some cases, excessive front-end shimmy or a severe tight turn and loose or

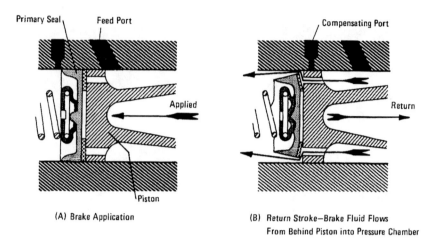

Figure 5-22. Primary seal operation.

worn suspension components may cause caliper piston pushback sufficient for the brakes to fail from excessive brake pedal travel. When the accident vehicle is examined, no excessive axial rotor runout may be observed with the true accident causation often remaining a mystery to many.

In the investigation of accidents involving four-wheel disc brake vehicles with the integrated disc parking brake using parking brake application to adjust the service brake pads, the parking brake should not be applied during the inspection since critical evidence may be destroyed (see Section 2.4.2).

2. *Brake Line Expansion*

When the basic equation for a pressurized cylinder is used, the volume increase $V_{BL}$ of a brake line may be determined by

$$V_{BL} = 0.79D^3Lp_\ell \, / \, tE \quad , \quad cm^3 \, (in.^3) \tag{5-25}$$

where   D  =  outer diameter of pipe, cm (in.)

   E  =  elastic modulus of pipe material, $N/cm^2$ (psi)

   L  =  length of brake line, cm (in.)

$p_\ell$ = brake line pressure, $N/cm^2$ (psi)

t = wall thickness of pipe, cm (in.)

For a brake line with D = 0.475 cm (0.187 in.), t = 0.0686 cm (0.027 in.), and E = $20.6 \times 10^6$ $N/cm^2$ ($30 \times 10^6$ psi), Eq. (5-25) yields the normalized brake line volume loss coefficient $k_{BL}$ of

$$k_{BL} = V_{BL} / p_\ell L = 0.044 \times 10^{-6} \quad , \quad cm^3/(N/cm^2)cm \quad (5\text{-}26)$$

$$[k_{BL} = V_{BL} / p_\ell L = 0.0064 \times 10^{-6} \quad , \quad in.^3/(psi)in.]$$

For a given vehicle with a specific brake line length L and a certain brake line pressure $p_\ell$, the volume loss due to brake line expansion is determined by

$$V_{BL} = k_{BL} L p_\ell \quad , \quad cm^3 (in.^3) \tag{5-27}$$

## 3. Brake Hose Expansion

Brake hose expansions have been measured. Typical values of brake hose expansion $V_H$ for vehicles in use today are computed by

$$V_H = k_H L_H p_\ell \quad , \quad cm^3 (in.^3) \tag{5-28}$$

with:

$$k_H = 4.39 \times 10^{-6} \quad , \quad cm^3/(N/cm^2)cm$$

$$[k_H = 0.47 \times 10^{-6} \quad , \quad in.^3/(psi)in.]$$

where $L_H$ = brake hose length, cm (in.)

## 4. *Master Cylinder Losses*

Volume losses for master cylinders in good mechanical condition generally vary with the size of the master cylinder diameter as indicated below.

| Diameter | $k_{mc}$ |
|----------|----------|
| 19.05 mm | $150 \times 10^{-6}$ cm$^3$/N/cm$^2$ |
| (3/4 in.) | ($6 \times 10^{-6}$ in.$^3$/psi) |
| 23.8 mm | $190 \times 10^{-6}$ cm$^3$/N/cm$^2$ |
| (15/16 in.) | ($8 \times 10^{-6}$ in.$^3$/psi) |
| 25.4 mm | $220 \times 10^{-6}$ cm$^3$/N/cm$^2$ |
| (1 in.) | ($9 \times 10^{-6}$ in.$^3$/psi) |
| 38.1 mm | $450 \times 10^{-6}$ cm$^3$/N/cm$^2$ |
| (1.5 in.) | ($19 \times 10^{-6}$ in.$^3$/psi) |

The volume loss $V_{mc}$ is determined by

$$V_{mc} = k_{mc}p_\ell \quad , \quad cm^3 \ (in.^3) \tag{5-29}$$

where $k_{mc}$ = specific master cylinder volume loss, cm$^3$/N/cm$^2$ (in.$^3$/psi)

## 5. *Caliper Deformation*

Caliper deformation is difficult to measure exactly, since residual pocket air and test fluid are compressed and cause a small fluid loss of their own. Furthermore, different caliper designs make it impossible to state one coefficient for all applications. However, tests conducted with "steel" pads and corrected for fluid compression show that the caliper volume loss coefficient for fixed caliper designs for one caliper may be approximated by:

$$V_c = k_c p_\ell \quad , \quad cm^3 \ (in.^3) \tag{5-30}$$

The values for $k_c$ are a function of the caliper size or the wheel cylinder. For wheel cylinder sizes between 38 and 60 mm (1.5 and 2.36 in.) $k_c$ is determined by

$$k_c = 482 \times 10^{-6} d_{wc} - 1.632 \times 10^{-6} \quad , \quad \text{cm}^3/\text{N/cm}^2 \qquad (5\text{-}31)$$

$$\left[ k_c = 52 \times 10^{-6} d_{wc} - 69 \times 10^{-6} \quad , \quad \text{in.}^3/\text{psi} \right]$$

where $d_{wc}$ = wheel cylinder diameter, cm (in.)

$V_c$ = volume loss in caliper, cm³ (in.³)

## 6. Drum Deformation

The hydraulic brake fluid volume loss $V_d$ due to mechanical drum deformation is computed by

$$V_d = k_d A_{wc}^2 p_\ell \quad , \quad \text{cm}^3 \text{ (in.}^3) \qquad (5\text{-}32)$$

with:

$k_d$ = (20 to 30) $10^{-6}$ cm/N

[$k_d$ = (35 to 53) $10^{-6}$ in./lb]

where $A_{wc}$ = wheel cylinder area, cm² (in.²)

## 7. Brake Pad Compression

For disc brakes, pad compression is an important factor in the selection of a proper material. A certain amount of compressibility or damping is essential for disc brakes to operate without undue noise.

For disc brake pads, the volume loss $V_p$ due to compression is determined by

$$V_p = 4\Sigma (A_{wc} C_s p_\ell)_i \quad , \quad \text{cm}^3 \text{ (in.}^3) \qquad (5\text{-}33)$$

where $A_{wc}$ = wheel cylinder area, cm² (in².)

$\quad$ $C_s$ = brake shoe compressibility, cm/(N/cm²) (in./psi)

$\quad$ i = identity of brake

$\quad$ $p_\ell$ = brake line pressure, N/cm² (psi)

For disc brake pads, a relatively well-damped pad material yields a compressibility factor $C_s = 11 \times 10^{-6}$ to $26 \times 10^{-6}$ cm/(N/cm²) $(3 \times 10^{-6}$ to $7 \times 10^{-6}$ in./psi) at normal (cold) brake temperature, $C_s = 15 \times 10^{-6}$ to $33 \times 10^{-6}$ cm/(N/cm²) $(4 \times 10^{-6}$ to $9 \times 10^{-6}$ in./psi) for hot brakes with a rotor temperature of approximately 672 K (750°F) with a backing plate temperature of approximately 380 K (225°F).

For example, for a passenger car with a four-wheel disc brake system with 5.71 and 3.81 cm (2.25 and 1.5 in.) wheel cylinder diameters, front and rear, hot brakes and a brake line pressure of 620 N/cm² (900 psi), Eq. (5-33) yields a maximum volumes loss $V_p$ due to pad compression of

$$V_p = 4C_s(A_{wcF} + A_{wcR})p_\ell$$

$$= 4(33)(10^{-6})(25.6 + 11.4)(620) = 3.02 \text{ cm}^2$$

$$[= 4(9)(10^{-6})(3.976 + 1.767)(900) = 0.186 \text{ in.}^3 )]$$

With a typical master cylinder size of 2.22 by 2.54 cm (7/8 × 1 in.) for a compact car, the fluid volume is approximately 11.5 cm³ (0.7 in.³). A pad compression loss of 3.02 cm³ (0.186 in.³) may account for nearly 30% of the pedal travel loss alone.

## 8. Brake Shoe and Lining Compression (Drum Brake)

The brake fluid volume loss resulting from two brake shoes including the apply mechanism is computed by

$$V_s = \frac{k_s A_{wc}^2 p_\ell}{dw} \quad , \quad \text{cm}^3 \text{ (in.}^3) \tag{5-34}$$

with:

$$k_s = (100 \text{ to } 150) \, 10^{-6} \, cm^3/N$$

$$[k_s = (271 \text{ to } 407) \, 10^{-6} \, in.^3/lb]$$

where   d = drum diameter, cm (in.)

   w = brake shoe width, cm (in.)

## 9. *Thermal Drum Expansion*

The brake fluid volume $V_T$ due to the expansion of the drum due to temperature is computed by

$$V_T = 3.14\alpha_T d T_d A_{wc} \quad , \quad cm^3 \, (in.^3) \tag{5-35}$$

where $A_{wc}$ = wheel cylinder area, $cm^2$ (in.$^2$)

   $T_d$ = drum temperature, K (°F)

   $\alpha_T$ = thermal expansion coefficient of cast iron drum material

   = $11 \times 10^{-6}$ cm/cmK [$6.55 \times 10^{-6}$ in./in. °F]

## 10. *Air in Drum Brake Hydraulics*

The brake fluid volume loss $V_a$ due to air inclusion is approximately

$$V_a = 0.035 \, A_{wc}, \, cm^3 \tag{5-36}$$

$$[V_a = 0.014 \, A_{wc}, \, in.^3]$$

## 11. *Brake Shoe/Drum Clearance*

The brake fluid volume $V_{c\ell}$ due to the clearance between shoes and drum for brakes with good automatic adjustment is

$$V_{c\ell} = (0.13 \text{ to } 0.15)A_{wc} \quad , \quad cm^3 \tag{5-37}$$

$$[V_{c\ell} = (0.05 \text{ to } 0.06)A_{wc} \quad , \quad in.^3]$$

## 12. Brake Fluid Compression

Volume losses resulting from the compression of brake fluid may have a significant effect on pedal travel as brake fluid temperatures and brake line pressures increase. Measured values of compressibility factor of different dry, gas-free fluids as a function of temperature are shown in Figure 5-23 for regular brake fluid based on polyglycolether, mineral oil, and silicone. Inspection of Fig. 5-23 reveals that regular brake fluid will double its compressibility factor when the brake fluid temperature increases from 294 to 477 K (70 to 400°F). Silicone-based brake fluids have the highest compressibility. For more details on brake fluids and their properties see Chapter 10.

The volume loss resulting from brake fluid compression is a function of the active volume $V_A$ in the brake system pressurized during the braking process.

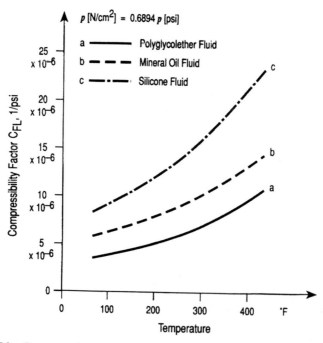

Figure 5-23. Compressibility factor $C_{FL}$ for dry brake fluids without gas content.

The active volume is determined by

$$V_A = V_o + 4\sum_{}^{n}(A_{wc}w)_i \quad , \quad cm^3 \ (in.^3) \tag{5-38}$$

where $A_{wc}$ = wheel cylinder area, $cm^2$ $(in.^2)$

$i$ = brake identity

$V_o$ = brake fluid volume with new shoes, $cm^3$ $(in.^3)$

$w$ = wear travel of shoes, cm (in.)

The volume loss $V_{FL}$ due to fluid compression is computed by

$$V_{FL} = V_A C_{FL} p_\ell \quad , \quad cm^3 \ (in.^3) \tag{5-39}$$

where $C_{FL}$ = brake fluid compressibility factor, $cm^2/N$ (1/psi)

For example, for a four-wheel disc brake vehicle with

$A_{wcF}$ = 25.6 $cm^2$ (3.976 $in.^2$)

$A_{wcR}$ = 11.4 $cm^2$ (1.767 $in.^2$)

$w_F$ = $w_R$ = 0.635 cm (0.25 in.)

$V_o$ = 164 $cm^3$ (10 $in.^3$)

$C_{FLF}$ = $C_{FLR}$ = $10 \times 10^{-6}$ $cm^2/N$ at 422 K ($7 \times 10^{-6}$ 1/psi at 300°F)

the active volume is computed by Eq. (5-38) as

$$V_A = 164 + 4(25.6 + 11.4) \times 0.635$$

$$= 258 \ cm^3 \ [10 + 4(3.976 + 1.767)(0.25) = 15.7 \ in.^3]$$

The volume loss at 620 $N/cm^2$ (900 psi) brake line pressure is determined from Eq. (5-39) as

$$V_{FL} = 258 \times 10 \times 10^{-6} \times 620$$

$$= 1.6 \ cm^3 \ (15.7 \times 7 \times 10^{-6} \times 900 = 0.099 \ in.^3)$$

For silicone-based fluids the volume loss would be approximately 3.6 cm³ (0.22 in.³), i.e., a significantly greater loss than that for normal brake fluid.

## 13. *Air or Gas in the Brake System*

Air can remain in the brake system when air pockets form which cannot be flushed out during the vacuum bleeding process at the factory. Small air bubbles will adhere to metal surfaces of springs and other parts. Due to surface tension, small-sized air bubbles will remain in the brake fluid, which can be removed only by ultrasound application. Typical residual air volumes in the entire brake system are approximately 3% of the active volume. The 3% includes the residual air in the front disc brake caliper of approximately 0.6%, and 0.4% in the rear disc brake caliper.

If one defines $V_G$ as the volume of the enclosed air at atmospheric pressure, and assumes an isothermal or constant temperature compression based on the valid assumption that the air temperature remains equal to the temperature of the brake fluid during the compression process, then with basic thermodynamics the decrease of air volume is determined by

$$V_{GL} = V_G T / T_0 [1 - p_o / (p_\ell + p_o)] \quad , \quad cm^3 \ (in.^3) \tag{5-40}$$

where $p_\ell$ = brake line pressure, N/cm² (psi)

$p_o$ = atmospheric pressure, N/cm² (psi)

$T_0$ = absolute temperature at initial conditions, K (°R)

$T$ = absolute temperature, K (°R)

$V_G$ = enclosed gas volume at ambient temperature, cm³ (in.³)

The absolute temperature T is computed by

$$T = T_{Celsius} + 273 \quad , \quad K \tag{5-41}$$

$$[T = T_{Fahrenheit} + 460 \quad , \quad °R]$$

At higher pressures, $p_\ell$, the square bracket in Eq. (5-40) goes to unity, indicating that the entire enclosed air volume will be compressed and cause a corresponding drop of the pedal height toward the floor. When the brake fluid is also heated, then the temperature increase will expand the air, thus making the volume loss stemming from the air greater.

For example, for $V_G = 4.9$ cm$^3$ (0.3 in.$^3$), and ambient and operating temperature of 294 and 394 K (70 and 250°F), respectively, Eqs. (5-40) and (5-41) yield:

$$T = 70 + 460 = 530° \text{ R}$$

and

$$T = 250 + 460 = 710° \text{ R}$$

and

$$V_{GL} = 4.9(394 \ / \ 294) = 6.57 \text{ cm}^3$$

$$[V_{GL} = 0.3(710 \ / \ 530) = 0.40 \text{ in.}^3\,]$$

### 14. *Fluid Loss in Hydrovac*

The hydraulic control of the hydrovac booster requires a small amount of brake fluid from the driver-operated master cylinder to actuate the vacuum unit. For most hydrovacs used in medium trucks, the volume loss is approximately 0.82 cm$^3$ (0.05 in.$^3$).

### 15. *Volume Loss in Valves*

Since the type of valve used in a hydraulic brake system varies with design, no specific loss coefficients can be stated. If a malfunctioning of a particular valve is suspected, special tests must be conducted with the accident and exemplar valves to clearly isolate any contribution to pedal travel loss.

After market valves installed in the brake system hydraulics will have an adverse effect on pedal travel. Depending on how "limited" the volume reserves are, they may cause increased pedal travels.

### 5.4.3.c Pedal Travel Computation

With the individual volume losses determined from the appropriate sections, the total volume loss can be computed. A brake line pressure sufficient to produce a deceleration of 0.9 g of the fully laden vehicle should be used for the volume loss calculations.

The pedal travel $S_p$ is determined by

$$S_p = [(\Sigma V_i / A_{mc}) + \ell_o]\ell_p \qquad (5\text{-}42)$$

where $A_{mc}$ = master cylinder cross-sectional area, cm$^2$ (in$^2$)

$\ell_o$ = master cylinder pushrod travel to overcome pushrod play and compensating port, cm (in.)

$\ell_p$ = pedal lever ratio

$V_i$ = volume loss for individual component, cm$^3$ (in.$^3$)

The pedal travel computed by Eq. (5-42) should not exceed 8.9 cm (3.5 in.) out of a maximum of approximately 15.2 cm (6 in.) or 60% of the maximum when the brakes are cold.

Example 5-1: For a deceleration of 0.9 g, determine the brake line pressure and pedal travel. Use the vehicle data that follow.

Weight: 16,458 N (3700 lb)

Wheel cylinder diameter, front: 5.71 cm (2.25 in.); rear: 3.81 cm (1.5 in.)

Effective rotor radius, front: 10.16 cm (4.0 in.); rear: 10.41 cm (4.1 in.)

Brake factor, front: 0.76; rear: 0.70

Tire radius: 31.75 cm (12.5 in.)

Pedal ratio: 4.2

Pedal travel: 12.7 cm (5.0 in.)

Pedal efficiency: 80%

Brake line length: 660 cm (260 in.)

Brake hose length: 152 cm (60 in.)

Wear pad travel: 0.38 cm (0.15 in.)

Initial brake system volume: 197 cm$^3$ (12 in.$^3$)

Residual air in brake fluid: 4.1 cm$^3$ (0.25 in.$^3$)

Proportioning valve: 207 N/cm$^2$ (300 psi) $\times$ 30%

Axial rotor runout front and rear: 0.00508 cm (0.002 in.)

Vacuum-boosted brakes

Solution: The master cylinder cross-sectional area is determined for the conditions of booster failure. As stated in Section 1.3.3, a pedal force of 100 lb must produce a deceleration of 0.3 g for the vehicle laden at GVW.

By use of Eq. (5-22), the cross-sectional area $A_{mc}$ of the master cylinder becomes

$$A_{mc} = \frac{2(445)(4.2)(0.8)[25.6(0.76)(10.16) + 11.41(0.7)(10.41)(0.3)](0.98)}{0.3(16,458)(31.75) - 2(11.4)(0.7)(10.41)(207)(1 - 0.3)(0.98)}$$

$$= 4.9 \text{ cm}^2$$

$$\left[ \begin{array}{l} A_{mc} = \dfrac{2(100)(4.2)(0.8)[3.976(0.76)(4) + 1.767(0.7)(4.1)(0.3)](0.98)}{0.3(3700)(12.5) - 2(1.767)(0.7)(4.1)(300)(1 - 0.3)(0.98)} \\[3mm] = 0.76 \text{ in.}^2 \end{array} \right]$$

A listing of master cylinder diameters usually available from the master cylinder manufacturers is shown in Table 5-1 in addition to the tolerance allowable for maximum bore diameter and minimum piston diameter. An area of 4.9 cm$^2$ (0.76 in.$^2$) requires a master cylinder size of 25.4 mm (1 in.) diameter. The associated cross-sectional area is 5.067 cm$^2$ (0.785 in.$^2$). The master cylinder piston travel $S_{mc}$ available from the pedal geometry is determined by Eq. (5-23) as

$$S_{mc} = 12.7 / 4.2 = 3.02 \text{ cm}$$

$$[S_{mc} = 5.0 / 4.2 = 1.19 \text{ in}]$$

### TABLE 5-1

### Master Cylinder Sizes

| Nominal Diameter | | Max. Bore Diameter | Min. Piston Diameter | Max. Allowable Tolerance |
|---|---|---|---|---|
| mm | in. | mm | mm | mm |
| 12.7 | 1/2 | 12.80 | 12.57 | 0.23 |
| 14.29 | 9/16 | 14.39 | 14.16 | 0.23 |
| 15.87 | 5/8 | 15.97 | 15.74 | 0.23 |
| 17.46 | 11/16 | 17.56 | 17.33 | 0.23 |
| 19.05 | 3/4 | 19.16 | 18.90 | 0.26 |
| 20.64 | 13/16 | 20.75 | 20.49 | 0.26 |
| 22.2 | 7/8 | 22.31 | 22.05 | 0.26 |
| 23.81 | 15/16 | 23.92 | 23.66 | 0.26 |
| 25.4 | 1 | 25.51 | 25.25 | 0.26 |
| 26.99 | 1- 1/16 | 27.10 | 26.84 | 0.26 |
| 27.78 | 1- 3/32 | 27.89 | 27.63 | 0.26 |
| 28.57 | 1- 1/8 | 28.68 | 28.42 | 0.26 |

## TABLE 5-1 (CONTINUED)

### Master Cylinder Sizes

| Nominal Diameter | | Max. Bore Diameter | Min. Piston Diameter | Max. Allowable Tolerance |
|---|---|---|---|---|
| mm | in. | mm | mm | mm |
| 31.75 | 1- 1/4 | 31.84 | 31.58 | 0.26 |
| 33.0 | 1.2992 | 33.09 | 32.83 | 0.26 |
| 34.92 | 1- 3/8 | 35.01 | 34.75 | 0.26 |
| 38.1 | 1- 1/2 | 38.19 | 37.93 | 0.26 |
| 41.27 | 1- 5/8 | 41.36 | 41.10 | 0.26 |
| 44.45 | 1- 3/4 | 44.54 | 44.28 | 0.26 |
| 46.83 | 1-27/32 | 46.92 | 46.66 | 0.26 |
| 48.42 | 1-29/32 | 48.51 | 48.25 | 0.26 |
| 50.8 | 2 | 50.90 | 50.60 | 0.3 |
| 54.0 | 2.1260 | 54.10 | 53.80 | 0.3 |
| 57.15 | 2- 1/4 | 57.25 | 56.95 | 0.3 |
| 65.0 | 2.5590 | 65.10 | 64.80 | 0.3 |
| 70.0 | 2.7559 | 70.10 | 69.80 | 0.3 |
| 75.0 | 2.9528 | 75.10 | 74.80 | 0.3 |

The effective master cylinder volume is $5.067 \times 2.82 = 14.3$ cm$^3$ ($0.785 \times 1.1 = 0.8635$ in.$^3$). The master cylinder travel actually used is approximately 2 mm (0.09 in.) less to account for pushrod play and compensating port travel.

The approximate brake fluid volume required by the disc brakes is determined next. The individual wheel cylinder piston travels $d_{minF}$ and $d_{minR}$ are obtained from Eq. (5-21) as

Front: $d_{\min F} = BF = 0.76$ mm $(0.76 / 25 = 0.0304$ in.$)$

Rear: $d_{\min R} = BF = 0.70$ mm $(0.70 / 25 = 0.028$ in.$)$

The brake fluid volume $V_F$ required by the disc brakes to provide sufficient pad displacement to overcome pad clearance, caliper deformation, and hose expansion is determined by

$$V_F = 4[25.65(0.076) + 11.4(0.07)] = 11 \text{ cm}^3$$

$$[V_F = 4[3.976(0.034) + 1.767(0.028)] = 0.681 \text{ in.}^3 ]$$

Because the master cylinder volume of 14.7 cm³ (0.8635 in.³) provided by the master cylinder exceeds the volume of 11 cm³ (0.681 in.³) required by the brake system, the master cylinder size is properly designed, provided the pedal travel calculations carried out next yield acceptable results.

The pedal travel computation is based on the individual volume loss calculations, which in turn require the brake line pressure as an input parameter.

The brake line pressure may be obtained by combining Eqs. (5-3) and (5-11) and eliminating rear brake line pressure:

$$p_\ell = \frac{(aWR / 2) - p_K(1 - SL)(A_{wc}BF\eta_c)_R}{(A_{wc}BF\eta_c)_F + (A_{wc}BF\eta_c)_R SL} \quad , \quad \text{N/cm}^2 \text{ (psi)} \quad (5\text{-}43)$$

where  $a$ = deceleration, g-units

$A_{wc}$ = wheel cylinder area, cm² (in.²)

BF = brake factor

$p_K$ = knee point pressure, N/cm² (psi)

$p_\ell$ = brake line pressure, N/cm² (psi)

R = tire radius, mm (in.)

SL = proportioning valve slope

W = vehicle weight, N (lb)

$\eta_c$ = wheel cylinder efficiency

Substitution of the appropriate data into Eq. (5-43) yields

$$p_\ell = \frac{[0.9(16,458)(31.75) \, / \, 2] - 207(1 - 0.3)(11.4)(0.7)(10.41)(0.98)}{25.6(0.76)(10.16)(0.98) + 11.4(0.7)(10.41)(0.98)(0.3)}$$

$$= 1024 \text{ N } / \text{ cm}^2$$

$$\left[ p_\ell = \frac{[0.9(3700)(12.5) \, / \, 2] - 300(1 - 0.3)(1.767)(0.7)(4.1)(0.98)}{3.976(0.76)(4.0)(0.98) + 1.767(0.7)(4.1)(0.98)(0.3)} \right.$$

$$\left. = 1482 \text{ psi} \right]$$

The individual volume losses are computed next.

1. Pad clearance:

$$V_{pc} = 4(0.0058)(25.6 + 11.4) = 0.752 \text{ cm}^3$$

$$[V_{pc} = 4(0.002)(3.976 + 1.767) = 0.046 \text{ in.}^3]$$

2. Brake line expansion (Eq. [5-27]):

$$V_{BL} = 0.06(10^{-6})(660)(1024) = 0.0405 \text{ cm}^3$$

$$[V_{BL} = 0.0064(10^{-6})(260)(1482) = 0.0018 \text{ in.}^3]$$

3. Brake hose expansion (Eq. [5-28]):

$$V_H = 4.39(10^{-6})(152)(1024) = 0.68 \text{ cm}^3$$

$$\left[ V_H = 0.47(10^{-6})(60)(1482) = 0.0417 \text{ in.}^3 \right]$$

4. Master cylinder (Eq. [5-29]):

$$V_{mc} = 220(10^{-6})(1024) = 0.2252 \text{ cm}^3$$

$$[V_{mc} = 9(10^{-6})(1482) = 0.0133 \text{ in.}^3 ]$$

5. Caliper deformation (Eq. [5-30]):

With Equation (5-31) we have on the front brakes

$$k_c = (482)(10^{-6})(5.71) - (1.632)(10^{-6})$$
$$= 1120.2 \text{ cm}^3/\text{N/cm}^2$$

$$\left[ \begin{array}{l} k_c = (52)(10^{-6})(2.25) - (69)(10^{-6}) \\ \quad = 48 \times 10^{-6} \text{ in}^3 / \text{psi} \end{array} \right]$$

The volume loss due to two front caliper deformations is (Equation [5-30])

$$V_c = (2)(1120.2)(10^{-6})(1024) = 2.29 \text{ cm}^3$$

$$\left[ V_c = (2)(48)(10^{-6})(1482) = 0.142 \text{ in.}^3 \right]$$

Similar computations for two rear calipers yield $V_c = 0.61$ cm$^3$ (0.038 in.$^3$).

6. Pad compression (Eq. [5-33]):

$$Cs = 18.5 \times 10^{-6} \text{ cm/(N/cm}^2)$$

$(5 \times 10^{-6}$ in./psi was chosen.)

Front:

$$V_{PF} = 4(25.6)(18.5)(10^{-6})(453) = 0.382 \text{ cm}^3$$

$$[V_{PF} = 4(3.976)(5)(10^{-6})(655) = 0.0231 \text{ in.}^3]$$

Rear:

$$V_{PR} = 4(11.4)(18.5)(10^{-6})(1024) = 0.864 \text{ cm}^3$$

$$[V_{PR} = 4(1.767)(5)(10^{-6})(1482) = 0.0523 \text{ in.}^3]$$

Total:

$$V_P = 2.322 \text{ cm}^3 (0.141 \text{ in.}^3)$$

7. Brake fluid compression (Eqs. [5-38] and [5-39]):

The active brake fluid volume $V_A$ is (Eq. [5-38])

$$V_A = 197 + 4(25.6 + 11.4)(0.38) = 253.2 \text{ cm}^3$$

$$[V_A = 12 + 4(3.976 + 1.767)(0.15) = 15.45 \text{ in.}^3]$$

The volume loss $V_F$ due to fluid compression is computed by Eq. (5-39) for a fluid temperature of 294 K (70°F) with $C_{FL}$ = $4.35 \times 10^{-6}$ cm$^3$/(N/cm$^2$) ($3 \times 10^{-6}$ in.$^3$/psi) from Fig. 5-23 as

$$V_F = (253.2)(4.35)(10^{-6})(1024) = 1.128 \text{ cm}^3$$

$$[V_F = (15.45)(3)(10^{-6})(1482) = 0.0687 \text{ in.}^3]$$

The difference in brake line pressures front and rear due to proportioning was ignored in computing fluid losses due to brake fluid compression.

8. Residual air in brake system (Eq. [5-40]):

The residual air of 4.1 cm$^3$ (0.25 in.$^3$) exists at atmospheric pressure. If no temperature increase of the air occurs, then no air volume increase has to be considered. Because these calculations are done for the "cold" brakes, the air volume is 4.1 cm$^3$ (0.25 in.$^3$).

With all individual volume losses established, the total volume loss $V_t$ experienced by the brake system at a brake line pressure of 1024 N/cm$^2$ (1482 psi) is $V_t = \Sigma V_i = 12.148$ cm$^3$ (0.743 in.$^3$).

The pedal travel $S_p$ required to produce a brake line pressure of 1024 N/cm$^2$ (1482 psi) as well as to provide sufficient master cylinder piston travel to "feed" the different brake fluid users is determined from Eq. (5-42) as

$$S_p = [(12.148 / 5.069) + 0.2](4.2) = 10.9 \text{ cm}$$

$$\left[ S_p = [(0.743 / 0.785) + 0.08](4.2) = 4.31 \text{ in.} \right]$$

The pedal travel falls outside of the requirements of Section 1.3.2 indicating that 75 to 90 mm (3 to 3.5 in.) pedal travel or approximately 60% out of a maximum of 150 mm (6 in.) should not be exceeded for cold brakes.

The brake system should be redesigned with the next larger master cylinder of 26.99 mm (27.10 in.). The pedal travel decreases to 3.85 in. (instead of 4.31 in.) The vacuum booster has to be designed based on the larger master cylinder diameter.

Any add-on devices "robbing" the brake system of fluid volume must not be used. In pedal-to-the-floor cases, the add-on device may have caused a brake failure and accident.

# 5.5 Dynamic Response of Hydraulic Brake Systems

## 5.5.1 Basic Considerations

A detailed analysis of the dynamic response of a hydraulic brake system is a complicated task. It involves the solution of several differential equations by means of computers. For valid conclusions to be reached, several input parameters must be measured for the specific brake system under consideration.

Generally, the response characteristics of hydraulic brake systems are such that the time lags between input and output variables are very small and are typically less than 0.1 to 0.2 s. With the increased use of ABS brakes, the dynamic response of hydraulic brake systems and individual brake components becomes important.

The dynamic response of a complete brake system consists of a quasi-static component and a transient component. The transient behavior is that associated with rapidly changing system variables, such as brake line pressure following a rapid pedal force input. The quasi-static behavior is associated with slowly changing variables, such as the change in coefficient of friction between lining and rotor due to a decrease in wheel speed during deceleration of the vehicle.

## 5.5.2 Brake Fluid Viscosity

The principal dynamic elements in a typical brake system are the brake lines and the vacuum booster. The flow of brake fluid from the master cylinder to the wheel cylinder is a function of fluid viscosity, cross-sectional flow area, and brake line length. The elements determining flow rate are the capacitance, resistance, and inertance of the section of brake line considered. The capacitance element accounts for fluid compressibility and wall compliance, while the resistance element introduces the pressure losses due to laminar or turbulent flow, i.e., frictional effects. Inertance effects result from the mass of the fluid in the brake lines. As fluid viscosity increases, the time interval between the application of force to the brake pedal and operation of the wheel brake increases. Similarly, there is an increase in brake release time.

On most vehicles, tubing length to the left front is shorter than that leading to the right front brake due to the location of the master cylinder. Because of the difference in tubing length in the front circuit, the left front brake is actuated before the right front brake. At normally low viscosity levels the difference is not perceptible. However, as viscosity increases with decreasing ambient temperatures, flow rates to each of the front brakes may become different and a noticeable unbalance in braking left to right may exist. The level of brake unbalance is affected by the rate of pedal force application, i.e., force application to the fluid. As fluid viscosity increases, the time required for fluid to return through the tubing from the brake to the master cylinder increases due to a slower flow rate, resulting in the brakes being applied for a longer time. Slow response due to viscosity effects will affect the performance of wheel anti-lock brake systems (ABS).

### 5.5.3  Brake Pedal Linkage

The dynamics of the brake pedal and linkage are of little importance to the dynamic response of a complete brake system. The weight and inertia of the foot and leg of the operator will greatly influence the dynamic behavior of the pedal.

### 5.5.4 Vacuum Booster

The vacuum booster contributes significantly to the response lag of the hydraulic brake system. The booster consists of several components such as pistons, valves, springs, and pushrod, all of which must be included in the mathematical expressions describing the dynamic behavior of the booster. In addition, certain assumptions must be made to simplify the thermodynamic relationships describing the pressure development in the vacuum and ambient pressure chambers (Ref. 20). Transient responses for the vacuum booster of a large domestic passenger car were measured and are illustrated in Figure 5-24. Inspection of Fig. 5-24 reveals the response characteristics for a slow pedal force application to be similar to the quasi-static behavior discussed in Section 5.2.2.a (Fig. 5-3). The response of the brake line pressure produced at the master cylinder outlet to a rapid pedal force application shows a significant lag compared with the response associated with a slow application. Only after the rapid pedal force has attained a value in excess of 1023 N (230 lb) does the resulting brake line pressure at the master cylinder reach the same level as that associated with the slow application. Generally, the lag of booster response during rapid brake applications is felt by the driver as a hard pedal similar to the pedal response with the engine turned off.

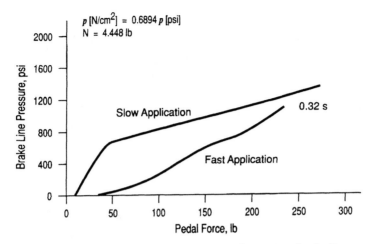

*Figure 5-24. Measured steady-state and transient brake line pressure/pedal force response.*

### 5.5.5 Master Cylinder

The dynamics of the master cylinder are relatively insignificant in comparison with those of the complete brake system. Major reasons are the small masses of the pistons and high geometrical stiffness of the cylinder. By application of the fundamentals of mechanics, the differential equations governing the dynamics of the master cylinder may be derived. In general, the equations describe the output flow of the master cylinder chambers as a function of the input force and the brake line pressure acting on the chambers.

### 5.5.6 Brake Line

In the past, hydraulic brake lines have been analyzed by means of the wave equation which describes the longitudinal vibrations of the fluid in the line. For small-diameter hydraulic lines, it has been found that the viscosity of the brake fluid has a significant effect on response time. Good correlation between theoretical analysis and experiment was obtained for a model consisting of a rod representing the fluid, a spring at one end representing the stiffness of the wheel brake, and a pressure input at the other end of the rod. Analysis and test data indicate that the brake line contributes significantly to the response lag of a complete brake system.

## 5.5.7 Wheel Brake

The dynamic behavior of the wheel brake may be analyzed by the use of several submodels, such as thermal submodel, friction coefficient submodel, static performance submodel, and dynamic performance submodel. The static performance submodel predicts the brake torque as a function of brake line pressure and coefficient of friction between lining and rotor. The brake torque of the submodel is based on expressions similar to Eq. (5-2) or (6-1). The dynamic performance submodel calculates a dynamic brake torque by treating the brake as a mass-spring-damper system. The thermal submodel considers the brake as an energy-conversion and heat-dissipation device in predicting brake temperature as discussed in Chapter 3. The friction material submodel considers the time-varying coefficient of friction between lining and rotor. The findings on the dynamic performance of wheel brakes indicate that typical wheel brakes are highly responsive brake system components. Furthermore, noticeable differences may exist from one brake design type to the next.

## 5.5.8 Hydraulic Boost Systems

Hydraulic boost systems without accumulators generally exhibit somewhat quicker response characteristics than those of vacuum-assisted brakes since they are not limited by the response lag of the vacuum booster. However, since they operate without accumulator, the operating boost pressure must be built up from zero, which contributes to response lag.

Hydraulic boost systems with accumulator have the assist pressure readily available for brake boosting. With hydraulic booster having the boost characteristics shown in Fig. 5-8, brake response time tests were conducted for a four-wheel disc brake vehicle (Ref. 21). The measured brake system transients are shown in Figure 5-25, where pedal force, master cylinder output pressure, and rear caliper wheel cylinder pressure are presented as a function of time. Inspection of the curves reveals that the pressure at the master cylinder outlet begins to rise approximately 0.06 s after pedal force begins. The average master cylinder brake line pressure lag is less than approximately 0.04 to 0.05 s for pressure up to 1034 N/cm (1500 psi). The corresponding brake line pressure at the rear brake lags the master cylinder pressure by only approximately 0.015 s.

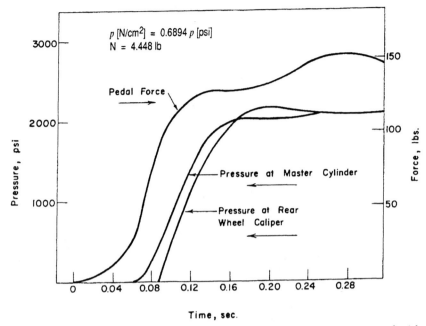

*Figure 5-25. Measured brake system transients for a Fiat XI/9 equipped with ATE master cylinder and hydraulic booster.*

The transient response behavior of the rear caliper brake line pressure in response to a near-instant accumulator pressure application electrically operated by a solenoid servo valve is shown in Figure 5-26. Both the human pedal force effects and the master cylinder/booster characteristics were bypassed in the experiment. Inspection of Fig. 5-26 reveals that the rear brake pressure response is not noticeably slowed by the master cylinder/booster system. The rotating wheel stopped approximately 0.06 s after the brake line pressure began to rise in the rear brake caliper.

$p \, [\text{N/cm}^2] = 0.6894 \, p \, [\text{psi}]$

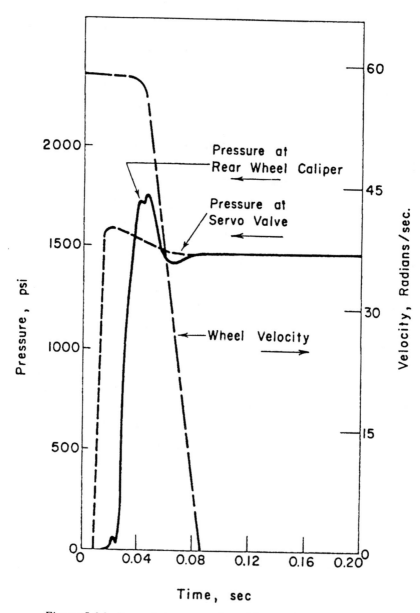

Figure 5-26. Recorded servo-activated brake system transients.

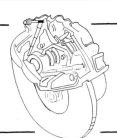

# CHAPTER 6

# Analysis of
# Air Brake Systems

*In this chapter the basic functions of air brake systems and their components are described. Equations for computing vehicle deceleration are presented for S-cam and wedge brakes. In-stop fade and the effects of brake adjustment on braking effectiveness are analyzed. Empirical relationships for predicting system response times are introduced. Air-over-hydraulic brakes are discussed.*

## 6.1 Basic Concepts

Air brakes are power brake systems in which compressed air is used as the energy medium. The brake pedal effort of the operator is used only to modulate the air pressure applied to the brake chambers. In the event of a failure of the energy source, no manual push-through by the driver is available and, hence, no brake line pressure is produced. Since 1975, all air brake systems have to comply with FMVSS 121. Based on US-DOT studies, FMVSS 121 was expected to prevent 29,000 truck related accidents, save 500 lives, and prevent 25,000 injuries. Brake systems must have a dual air brake system so that in the event of one circuit failure, emergency braking function is maintained. Although the energy source is compressed air, the transmission of the energy from the brake chambers to the friction surfaces involves lever arms, cams, and rollers, or wedges. In the case of air-over-hydraulic brakes, the air pressure is converted into hydraulic pressure, which is used to press the shoes against the drums. All air brake systems have certain components in common. Differences exist between straight trucks, tractors, and trailers. Differences also exist in some design details, particularly with respect to the number of air tanks. In some cases, the functions of two air tanks are combined into one larger air tank. In addition, customer preference may introduce slight changes in terms of valves used. Design details vary also between conventional and cab-over vehicles (tilt cab).

The basic tractor trailer system functions as follows: The compressor charges a wet supply reservoir from which two tractor (or truck) reservoirs are fed, namely one front and one rear circuit reservoir. The compressor also charges two reservoirs of the trailer: the trailer service reservoir and the spring brake reservoir. The dual air brake system is modulated by the driver through the dual brake application valve. When the brake application valve is released, all brake chambers exhaust through their respective quick release valves. When the front brake circuit fails, a double check valve and reservoir single check valve immediately close off the front circuit to protect the rear circuit, which continues to function normally. A similar protection is installed in the event of a rear brake circuit failure. Because of the double check valve, air is supplied to the tractor and trailer spring brakes and the trailer service brakes if the tractor rear system becomes inoperative. If both the front and rear brake systems become inoperative, spring brakes will apply automatically when the air pressure drops below approximately 40 psi.

Some vehicle are also equipped with a spring brake control valve. The valve allows the operator to modulate spring brake application by using the brake application valve in the event of a rear circuit failure. This allows the rear spring brakes to operate like service brakes up to the capacity of the spring force equivalent to approximately 60 psi brake line pressure. The spring brake control valve also eliminates the need to activate dash control valves in the case of an emergency and provides normal application and release of the spring brakes.

In the event of a complete trailer breakaway the trailer brakes are applied automatically.

## 6.2 Brake System Plumbing Schematics

### 6.2.1 Straight Truck Without Trailer

The basic schematic of a piping diagram is shown in Figure 6-1. The brake system of the truck is not designed to supply air pressure to a trailer.

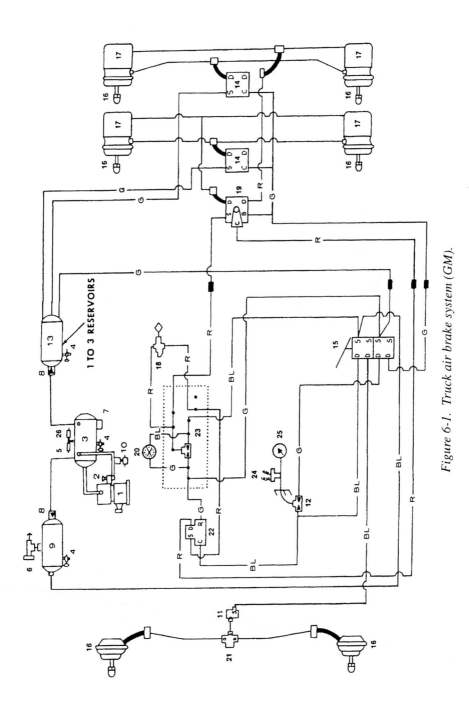

*Figure 6-1. Truck air brake system (GM).*

The numbers shown identify different brake system components. The letters written in the various brake lines designate color to assist in following lines from one brake component to the next: BK = black, BL = blue, G = green, R = red, and Y = yellow. The letters written at the different valves identify inlet or exit port functions: B = balance, C = control, D = delivery, R = rear pressure, S = supply.

1. Air compressor

2. Compressor governor

3. Wet supply reservoir

4. Drain cock

5. Safety pressure valve

6. Pressure protection valve

7. Automatic drain valve

8. One-way check valve

9. Front system reservoir

10. Low pressure switch

11. Automatic front brake limiting valve (ratio valve)

12. Double check valve

13. Rear system reservoir

14. Service relay valve (if ABS wheel lock control modulator)

15. Dual application valve (also called foot valve or treadle valve)

16. Service brake chamber

17. Spring brake chamber

18. System park control valve

19. Spring brake relay valve

20. Dual air gage

21. Quick release valve

22. Spring brake control valve

23. Instrument package manifold valve

24. Stoplight switch

25. Application pressure air gage

26. Filler valve

## 6.2.2. Tractor Brake Schematic

A typical plumbing schematic for a tractor designed to tow a semitrailer is illustrated in Figure 6-2. The letter codes are identical to those of Section 6.2.1. The numbered components are as follows:

1. Air compressor

2. Compressor governor

3. Wet supply reservoir

4. Drain cock

5. Safety pressure valve

6. Pressure protection valve

7. Automatic drain valve

8. One-way check valve

9. Front system reservoir

10. Low pressure switch

11. Automatic front brake limiting valve (ratio valve)

12. Trailer couplings

13. Rear system reservoir

14. Service relay valve (if ABS wheel lock control modulator)

15. Dual application valve (also called foot or treadle valve)

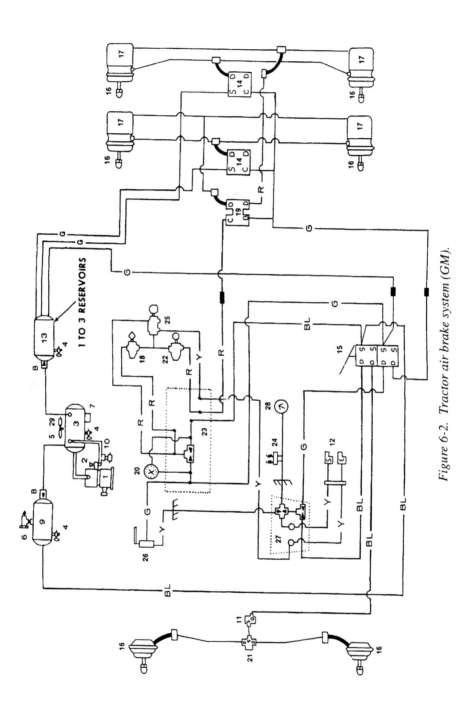

Figure 6-2. Tractor air brake system (GM).

16. Service brake chamber

17. Spring brake chamber

18. System park brake control valve

19. Combination quick release/double check valve

20. Dual air gage

21. Quick release valve

22. Tractor-only parking brake control valve

23. Instrument package manifold valve

24. Stoplight switch

25. Trailer supply valve

26. Trailer brake hand control valve

27. Combination tractor protection/quick release valve

28. Application pressure air gage

29. Filler valve

## 6.2.3 Trailer Brake Schematic

The basic plumbing of a trailer brake system is illustrated in Figure 6-3. Compressed air comes through the supply line from the tractor protection valve into the spring brake control valve to charge both the trailer service reservoir and the spring brake reservoir. The control of the trailer service brake is accomplished by the service relay valve or, in the case of ABS, by the wheel lock control valve. Differences may exist for trailer systems using two service reservoirs, or for dollies employing speed-up valves.

Older trailers manufactured before 1975 when FMVSS 121 became effective used a relay/emergency valve. The relay portion is used for the normal driver modulated trailer service brake application. The emergency portion of the valve is used to apply all trailer service reservoir air pressure to the trailer brakes in the event of an air loss in the supply (also called emergency) line. The emergency portion of the valve is also used as a trailer

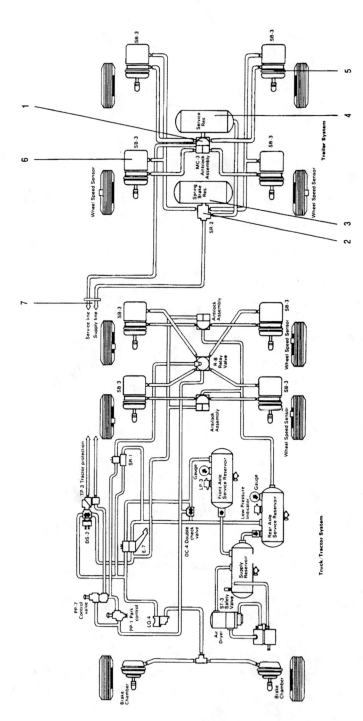

*Figure 6-3. Typical "121" dual tractor/trailer air brake system.*

parking brake since no mechanical spring brakes were used. The parking brake effectiveness would usually diminish because air leaks over time would deplete air pressure.

1. Service relay valve

2. Trailer spring brake valve

3. Spring brake reservoir

4. Service brake reservoir

5. Brake chamber

6. Spring brake

7. Glad hands

# 6.3 Brake System Components

## 6.3.1 Air Compressor and Governor

The dual-piston air compressor provides the energy source for the brake system. It is driven by the truck engine and runs whenever the engine is running. Compressors are either belt or gear driven and generally lubricated with engine oil. Worn compressors may introduce a substantial amount of oily sludge-type contamination into the air brake system, easily seen when the wet tank is drained.

When the air pressure in the system reaches a preset level of approximately 83 N/cm$^2$ (120 psi), the governor high pressure setting, the governor opens by holding the inlet valves off their seats. As long as this condition exists, air is merely pumped between the two cylinders of the compressor without increasing reservoir pressure.

When the air pressure drops in the system to approximately 72 N/cm$^2$ (105 psi), the low pressure setting, the governor closes, the intake valves return to their seats, and compression resumes.

The proper functioning of the compressor and governor can be tested by completely draining the reservoir, and by measuring the time required to build pressure. With the engine at fast idle of approximately 1000 rpm, the compressor cutout pressure should be reached in less than five minutes. With the engine at 2100 rpm, compressors in good mechanical condition will reach maximum system pressure in one to two minutes when both tractor and trailer are connected.

## 6.3.2 Air Pressure Reservoir

The basic function of the reservoir is to store compressed air and hold it available when the operator makes a brake application. Another purpose is to provide a place for the air heated during compression to cool and for water vapor to condense and collect.

The compressed air coming from the compressor first enters the supply reservoir. Since the air contains water due to the moisture drawn with the air into the compressor, water will accumulate in the supply reservoir. This is the reason why the first reservoir is also called wet tank or wet reservoir. In some cases the wet supply reservoir is combined with the regular reservoir into one unit with a partition separating the two. Reservoirs consist of rolled welded steel plates with the inside painted in a special process to minimize corrosion. The moisture in the compressed air will be a safety hazard when the vehicle is operating in below-freezing temperatures. Under those conditions the proper functioning of valves may be impaired due to the freezing and stiffening of seals or lack of movement of plungers. The wet supply reservoir must be located at the lowest point of the air brake system to allow moisture to drain toward the wet tank. The fitting for the drain cock or moisture ejection valve must be located at the bottom of the tank in a location where it can be reached easily. Operators or maintenance personnel must drain the wet supply reservoir daily. Automatic ejection valves ensure that the system is drained on a regular basis.

From the wet supply reservoir the air is discharged into several "dry" service reservoirs. All pressure for operating the brakes is taken from the dry tanks.

The air tanks are mounted against the frame by straps or support bands, or bolted against the frame by brackets welded against the tanks. When brackets are used, potential for fatigue fracture exists.

### 6.3.3  Safety Valve

A safety valve is installed in the wet air tank to prevent air pressure buildup beyond a certain safe limit in the event the air compressor governor fails to function properly.

When the system pressure rises to approximately 150 psi, air pressure forces a spring-loaded ball off its seat and vents the compressed air to the outside atmosphere. When the air pressure has decreased sufficiently low, the spring will close the valve and seal off the reservoir. In most cases the pressure setting is stamped on the valve body.

### 6.3.4  Filler Valve

A filler valve is located on the supply reservoir to provide an efficient means of filling of the air system from an outside source of compressed air. It should be checked periodically with a soap solution to ensure no leakage.

### 6.3.5  Automatic Drain Valve

The automatic ejector valve drains moisture and contaminants from the wet supply reservoir. When the air compressor is operating to increase reservoir pressure, the inlet valve opens allowing moisture and contaminants to collect in the sump of the ejector.

When the reservoir pressure drops slightly by approximately $1.4 \text{ N/cm}^2$ (2 psi) when air is used from the system, air pressure in the sump cavity opens the exhaust valve and allows moisture and contaminants to be ejected. Manual draining can be done. Maintenance at regular intervals is required.

### 6.3.6  One-Way Check Valve

A one-way check valve is installed in each air line leading from the wet supply reservoir to the service reservoirs. Installed at the inlet to the wet air tank, the one-way check valve prevents loss of air from the service reservoir in the event of a leakage in the supply reservoir, air compressor, or connecting lines or hardware.

Air flow in the normal direction moves the check valve disc from its seat allowing unobstructed air flow into the service reservoir. Flow in the opposite direction is prevented by the disc closing off the valve.

### 6.3.7 Pressure Protection Valve

This valve shuts off the air line to accessory components such as air horn, air seat, transmission shift control and others when the air pressure in the main system has fallen to approximately 65 psi. Eliminating other nonessential air users ensures that sufficient air pressure remains to apply the service brake or to release the spring brakes.

The valve may be of the diaphragm or piston-type design. It is located at one of the service reservoir tanks at the outlet to the accessory air line.

### 6.3.8 Low Air Pressure Switch

The low air pressure switch activates when the air system pressure in the supply tank has decreased to approximately 45 to 52 $N/cm^2$ (65 to 75 psi). The switch closes an electrical circuit which sounds an audio buzzer and illuminates a low pressure warning light. In some older models a red metal "flag" would appear to visually alert the driver of the unsafe conditions of the air pressure system.

### 6.3.9 Brake Application Valve

The brake application valve controls the level of braking of the truck or tractor through the pedal force and displacement by the driver. Foot force on the treadle pushes a plunger down, moving internal valve parts and allowing air flow to the exits or delivery ports of the valve and into the system. The application valve also controls the service brake system of the trailer by way of the trailer control valve.

In the past, single-circuit application valves were used. Since the mid-'70s dual-circuit application valves are used to meet the dual-circuit safety requirements of air-brake-equipped vehicles (Figs. 6-1 and 6-2).

Application valves come in basically two designs, namely floor- and fire-wall-mounted (suspended) configurations. The working principles are identical for either design.

Air at reservoir pressure is constantly supplied to the brake application at its inlet port. Brake lines, running from the application valve to the various valves and wheel brakes, contain air only when the brakes are applied, and then only at the pressure demanded by the driver. From the pushrod, pedal force is transferred to the metering spring, which strokes the piston against its return spring. During this stroke, the inlet-exhaust valve cartridge closes off the exhaust port. The continued stroke unseats the inlet poppet, permitting compressed air to flow through the valve delivery ports into the brake system. Compressed air also is bypassed to the piston through an equalization orifice, and pressure beneath the piston forces it to move to compress the metering spring. The piston reaches a balanced position between these two opposing forces. Further movement of the pushrod unbalances the forces and admits higher air pressure to increase brake force. Brake pedal release exhausts the system.

Dual-circuit application valves utilize two separate supply and delivery circuits for each of the individual brake circuits. Valve designs may vary between manufacturers.

Bendix operates the number one or primary circuit, which usually provides air pressure to the rear brakes, through mechanical action of the pedal/treadle/plunger system. The number two or secondary circuit operates similarly to a relay valve. Control air is delivered from the number one circuit, which allows a corresponding air flow in the number two circuit. In the event of a number one circuit air failure the secondary inlet valve is opened mechanically by a push-through of the mechanical force from the driver's foot via the treadle/plunger and primary piston.

The Bosch service-brake valve shown in Figure 6-4 consists of two separate brake valves arranged one behind the other and actuated by a common control. The brake pressure can be modulated delicately in both circuits. A short activation distance is achieved by using a preloaded spring. If one circuit fails, the other one remains fully functioning.

The operation of the Bosch service brake valve is explained next.

No Brake Application (Fig. 6-4): When the brakes are not applied, the cup seals (7) and (14) contact the inlet valve seats (8) and (13). The compressed air cannot flow into the brake circuits 1 and 2 by way of ports 21 and 22. Ports 21 and 22 are connected to the atmosphere vent 3 so that both brake circuits are vented.

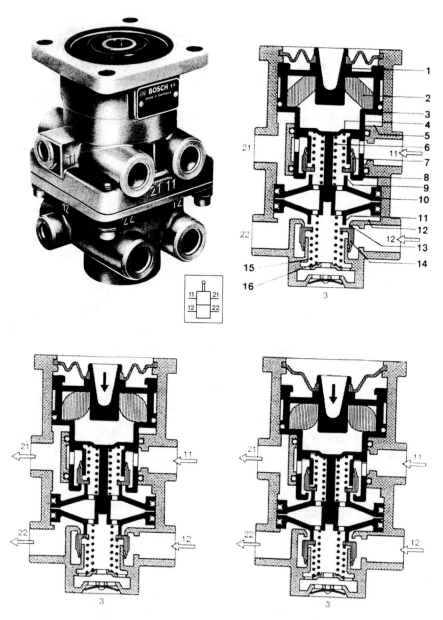

*Figure 6-4. Brake application valve (Bosch).*

Partial Brake Application: The brake pedal is partially depressed. The tappet (1) pushes against the reaction piston (3) downward by way of the travel-limiting spring (2) until the discharge valve seat (9) closes. The rocking piston (10) is pushed downward by the plunger spring (6) so that the discharge valve seat (11) also closes and then the inlet valve seats (8) and (13) open. The inlet valve seats remain open until the compressed air entering through port 11 has built up sufficient pressure beneath the reaction piston (3) to push the reaction piston upward against the force of the travel-limiting spring (2) and closes the inlet valve seat (8) again. The inlet and outlet of brake circuit 1 are closed, the valve is in the center position.

Together with the reaction piston (3), the rocking piston (10) moves upward and closes the inlet valve seat (13) so that the brake pressures in circuit 1 and 2 are equal. The level of brake line pressure in both circuits is a function of the actuation force. The more the brake pedal is depressed, the higher the brake line pressures will be.

Full Brake Application: During full brake application, the brake pedal is completely depressed. Tappet (1) is pressed down to such an extent that, after overcoming the force of the travel-limiting spring (2), the reaction piston (3) is pushed downward until it reaches stop (5). The rocking piston (10) is also pressed downward by the helical compression springs (4) and (6) until it reaches stop (12). During the downward motion of these two pistons the two discharge valve seats (9) and (11) close first, then the two inlet valve seats (8) and (13) open and remain open as long as the brake pedal is depressed. During full brake application the brake line pressure in both circuits is equal to the pressure of the reservoir.

Failure on One Brake Circuit: If one brake circuit fails, the pressure in the unfailed brake circuit can still be controlled in small modulating steps by the driver. If brake circuit 1 (port 11) fails so that piston (3) cannot act as the reaction piston during partial brake application, the rocking piston (10) performs the function of the reaction piston and closes off the inlet valve seat (13) as soon as sufficient pressure has built up beneath the piston during partial brake application. Furthermore, regardless of which circuit fails in the tractor, the trailer brakes are actuated by the remaining unfailed brake circuit of the tractor or truck.

Typical application valves in use on domestic air-braked vehicles require only small plunger displacements to actuate air flow to the delivery ports. Depending on the metering spring and other parts used, a soft or hard pedal feel may be obtained. A soft pedal is achieved by requiring large plunger travel to obtain large pressure output. For example, when expressed in terms of brake pedal angular rotation, the first 8 to 10 degrees may not cause any air flow, while a pedal rotation from 10 to 20 degrees will deliver a linear increase of air pressure at the delivery ports of approximately 52 N/cm² (75 psi) with a rapid increase of pressure to its maximum value of approximately 79 to 83 N/cm² (115 to 120 psi) during the remaining 2 or 3 degrees of pedal rotation. Typically, not more than 10 to 15 degrees of pedal rotation are required from initial pressure rise to maximum reservoir pressure at the exit or delivery ports of the application valve.

In general, a pedal force of 180 to 269 N (40 to 60 lb) applied approximately 203 mm (8 in.) from the fulcrum pin produces a brake line pressure of 41 to 48 N/cm² (60 to 70 psi), whereas a force range of 269 to 356 N (60 to 80 lb) produces a brake line pressure of approximately 62 N/cm² (90 psi).

## 6.3.10 Automatic Front Brake Limiting Valve

The valve reduces the exit pressures to the front brakes in relationship to the inlet pressures. The purpose of the automatic limiting valve is to limit the level of braking done by the front brakes for low- to medium-pressure applications to ensure steerability of the vehicle on low-friction road surfaces. Since the ratios between delivery or exit port and inlet pressures are fixed by design, i.e., they cannot be selected by the driver, automatic limiting valves are also called ratio valves. In the past, variable ratio valves, allowing the driver to select a slippery road mode with low exit port pressures or a dry road mode for high exit port pressures, had been in use.

Typically, automatic limiting valves reduce the supply pressure delivered to the valve by approximately 50% as it passes through the valve for supply pressures between 0 to 28 N/cm² (0 to 40 psi). Supply pressures between 28 and 41 N/cm² (40 and 60 psi) are reduced by less than 50% with no reduction for supply pressures above 41 N/cm² (60 psi).

The advantage of the automatic limiting valve is that full braking is available on the front axle with maximum pedal force while inadvertent lockup is less likely on slippery road surfaces. With the manually operated valve, inadvertent setting of the valve to the slippery road mode could result in longer stopping distances while actually braking on a dry road.

A major disadvantage of the valve is that the front brakes are not fully braked, thus allowing the rear brakes of a straight truck or tractor to lock first. Locking of the truck rear axle while the front wheels are still rolling causes directional instability with the truck spinning out of control if it is traveling at a sufficiently high speed. Similarly, for tractor-trailer combinations, locking of the tractor rear brakes first causes jackknifing (see Chapter 8).

Use of automatic limiting valves on axles other than the front axle will also minimize the potential for inadvertent lockup on that axle. For example, installing an automatic limiting valve on the tractor rear axle will improve directional stability during braking and pedal control, since premature rear axle locking is largely prevented. Installing automatic limiting valves on rear axles will affect brake wear balance among axles and may increase the basic response time of the brake system. Advantages and disadvantages must be evaluated carefully by means of a design solution selection table (see Chapter 1) and testing.

## 6.3.11 Quick Release Valve

The purpose of the quick release valve is to reduce the time required to release the brakes. When the brakes are being released, the exhaust of the quick release valve opens and the air pressure from the line is exhausted through the quick release valve rather than going all the way back through the brake application valve.

A basic schematic of a quick release valve is shown in Figure 6-5. Air pressure from the brake application valve enters the valve through the port above the diaphragm and forces the center of the diaphragm to seat tightly against the exhaust port. Air pressure also overcomes diaphragm cup tension to deflect the outer edges of the diaphragm, and air flows through the side ports to the brakes. During release the pressure above the diaphragm is released quickly and the brake line pressure coming from the brakes lifts the center of the diaphragm from the exhaust port and permits air to escape directly to the atmosphere.

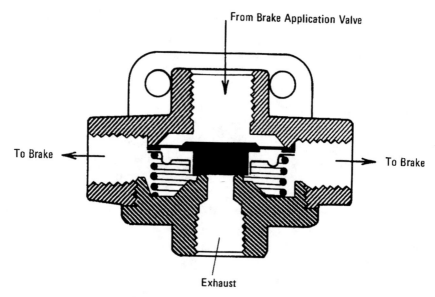

*Figure 6-5. Quick release valve.*

## 6.3.12 Relay Quick Release Valve

The relay portion of the valve acts as a remote control relay station to speed up the application and release of the brakes. The relay valve is usually located at the rear of the vehicle, closest to the brake chamber(s) it serves.

A basic schematic of a relay quick release valve is shown in Figure 6-6. Typically, the relay valve will have a supply or reservoir port which provides reservoir pressure to the valve from a reservoir located close to the axle(s) braked, a service port which receives the signal pressure from the application valve, one or more delivery ports to permit air pressure to the brake chambers, and an exhaust port which permits air to be exhausted directly to the atmosphere when the brakes are released.

Until the brake application cycle starts, the relay inlet valve is closed and the exhaust is open to the atmosphere. When the brakes are applied, the metered air pressure from the brake application valve forces the relay piston down, and closes the exhaust port. Further movement of the piston opens the inlet

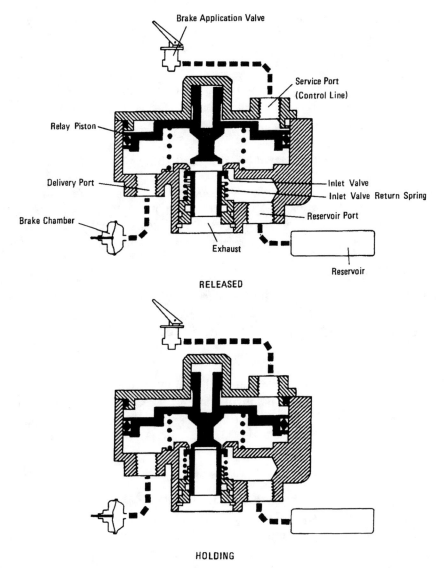

RELEASED

HOLDING

*Figure 6-6. Relay quick release valve.*

valve, allowing air to pass from the reservoir to the valve, pass through the delivery ports, and on to the brake chambers. As brake line pressure underneath the piston equals the controlling pressure above, the piston balances and allows the inlet valve return spring to close the inlet valve. When the brakes are released, the decrease in controlling pressure unbalances the relay piston and permits the exhaust port to open, thus releasing brake line pressure directly to the atmosphere.

Spring brake relay valves are used in straight trucks to control the operation of the emergency spring brakes of the truck. Tractors do not use this valve for spring brake operation.

## 6.3.13 Wheel Lock Control Modulator (ABS)

For air-braked vehicles manufactured between 1975 and late 1978, FMVSS 121 had certain no-lock brake lock requirements. Optional anti-lock brake systems (ABS) were available during later model years. Consequently, properly functioning or disabled ABS systems may be found on existing vehicles.

The wheel lock control modulator takes the place of the relay valve. It is located near the axle(s) it controls. The modulator is connected to the cab and to wheel speed sensors by cables (see Chapter 9). The valve contains the computer or logic center, the solenoid to operate the valve, and a relay valve.

The valve performs as a standard relay quick release valve during all normal braking applications not involving any brake lockup. During impending wheel lockup the valve functions as a solenoid-operated air release valve.

## 6.3.14 Spring Brake Control Valve

The valve indirectly ties together the front and rear service brake systems. The purpose of the spring brake control valve is to provide a modulated spring brake application in the event of inadequate air pressure in the rear (primary) service brake system due to a brake failure. The valve eliminates the need to activate dash-mounted control valves in an emergency situation and provides normal application and release of the spring brakes. When both circuits are working properly, the valve is balanced by air pressure from the front and rear circuit. When balanced, the spring brake control valve maintains system pressure in the spring brake chamber to keep the springs compressed, i.e., the spring brakes released.

### 6.3.15 System Parking Brake Control Valve

All vehicles equipped with air brakes, whether they are straight trucks or tractors, have a parking brake control located in the cab to apply and release the parking brake of the vehicle. For a truck, the brakes equipped with spring brakes are the parking brake; for a tractor-trailer combination the spring brakes on the tractor and trailer are the parking brake.

The shape of the knob is square, usually painted yellow. The valve is basically a pull-push on-off type semi-automatic valve. When the knob is pulled out, the parking brakes are applied. With 50 psi or more air pressure in the supply or reservoir system the control knob will stay in. When the system air pressure falls below approximately 28 N/cm² (40 psi), the valve will automatically pop out the control valve knob and the delivery line air pressure to the spring brake chambers will be released and the brakes will be applied.

For tractors coupled to trailers the valve must be in the released position, i.e., the knob must be pushed in, to charge the trailer brake reservoir.

### 6.3.16 Tractor-Only Parking Control Valve

The valve is used optionally to apply the tractor parking brakes only. The knob is round and usually painted blue. It functions in the same fashion as the system parking brake valve.

### 6.3.17 Tractor Protection Valve

The primary function of the tractor protection valve is the protection of the tractor air brake system under trailer breakaway conditions and/or when severe air leakage develops in the tractor or trailer. The valve operates automatically or can be operated by the cab-mounted trailer air supply valve. When the air system pressure in the trailer falls below approximately 31 N/cm² (45 psi), the valve automatically disconnects the air lines to the trailer. In pre-1970 vehicles the relief pressure setting may be as high as 55 N/cm² (80 psi). In addition, in every use the valve is used to shut off the trailer service and supply lines before disconnecting the tractor from the trailer by discoupling of the glad hands. Over the years changes have been made in response to pre- or post-FMVSS 121 requirements. Both older and newer models use dash-mounted control valves to remotely control the tractor protection valve.

## 6.3.18 Trailer Supply Valve

The trailer supply valve is located in the cab generally next to the parking brake control valves. Its knob is octagonal and usually painted red. The knob must be pushed in for the trailer reservoir to be charged. When the system air pressure falls below approximately 28 N/cm$^2$ (40 psi), the exhaust of the valve will open and automatically vent the trailer supply air to the atmosphere. Venting of the trailer supply air will activate the tractor protection valve. The trailer supply valve provides in-cab control for the tractor protection system and functions in conjunction with the tractor protection valve.

## 6.3.19 Trailer Brake Control Valve

The trailer brake control valve, either dash- or steering-column-mounted, provides graduated brake line pressure to the trailer brakes. The level of brake line pressure is a function of the angle through which the lever is rotated. The handle will remain in the applied position until manually moved.

When a foot brake is also applied while the trailer brake is applied, the valve releasing the greater amount of brake line pressure will control the brakes to which it is connected. Due to the design of the double check valve (see Section 6.3.20), the line with the higher pressure will pass through the exit port of the double check valve. If, for example, the foot brake application provides a lower pressure than the trailer hand control valve, then the double check valve will allow the higher trailer pressure to go to the trailer brakes, while the lower tractor service brake pressure goes to the tractor brakes.

The trailer hand control valve operates only the service brakes of the trailer and does not actuate the parking or spring brakes of the trailer. Consequently, the trailer hand control valve must not be used as a parking brake. Air leakage in the trailer brake system may render the trailer service brake without air line pressure and, hence, without brakes.

## 6.3.20 Double Check Valve

The double check valve is used to direct air pressure from two different sources into a common line. For example, it is used between the brake application valve and trailer hand control valve with the common line running to the rear brakes of the tractor and brakes of the trailer.

A movable shuttle piston on the inside seals off the inlet of one of the inlet ports depending on the air pressure against it from the opposite end. Due to the geometry of the valve and the size of the shuttle piston, air pressure is never cut off from the outlet port.

A combination double check valve/stoplight switch is used in some brake systems equipped with trailer controls. This check valve is used in the normal fashion to connect both the foot brake application valve and the trailer hand control to the trailer brake service line. In addition, every time brake line pressure of more than $2.75$ $N/cm^2$ (4 psi) exits at the outlet port to apply the brakes, the stoplight switch closes to illuminate the stoplights of the vehicle.

A combination quick release/double check valve is used on vehicles with the trailer brake controls. It is plumbed into the parking brake circuit near the rear axle. Its purpose is to prevent brake pressure compounding and to provide a rapid air release location for the spring brake chambers for parking brake apply.

## 6.3.21 Dual Air Pressure Gage

The dual air pressure gage is located in the instrument panel within easy view by the driver. The dual gage has two needles of contrasting colors. One needle indicates reservoir air pressure in the front brake system, while the other needle indicates rear system air pressure.

The dual air pressure gage provides a simple means by which the driver can determine the adjustment condition of the foundation brakes. With the air reservoirs fully charged and the engine stopped, a full brake pedal application causes a certain amount of compressed air to apply the brake chambers. The travel of the diaphragm and pushrod in the brake chambers is a function of the adjustment of the brakes and will require more air volume for out-of-adjustment brakes. However, more air volume used results in a greater air pressure drop indicated on the dual air gage. For brakes in good adjustment a pressure drop between $5.5$ and $8.3$ $N/cm^2$ (8 and 12 psi) is permissible. When the air pressure drops between 14 to 17 $N/cm^2$ (20 to 25 psi) or approximately 20% of maximum reservoir pressure brake adjustment must be made.

The vehicle should never be driven with a pressure reading of less than $45$ $N/cm^2$ (65 psi).

## 6.3.22 Application Air Pressure Gage

The application air pressure gage measures the air pressure going from the brake application valve to the brakes. It provides a means by which the driver may know how hard the brakes are being applied.

## 6.3.23 Air Dryer and Aftercooler

The aftercooler is designed to remove water and other contaminants from the air coming from the compressor. It is a heat transfer device with the heat coming from the compressed air. It requires that cooling air flows over it as the vehicle moves. The cooler should not be obstructed and no after-market changes should be made to it.

## 6.3.24 Alcohol Evaporator

The alcohol evaporator provides a means of putting vaporized alcohol into the air system of the vehicle as an aid to prevent freezing up at low ambient temperatures. When used, it is installed in the charge line coming from the compressor to the wet supply reservoir. The evaporator functions automatically, but still requires daily reservoir drainage and alcohol level checking.

## 6.3.25 Air Lines

Air lines connect the different components to form the brake system. Metal tubing, nylon tubing, and flexible hoses are used. In general, all signal lines carrying variable pressure have a diameter of 9.5 mm (3/8 in.), and all supply and constant pressure lines have a diameter of 12.7 mm (1/2 in.).

Metal tubings are annealed copper with three-piece compression fittings. Flared-type fittings should not be used.

Nylon tubing is used for air lines in areas of the vehicle where this material can be used safely, i.e., away from heat sources exceeding 366 K (200°F). Nylon lines are color coded for easy identification of subsystem functions:

Black ..................... accessory

Blue ................... front system

Green ................. rear system

Red ................. parking brake

Yellow ............ trailer system

Flexible hoses are used at each brake chamber, between frame and axle, and at trailer connection where constant flexing prevents the use of rigid metal tubing. Wire braid air lines are heavier and generally only available as an option.

## 6.3.26 Brake Chambers

Compressed air entering the brake chamber is converted into mechanical work in terms of pushrod force multiplied by pushrod travel. In the case of cam brakes the pushrod connects to a slack adjuster, which turns a camshaft and cam, which applies the brake shoes (see Chapter 2). A standard brake chamber, pushrod, and slack adjuster assembly is shown in Figure 6-7. A combination service brake/spring brake is shown in Figure 6-8. It is used as a service brake chamber, an emergency brake in case of air pressure loss, and a spring-applied parking brake. The regular service brake chamber located closest to the slack adjuster applies the service brake in a normal

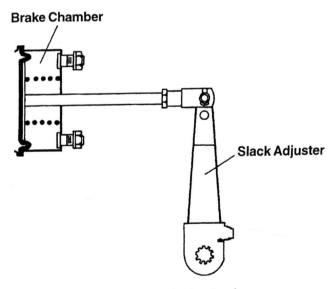

*Figure 6-7. Air brake chamber.*

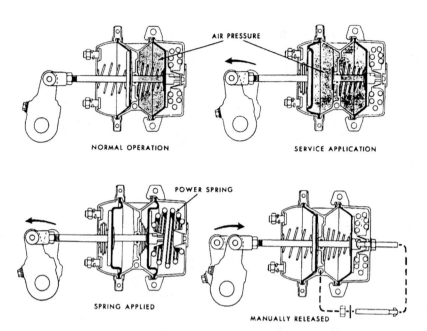

NORMAL OPERATION          SERVICE APPLICATION

SPRING APPLIED

MANUALLY RELEASED

*Figure 6-8. "Anchorlok" spring brake chamber.*

fashion. The spring-loaded parking/emergency brake (power spring) located behind the service brake chamber moves the pushrod, i.e., applies the brake as the spring is allowed to expand by releasing the hold-off air pressure in the spring brake chamber.

The brake chambers illustrated in Figs. 6-7 and 6-8 use a diaphragm. The size of the brake chamber is identified by the nominal area size of the diaphragm. For example, a brake chamber type 30 has a nominal area of $194 \text{ cm}^2$ ($30 \text{ in.}^2$) against which air pressure acts. The actual effective area involved in pushrod force production is approximately 5 to 10% smaller due to the rolling action of the diaphragm. Standard brake chamber sizes, maximum and adjustment stroke are shown in Table 6-1.

## TABLE 6-1

### Pushrod Adjustments

| | Chamber Type | Overall Diameter in. | Overall Diameter cm | Maximum Stroke at Which Brakes Should Be Readjusted (for disc brakes, consult manufacturer) in. | cm |
|---|---|---|---|---|---|
| Bolted Flange | A (12) | 6-15/16 | 17.62 | 1-3/8 | 3.49 |
| Brake Chambers | B (24) | 9- 3/16 | 23.34 | 1-3/4 | 4.45 |
| | C (16) | 8- 1/16 | 20.48 | 1-3/4 | 4.45 |
| | D (6) | 5- 1/4 | 13.34 | 1-1/4 | 3.18 |
| | E (9) | 6- 3/16 | 15.72 | 1-3/8 | 3.49 |
| | F (36) | 11 | 27.94 | 2-1/4 | 5.72 |
| | G (30) | 9- 7/8 | 25.1 | 2 | 5.08 |
| Clamp Ring | 9 | 5- 1/4 | 13.34 | 1-3/8 | 3.49 |
| | 12 | 5-11/16 | 14.45 | 1-3/8 | 3.49 |
| | 16 | 6- 3/8 | 16.19 | 1-3/4 | 4.45 |
| | 20 | 6-13/16 | 17.30 | 1-3/4 | 4.45 |
| | 24 | 7- 1/4 | 18.42 | 1-3/4 | 4.45 |
| | 30 | 8- 1/8 | 20.64 | 2 | 5.08 |
| | 36 | 9 | 22.86 | 2-1/4 | 5.72 |
| Rotochambers | 9 | 4- 9/32 | 10.87 | 1-1/2 | 3.81 |
| | 12 | 4-13/32 | 11.19 | 1-1/2 | 3.81 |
| | 16 | 5-13/32 | 13.73 | 1-7/8 | 4.76 |
| | 20 | 5-15/16 | 15.08 | 1-7/8 | 4.76 |
| | 24 | 6-13/32 | 16.27 | 1-7/8 | 4.76 |
| | 30 | 7- 1/16 | 17.94 | 2-1/4 | 5.72 |
| | 36 | 7- 5/8 | 19.37 | 2-5/8 | 6.67 |
| | 50 | 8- 7/8 | 22.54 | 3 | 7.62 |

A typical diaphragm-type brake chamber pushrod force curve as a function of pushrod travel is shown in Figure 6-9. Inspection of Fig. 6-9 reveals that the pushrod force decreases for pushrod travels exceeding certain critical values making adjustment as shown in Table 6-1 necessary.

Roto chambers use a piston-like design rather than a diaphragm to seal the chamber, resulting in a linear pushrod force output as pushrod travel increases.

The angle formed between the pushrod and slack adjuster should be 90 degrees when one-half or less of the pushrod travel is used.

In selecting an air brake chamber the following factors must be considered:

For a diaphragm chamber:

    Advantages:          *   Compact size requires less space

                                    *   Less sensitive to dirt

                                    *   Low wear

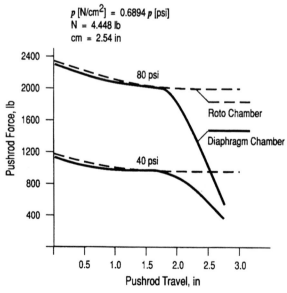

*Figure 6-9. Diaphragm and roto chamber pushrod force vs. travel.*

Disadvantages:          * Relatively small pushrod travel, therefore
                          high adjustment frequency

                        * Decreasing pushrod force


For a roto air chamber:

Advantages:             * Relatively long pushrod travel, therefore
                          low adjustment frequency

                        * Constant pushrod force over pushrod
                          travel

Disadvantages:          * Installation length longer

                        * Dirt and wear sensitive

                        * Frequent checking of dust boot damage
                          to avoid entry of contaminants


# 6.4  Air-Over-Hydraulic Brake Systems

Standard air brakes transmit the force produced in the brake chamber to the brake shoes by mechanical means in the form of rods, slack adjusters, shafts, cams, and rollers in the case of S-cam brakes, or wedges and rollers in the case of wedge brakes. Air-over-hydraulic brakes use standard hydraulic brake system components such as brake lines, wheel cylinders, and a slave cylinder similar to a master cylinder to transmit the air-pressure-produced braking energy to the wheel brakes. Air-over-hydraulic systems are used frequently when greater braking capacity is required but space limitations make the use of high-pressure hydraulic apply components necessary. For example, front brakes on three-axle heavy trucks may be equipped with hydraulically actuated disc brakes while the rear are standard air brakes.

A typical air-to-hydraulic conversion cylinder is illustrated in Figure 6-10. The larger assist cylinder is controlled by the brake application valve of the air brake system. The pushrod force applies the master cylinder piston to pressurize hydraulic brake fluid. The pressure ratio is a function of the ratios of the cross-sectional areas of both pistons.

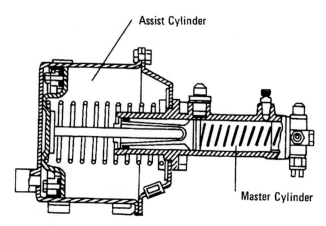

*Figure 6-10. Air-over-hydraulic brake unit.*

An integral air-over-hydraulic unit is shown in Figure 6-11. The released position of the brakes is illustrated in Figure 6-11(A). Compressed air is separated from the booster section. Both sides of the diaphragm (1) are exhausted to ambient air and the return spring (2) forces the piston to the far left position. The maximum brake application is illustrated in Figure 6-11(B). A brake pedal application causes the chambers located to the left and right side of the diaphragm (1) to be separated by the small valve (3). Further movement of the brake pedal causes the compressed air to be applied to the diaphragm by opening of the compressed air valve (4). The result is a booster application to the master cylinder. In the event of a medium brake application, the compressed air valve (4) closes and a constant air pressure is applied to the booster piston. In a full application the compressed air valve remains open as illustrated in Figure 6-11(B).

## 6.5 Brake Torque

The brake torque $T_B$ of a drum brake is determined by use of the definition of brake factor (see Section 2.8.1). It may be computed by the following expression:

$$T_B = (p_\ell - p_o)A_c BF\eta_m \rho k_A k_T r \quad , \quad \text{Nm (lb-in.)} \tag{6-1}$$

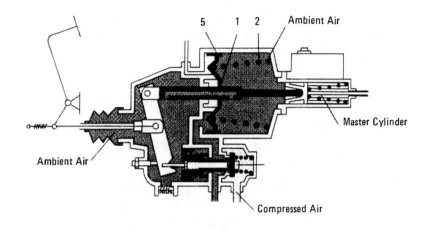

(A) Released

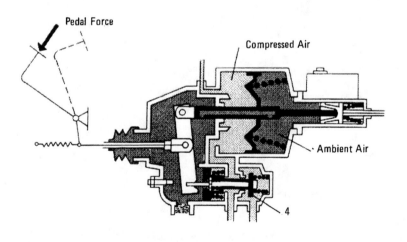

(B) Applied

*Figure 6-11. Air-over-hydraulic brake operation.*

279

where  $A_c$ = brake chamber area, cm² (in.²)

   BF = brake factor

   $k_A$ = brake adjustment reduction factor

   $k_T$ = brake temperature reduction factor

   $\ell_1$ = brake line pressure, N/cm² (psi)

   $p_o$ = pushout pressure, N/cm² (psi)

   r = drum or effective rotor radius, m (in.)

   $\eta_m$ = mechanical efficiency

   $\rho$ = lever ratio

For S-cam brakes, the lever ratio $\rho$ is given by

$$\rho = \ell_s \, / \, 2\ell_c \qquad\qquad (6\text{-}2)$$

where  $\ell_c$ = effective cam radius, cm (in.) (standard S-cam brakes use
1.27 mm [0.5 in.] cam radius)

   $\ell_s$ = effective slack adjuster length, cm (in.)

For wedge brakes the lever ratio is given by

$$\rho = 1 \, / \, [2\tan(\alpha \, / \, 2)] \qquad\qquad (6\text{-}3)$$

where  $\alpha$ = wedge angle, deg

Wedge angles vary in increments of two degrees, generally between 10 and 18 degrees.

The brake factor is obtained from Chapter 2. The brake factor of S-cam brakes is determined by Eq. (2-26). Brake factors for wedge brakes are obtained from the appropriate equation of Section 2.8.2.

The brake factor of a typical leading-trailing type S-cam brake is approximately 1.6, indicating an average coefficient of approximately 0.35. The typical dual-chamber wedge brake factor is approximately 4 to 4.5 with an average lining friction coefficient of approximately 0.4 or higher.

For brakes in good mechanical condition, the mechanical efficiencies exhibited by S-cam and wedge brakes range from 0.7 to 0.75, and 0.8 to 0.88, respectively (Ref. 22).

Inspection of Fig. 6-9 reveals the decrease of pushrod force with increasing travel of the pushrod. A type 30 brake chamber will produce a pushrod travel of approximately 70 mm (2.75 in.) when actuated by 69 N/cm$^2$ (100 psi). Brake chambers of size 20 and greater begin to experience an initial pushrod force decrease near a travel limit of 44 mm (1.75 in.) (Ref. 22).

If one idealizes the pushrod force as unaffected for pushrod travels less than 44.5 mm (1.75 in.), and assumes a linear decrease to zero for greater pushrod travel values, then effective pushrod force reduction factor $k_A$ as a function of brake adjustment for adjustments greater than the limit travel of 44.5 mm (1.75 in.) may be determined by an empirical relationship of the form

$$k_A = 1 - 0.5\Delta d \qquad (6\text{-}4)$$

$$[k_a = 1 - 0.4\Delta d]$$

where $\Delta d$ = pushrod travel beyond limit travel, mm (in.)

Eq. (6-4) can be used for any size brake chamber. The limit drop-off pushrod travel is 44 mm (1.75 in.) for type 20 or greater, 38 mm (1.5 in.) for type 16, and 32 mm (1.25 in.) for type 12 brake chambers.

For brakes at elevated temperatures, the pushrod travel increases by approximately 0.036 mm/K (0.0008 in./°F). For example, a brake temperature increase of 222 K (400°F) increases the pushrod travel by 8 mm (0.31 in.) solely due to temperature.

In the examination of accident vehicles, investigators frequently determine the brake adjustment by counting the number of turns or clicks of the adjustment nut as they tighten the shoes against the drum. This should never be done since it will destroy evidence, even if the slack adjusters are backed off by the exact amount to "restore the evidence." The tightening of the brakes will force the shoe rollers to operate over a different cam surface, thus eliminating the possibility of determining brake adjustment from a detailed cam surface/roller displacement analysis. The rule of thumb that three clicks correspond to 12.7 mm (0.5 in.) of pushrod travel may not be accurate enough.

For a standard type 30 brake chamber/drum assembly, a 0.04 in. clearance between lining and drum equates to approximately 0.3 in. of pushrod travel.

When brake temperature exceeds a certain level, braking effectiveness will be reduced. More specifically, the lining to drum friction coefficient will decrease for higher brake temperatures, resulting in a reduced brake factor and, hence, brake torque. For brake temperatures greater than 589 K (600°F), the brake torque will drop significantly. In general, with the slack adjuster adjusted at 38 mm (1.5 in.) or better, the brake torque will decrease by approximately 30% of its cold value when heated to 589 K (600°F), or by 50% when adjusted to 64 mm (2.5 in.).

In-stop temperature increase will have a similar effect on brake torque. For example, a wedge brake when tested at 32.2 km/h (20 mph) and 41 N/cm² (60 psi) application pressure may produce a brake torque of approximately 13,000 Nm (115,000 in.-lb) as compared with 8500 Nm (75,000 in.-lb) when tested at 97 km/h (60 mph). It appears that the greater friction surface temperature causes in-stop fade, which may significantly reduce brake torque. It is important to recognize that in-stop fade is a function of the brake lining material used.

Based on road test and dynamometer data, the average brake torque reduction factor $k_T$ as a function of brake temperature for brake temperatures exceeding 366 K (200°F) may be determined by the following empirical expression:

$$k_T = 1.2 - 0.0018(T - 255) \tag{6-5}$$

$$[k_T = 1.2 - 0.001T]$$

where   T = brake temperature, K (°F)

The spring force $F_{spring}$ actuation of a 30/30 type S-cam brake against the pushrod is approximately 8896 N (2000 lb) when the spring is fully compressed, and approximately 3111 N (700 lb) at 64 mm (2.5 in.) adjustment. From these data a reduction similar to the service brake chamber may be determined as

$$F_{spring} = 91(98 - d) \quad , \quad N \qquad (6\text{-}6)$$

$$[F_{spring} = 2000 - 520(d) \quad , \quad lb]$$

where   d = brake adjustment, mm (in.)

For example, for an S-cam brake having a pushrod travel of 57 mm (2.25 in.) the spring force pushing against the slack adjuster arm will be only $91(98 - 57)$ = 3731 N $[2000 - (520)(2.25) = 830$ lb$]$.

## 6.6  Vehicle Deceleration

For air brake vehicles an equation similar to that for vehicles equipped with hydraulic brakes (Eq. [5-3]) is used. Division of brake torque (Eq. [6-1]) by tire radius R yields the braking force acting on the circumference of the tire, and multiplication by two yields the braking force produced by one axle equipped with two brakes. Summation of the braking forces on all axles yields the deceleration a as

$$a = (2 \, / \, RW) \sum_{}^{n} [(p_\ell - p_o) A_c BF \eta_m \rho k_A k_T r]_i \quad , \quad \text{g-units} \qquad (6\text{-}7)$$

where   i = identity of axle braked

n = number of axles braked

R = tire radius, mm (in.)

W = vehicle weight, N (lb)

For example, for a three-axle truck equipped with an automatic limiting valve on the front axle, Eq. (6-7) is expressed as

$$a = (2 / RW)\{(p_\ell - p_o)A_cBF\eta_m\rho k_A k_T r]_1$$

$$+(p_\ell - p_o)r[(A_cBF\eta_m\rho k_A k_T)_2$$

$$+(A_cBF\eta_m\rho k_A k_T)_3]\} \quad , \quad \text{g-units} \quad (6-8)$$

where subscripts 1, 2, and 3 identify front, second, and third vehicle axle, respectively.

The summation term for each axle having a brake line pressure reducing valve is expressed separately such as the front axle of the truck. If proportioning or limiting valves are used, expressions for the brake line pressures have to be derived for the particular axle served by the proportioning valve.

European safety standards require the use of brake line pressure reducing valves on trucks and trailers exceeding a certain weight. Automatic reducing valves are manufactured by different manufacturers including Bosch and Wabco.

## 6.7 Response Time of Air Brake Systems

Air brakes have a relatively long response time and high pressure losses when compared with hydraulic systems. The time lag can be minimized through adequate design of the plumbing system.

An investigation into the dynamic behavior of air brake systems indicates that the time required to overcome clearance between linings and drum becomes smaller with increased brake line pressure and decreased brake chamber piston travel. The time required to build up brake torque also decreases with increasing brake line pressure, increased reservoir pressure, decreased piston travel, and decreased brake line length. For example, increasing the brake line length between the brake application valve and the brake chamber from 1.98 to 10.7 m (6.5 to 35 ft) increases the application time only a little, while the brake line pressure buildup time is nearly doubled. This is the major reason why air reservoirs are located near the brake chambers they serve. In general, brake application time is defined as the time elapsed between the instant of the first brake pedal movement and the instant the brake shoes contact the drum. The buildup time is defined as the time elapsed between the instant the brake shoes contact the drum and the instant a specified brake line pressure is obtained at the brake chambers. FMVSS 121 requires that for trucks and buses a brake line pressure of 41 N/cm$^2$ (60 psi) be reached at the farthest brake chamber within 0.45 s or less. The optimum result should therefore be achieved with a minimum volume, i.e., tight brake adjustment, and maximum brake line pressure.

Experiments have shown that considerable time delays are associated with the control and flow processes in the brake application valve (Ref. 24). The time lag of the application valve varies slightly from design to design and depends also on the volume to be pressurized. Typical time delays for application valves range from 0.05 s for a volume of $9.9 \times 10^{-4}$ m$^3$ (0.035 ft$^3$) to 0.25 s for a volume of $3.54 \times 10^{-3}$ m$^3$ (0.125 ft$^3$).

Brake response time tests are conducted to determine the time required by the brake chamber pressures to reach a specified value. To measure the brake pressure response time of a brake system, pressure transducers are fitted to the output port of the application valve and at the brake chambers to be measured. Typical response times measured on a tractor-semitrailer are presented in Figure 6-12 (Ref. 23). Close inspection of the pressure

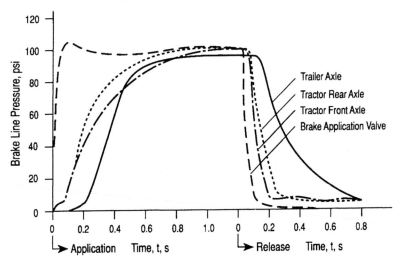

*Figure 6-12. Brake response times for tractor-semitrailer combination.*

curves reveals that approximately 0.55 to 0.6 s is required for the pressure to reach 90% of the maximum system pressure.

Inspection of the pressure rise curves of Fig. 6-12 and studies with scaled physical models representing actual pneumatic brake systems have shown that brake system response times may be composed of three parts as illustrated in Figure 6-13, each influenced by different factors (Ref. 24).

In the first part, a time lag $t_1$ derives from the speed with which the pressure wave travels through a brake line of given length. This time delay indicates how long it will take for the pressure signal to travel from the brake application valve to the relay valve.

The second time lag $t_2$ derives from the motion of the brake chamber piston required to overcome lining to drum clearance. This time lag is proportional to the volume of the brake chamber displaced as the shoes move against the drum.

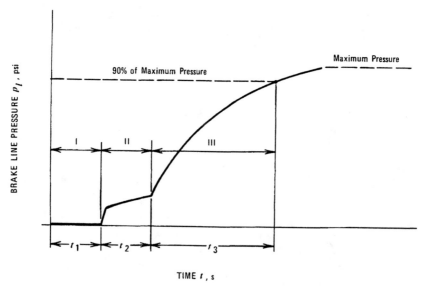

*Figure 6-13. Schematic of pressure rise in air brake system.*

The third time lag $t_3$ consists of the time required for the brake line pressure to reach a specified maximum value, typically 90% of the reservoir pressure. This lag is proportional to both the total volume and the flow resistance of the brake system.

To simplify the analysis, a basic air brake system schematic as illustrated in Figure 6-14 is used.

The time $t_1$ required for the pressure wave to travel between the brake application valve and the brake chamber is determined by

$$t_1 = \ell_2 / c \quad , \quad s \tag{6-9}$$

where  $c$ = speed of sound in air, m/s (ft/s)

$\ell_2$ = length of brake line between application valve and brake chamber, m (ft)

The speed of sound is a function of the density of the air. For atmospheric conditions, $c \approx 333$ m/s (1000 ft/s). The time is little affected by typical curves and fittings found in air brake systems. The time required by the pressure wave to travel between the application valve and the brake chambers located the farthest away is approximately 0.02 to 0.05 s for a tractor-trailer combination.

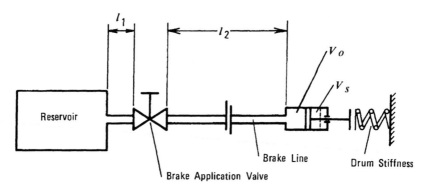

*Figure 6-14. Air brake system schematic.*

The time lag $t_2$, required by the brake line pressure of a typical air brake system in good mechanical condition to overcome brake chamber piston slack and shoe return springs, is determined by the volume $V_o$ to be filled prior to any brake chamber piston movement, the volume $V_s$ to be filled to overcome brake chamber piston slack, and the brake line length $\ell_1$ between reservoir and application valve, as well as the line length $\ell_2$ between application valve and brake chamber(s). An approximate expression determined from experiment for a typical air brake line is (Ref. 10)

$$t_2 = 35.3(V_o + V_s)(0.023\ell_1 + 0.082\ell_2) \quad , \quad s \qquad (6\text{-}10)$$

$$[t_2 = (V_o + V_s)(0.007\ell_1 + 0.025\ell_2) \quad , \quad s]$$

where $\ell_1$ = brake line length between reservoir and brake application valve, m (ft)

$\ell_2$ = brake line length between brake application valve and brake chamber, m (ft)

$V_o$ = brake chamber volume to be filled prior to any piston displacement, m³ (ft³)

$V_s$ = brake chamber volume to be filled to take up slack, m³ (ft³)

The time lag $t_3$, required for the brake line pressure in the brake chambers to attain 90% of the maximum reservoir pressure, is determined by the total volume between the brake application valve and brake chamber including the brake line, and is given by an empirical expression

$$t_3 = 4.87(\ell_1 + \ell_2)(V_s + V_o + V_2) \quad , \quad s \qquad (6\text{-}11)$$

$$[t_3 = 0.042(\ell_1 + \ell_2)(V_s + V_o + V_2) \quad , \quad s]$$

where $V_2$ = volume of brake line connecting brake application valve and brake chamber, m³ (ft³)

The total time lag is increased by the time lags of the brake application valve associated with each of the three pressure phases. The valve time lag can only be determined conveniently by experiment for the particular brake application valve installed in the brake system.

The total time lag $t_{total}$ is

$$t_{total} = t_1 + t_2 + t_3 + t_v \quad , \quad s \qquad (6\text{-}12)$$

where $t_v$ = time lag of brake application valve, s

For example, a tractor-semitrailer combination may have the brake system data that follow:

$$\ell_1 = 3.05 \text{ m (10 ft)}$$

$$\ell_2 = 9.14 \text{ m (30 ft)}$$

$$V_2 = 0.00085 \text{ m}^3 \text{ (0.03 ft}^3\text{)}$$

$$V_s = 0.0028 \text{ m}^3 \text{ (0.10 ft}^3\text{)}$$

The total time lag computed by Eq. (6-12) is

$$t_{total} = 0.010 + 0.082 + 0.220 + 0.25 = 0.562 \text{ s}$$

In the calculations $t_1 = 0.01$ s, $t_2 = 0.082$ s, $t_3 = 0.220$ s, and $t_v = 0.25$ s were either assumed or computed.

The time lag due to long brake line length associated with the brakes of articulated vehicles can become critical at higher speeds. Studies have shown that a time lag of one second or more between the brakes of the empty semitrailer and the tractor brakes may cause instability for speeds in excess of 97 km/h (60 mph) due to the increased horizontal forces at the kingpin of the fifth wheel, and the premature brake lockup associated with an empty combination.

Improvements of brake response times of pneumatic brake systems can be achieved through the use of larger cross section hoses and pipes, improved connectors and fittings, quick release valves, relay valves on tractors and trailers, and trailer brake synchronization. Larger cross section losses are not always beneficial since the increased air mass increases response times.

In articulated vehicles it is essential that all brakes of the vehicle combination are applied at the same time. If that is not possible, then those of the rear axle of the last trailer are applied first and then the brakes are applied progressively forward to the front axle of the tractor in order to avoid too large fifth wheel kingpin or hitch forces. The proper application of brakes is accomplished by adjusting the relay valve crack pressures so that the lowest value is associated with the relay valve located the farthest from the brake application valve. The relay valve located closest to the brake application valve has the highest crack pressure. Electrically operated solenoid valves (synchronizing valves) have been used to improve sequencing and rapid brake application. Tests have shown that brake synchronization improves (decreases) trailer brake application time by about 25% and the release time by more than 40% (Ref. 22). The installation of a pressure reducer or proportioning valves and/or wheel anti-lock (ABS) brake systems does not seem to affect either application or release times.

Brake application sequence must not be confused with brake lockup sequence, which, for stability reasons should always be tractor front axle first, semi-trailer axle second, and tractor rear axle last (see Chapter 8).

In the United States it is common practice to define the response time of an air brake system as the time required from the instant of brake pedal movement until a pressure of 41 $N/cm^2$ (60 psi) is attained in the brake chambers. Results of road tests have shown that it takes considerably more time to reach maximum brake line pressure and, hence, maximum deceleration than that required to reach 41 $N/cm^2$ (60 psi).

## 6.8 Braking by Wire

In standard air brake systems air pressure is used to control the braking process. In the electric brake system (EBS) electronic and electric components control the brakes. Compressed air serves only to press the shoes against the drum or pads against the rotor.

The electrical/electronic system consists of the following components: EBS control unit, brake valve with pedal travel sensor, load sensor, trailer control valve, pressure control module per wheel including control for ABS and/or drive traction, brake pad wear sensor per wheel, wheel speed sensor per wheel, and wire cable from each brake to control unit.

The mechanical components are disc brakes (or drum brakes if used), air pressure producing and storage system, as well as mechanical parking brake system.

The driver communicates braking from the brake pedal and sensor electronically to the EBS unit. The EBS also receives signals from the individual brake modules (one per wheel) and the load sensing system (measuring vehicle weight and distribution). The EBS analyzes the information and sends brake pressure signals to the individual brake pressure modulators, as well as to the EBS of the trailer (if any). The pressure modulators cause the appropriate air pressure to be delivered to the brakes from the air pressure reservoirs. The ABS function is improved by knowing the "intended" brake pressure on each brake.

Advantages include good pedal feel, modulation of braking forces on each wheel as a function of loading and deceleration, and optimization of braking between tractors and trailers. Scania of Sweden is currently the first manufacturer of commercial vehicles offering braking by wire. Expected improvement are rapid response times, balanced braking among all axles, high decelerations, and simpler maintenance than air brakes.

# CHAPTER 7

# Single Vehicle Braking Dynamics

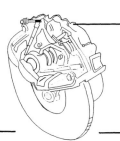

*This chapter analyzes the response of the vehicle to the forces produced by the braking system. The optimum braking forces for the straight-line, level-surface braking process are presented. Optimum and actual braking forces are compared and the concepts of braking efficiency and friction utilization are discussed. Practical methods for brake balance analysis including proportioning valves are presented. The equations describing vehicle directional stability during braking are shown. The effects of inter-axle load transfer of tandem axle trucks on optimum braking are analyzed. The basic relationships for braking while turning are discussed.*

## 7.1 Static Axle Loads

The forces acting on a non-decelerating vehicle, either stationary or traveling at constant velocity on a level roadway are illustrated in Figure 7-1. Due to the front-to-rear weight distribution, the front and rear axle may carry significantly different static axle loads.

The static axle load distribution is defined by the ratio of static rear axle load to the total vehicle weight, designated by the Greek letter $\Psi$ as

$$\Psi = F_{zR} / W \tag{7-1}$$

where $F_{zR}$ = static rear axle load, N (lb)

$W$ = vehicle weight, N (lb)

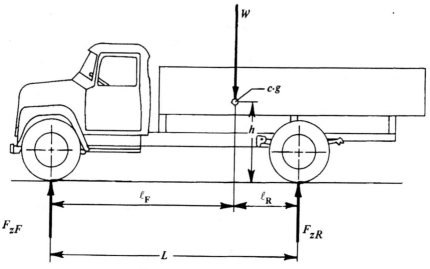

*Figure 7-1. Static axle forces.*

The relative static front axle load is given by

$$1 - \Psi = F_{zF} / W \qquad (7-2)$$

where   $F_{zF}$ = static front axle load, N (lb)

Modern front-wheel-driven cars have $\Psi$-values for the empty conditions as low as 0.35, indicating that only 35% of the total weight is carried by the rear axle. This relatively low static rear axle load of a car or pickup truck when lightly loaded is one of the major reasons that a careful brake balance analysis to avoid premature rear brake lockup is required.

Application of moment balance about the front axle of the stationary vehicle shown in Fig. 7-1 yields

$$W\ell_F = F_{zR}L$$

where   L = wheelbase, m (ft)

$\quad\quad\ \ \ell_F$ = horizontal distance from center of gravity to front axle, m (ft)

Solved for the horizontal distance $\ell_F$ between front axle and center of gravity

$$\ell_F = F_{zR}L / W = \Psi L \quad , \quad m \text{ (ft)}$$

Similarly, for the horizontal distance $\ell_R$ between the rear axle and the center of gravity

$$\ell_R = (1 - \Psi)L \quad , \quad m \text{ (ft)}$$

## 7.2  Dynamic Axle Loads

When the brakes are applied, the torque developed by the wheel brakes is resisted by the tire circumference where it comes in contact with the ground. Prior to brake lockup, the magnitude of the braking forces is a direct function of the torque produced by the wheel brake. For hydraulic brakes, Eq. (5-2) is used for determining the actual braking forces; for air brakes an equation similar to Eq. (6-1) is used.

The forces acting on a two-axle vehicle decelerating on a level road are illustrated in Figure 7-2. Application of moment balance about the rear tire-to-ground contact point yields the dynamic normal force $F_{zF,dyn}$ on the front axle:

$$F_{zF,dyn} = (1 - \Psi + \chi a)W \quad , \quad N \text{ (lb)} \tag{7-3a}$$

where  $a = F_{x,total}/W = \text{deceleration, g-units}$

$F_{x,total} = \text{total braking force, N (lb)}$

$F_{zR,static} = \text{normal rear axle load without braking, N (lb)}$

$W = \text{vehicle weight, N (lb)}$

$\chi = \text{center of gravity height (h) divided by wheelbase (L)}$

$\Psi = F_{zR,static}/W$

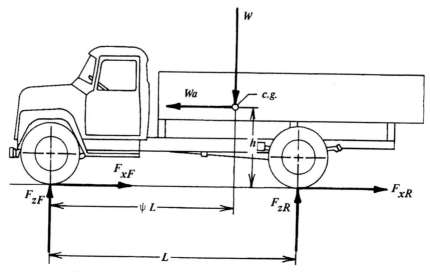

*Figure 7-2. Forces acting on a decelerating vehicle.*

Similarly, moment balance about the front tire-to-ground contact point yields the dynamic rear axle normal force $F_{zR,dyn}$:

$$F_{zR,dyn} = (\Psi - \chi a)W \quad , \quad N \text{ (lb)} \qquad (7\text{-}3b)$$

Inspection of Eqs. (7-3a) and (7-3b) reveals that the dynamic normal axle forces are linear functions of deceleration a, i.e., straight-line relationships. The amount of load transfer off the rear axle (and onto the front axle) is given by the term $\chi aW$ in Eqs. (7-3a) and (7-3b). The normal axle loads of a typical front-wheel-driven car are illustrated in Figure 7-3 for the driver-only and fully laden cases. Inspection of the axle loads reveals that the rear axle load is significantly less at higher decelerations than that associated with the front axle. For example, the rear axle load has decreased from a static load of 3114 N (700 lb) to only 1334 N (300 lb) for a 1 g stop, while the front axle load has increased from 5782 to 7562 N (1300 to 1700 lb).

The relative center-of-gravity height $\chi$ of a typical passenger car does not change significantly, if at all, from the driver-only to the fully laden condition.

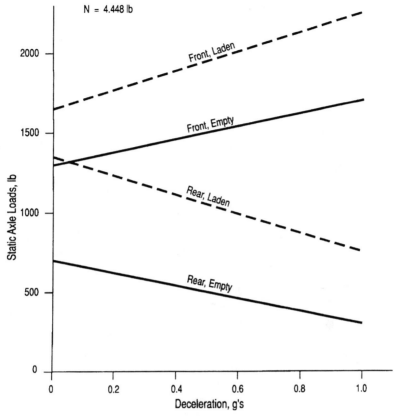

*Figure 7-3. Dynamic axle loads for empty and fully laden vehicle (2000 and 3000 lb, respectively).*

## 7.3   Optimum Braking Forces

### 7.3.1  Braking Traction Coefficient

The wheel brake torques generate braking or traction forces between the tire and the ground. The ratio of braking force to dynamic axle load is defined as the traction coefficient $\mu_{Ti}$

$$\mu_{Ti} = F_{xi} / F_{zi,dyn} \tag{7-4}$$

where $F_{xi}$ = axle braking force, N (lb)

$F_{zi,dyn}$ = dynamic axle load, N (lb)

i = designates front or rear axle

The traction coefficient is the level of tire-road friction needed by the braked tire so that it will just not lock up. The traction coefficient varies as either braking force or dynamic axle load change and, consequently, is a vehicle and deceleration-dependent parameter. In general, the front and rear axle traction coefficients will be different. The traction coefficient must not be confused with the tire-road friction coefficient. Only when the numerical values of traction and tire-road friction coefficient are equal does the tire lock up.

## 7.3.2 Dynamic Braking Forces

Multiplication of the dynamic axle loads by the traction coefficients yields the dynamic braking forces $F_{xF}$ for the front axle:

$$F_{xF} = (1 - \Psi + \chi a)W\mu_{TF} \quad , \quad N \text{ (lb)} \qquad (7\text{-}5a)$$

Similarly, the dynamic braking force $F_{xR}$ for the rear axle is:

$$F_{xR} = (\Psi - \chi a)W\mu_{TR} \quad , \quad N \text{ (lb)} \qquad (7\text{-}5b)$$

where $\mu_{TF}$ = front traction coefficient

$\mu_{TR}$ = rear traction coefficient

The tire-road friction coefficients $\mu_F$ or $\mu_R$, existing at either the front or rear tires, are indicators of the ability of a road surface to allow traction to be produced for a given tire and, as such, are fixed numbers. A braked tire will continue to rotate as long as the traction coefficient computed by Eq. (7-4) is less than the tire-road friction coefficient, otherwise it will lock up. At the moment of incipient tire lockup, the traction coefficient equals the tire-road friction coefficient. When both axles are braked at sufficient levels so that the front and rear wheels are operating at incipient or peak friction conditions, then the maximum traction capacity between the tire-road system is utilized.

Under these conditions the vehicle deceleration will be a maximum, since the traction coefficients front and rear are equal, and are also equal to the vehicle deceleration measured in g-units.

### 7.3.3 Optimum Straight-Line Braking

For straight-line braking on a level surface in the absence of any aerodynamic effects, optimum braking in terms of maximizing vehicle deceleration is defined by

$$\mu_F = \mu_R = a \tag{7-6}$$

where  $a$ = vehicle deceleration, g-units

  $\mu_F$ = front tire-road friction coefficient

  $\mu_R$ = rear tire-road friction coefficient

The optimum condition expressed by Eq. (7-6) must not be confused with "ideal" conditions since there are a variety of operational conditions under which Eq. (7-6) does not yield satisfactory braking results. For example, for braking-in-a-turn maneuvers, not all tire-road friction can be utilized for braking because lateral tire forces must share in the total traction available with the braking forces. Furthermore, simultaneous front and rear wheel lockup produces a different vehicle response which may exhibit gentle vehicle rotation, while front wheel lockup first will not.

The optimum braking forces may be determined by setting the traction coefficients equal to vehicle deceleration in Eqs. (7-5a) and (7-5b), resulting in the optimum braking forces $F_{xF,opt}$ on the front axle

$$F_{xF,opt} = (1 - \Psi + \chi a)aW \quad , \quad N \text{ (lb)} \tag{7-7a}$$

and the optimum braking forces $F_{xR,opt}$ on the rear axle

$$F_{xR,opt} = (\Psi - \chi a)aW \quad , \quad N \text{ (lb)} \tag{7-7b}$$

Inspection of Eqs. (7-7a) and (7-7b) reveals a quadratic relationship relative to deceleration a. The graphical representation is a parabola as illustrated in Figure 7-4 for a pickup truck. Any point on the curve represents an optimum point identified by deceleration equal to friction coefficient. Inspection of the empty case reveals that for decelerations greater than 0.8 to 1 g, optimum rear braking begins to decrease due to the significant load transfer onto the front axle. For the loaded case, the higher static rear axle load yields an increasing optimum rear brake force with higher deceleration.

The deceleration scales for the empty and loaded vehicle are different as shown in Fig. 7-4. A simplification can be obtained by expressing the optimum braking forces relative to vehicle weight, or per one pound of weight, by dividing the optimum braking forces by vehicle weight. Consequently, the optimum braking forces $F_{xF}$ and $F_{xR}$ are

Front:    $F_{xF} / W)_{opt} = (1 - \Psi + \chi a)a$    (7-8a)

Rear:    $F_{xR} / W)_{opt} = (\Psi - \chi a)a$    (7-8b)

The graphical representation of Eqs. (7-8a) and (7-8b) is illustrated in Figure 7-5 for the empty and laden cases. Any point on the optimum braking force curve represents optimum braking, i.e., a condition under which the tire-road friction coefficient for front and rear equals vehicle deceleration. For example, for the empty vehicle and for a deceleration of 0.61 g, the relative optimum front braking force is 0.41 and the relative rear optimum braking force is 0.2. The corresponding approximate values for the laden condition are 0.3 and 0.3, respectively. If the actual brake torque balance front to rear were distributed according to the optimum ratios indicated, simultaneous front and rear brake lockup would occur, yielding minimum stopping distance.

Only one deceleration scale is used in Fig. 7-5 for both loading conditions. The lines of constant deceleration run under an angle of 45 deg, assuming equal scales are used for front and rear braking forces. The reason for the 45-deg angle follows from Newton's Second Law, expressed by the relative optimum braking forces as

$$F_{xF} / W + F_{xR} / W = a$$    (7-9)

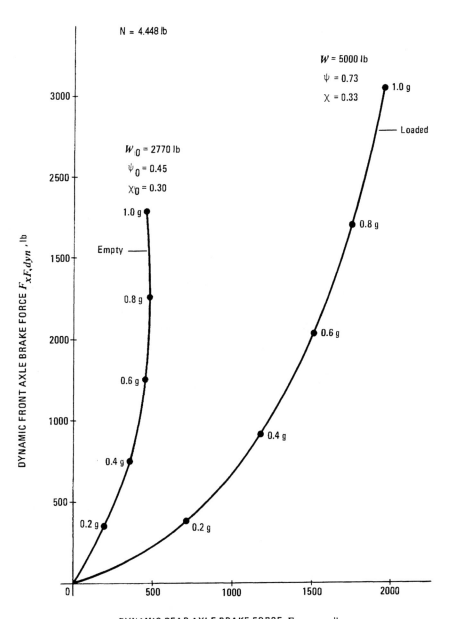

*Figure 7-4. Dynamic braking forces.*

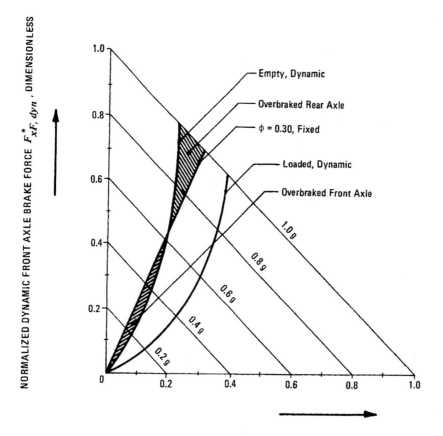

Figure 7-5. *Normalized dynamic brake forces.*

Inspection of Fig. 7-5 reveals, for example, for the 0.8 g-deceleration line substituted into Eq. (7-9), that for the empty case: $0.58 + 0.22 = 0.8$; and for the laden case: $0.45 + 0.35 = 0.8$. The numerical values indicate that the front braking force decreased by $0.58 - 0.45 = 0.13$ on the vertical axis, which is the increase experienced by the rear braking force on the horizontal axis. The result is a right-angle triangle with two sides of equal distance, namely the vertical and horizontal components, thus yielding a 45-deg slope for the deceleration lines.

Inspection of Eqs. (7-8a) and (7-8b) reveals that the optimum braking forces are only a function of the particular vehicle geometry and weight data, i.e., $\Psi$ and $\chi$, and vehicle deceleration a. They are not a function of the brake system hardware installed.

To better match actual with optimum braking forces, it becomes convenient to eliminate vehicle deceleration by solving Eq. (7-8a) for deceleration a, and substituting into Eq. (7-9). The result is the general optimum braking forces equation

$$F_{xR} \, / \, W)_{opt} = \sqrt{\frac{(1 - \Psi)^2}{4\chi^2} + \left(\frac{1}{\chi}\right)\left(\frac{F_{xF}}{W}\right)} - \frac{1 - \Psi}{2\chi} - \frac{F_{xF}}{W} \qquad (7\text{-}10)$$

Eq. (7-10) allows computation of the appropriate optimum rear braking force associated with an arbitrarily specified (optimum) front braking force.

The graphical representation of Eq. (7-10) is that of a parabola illustrated in Figure 7-6. The optimum curve located in the upper right quadrant represents braking, the lower left acceleration. Only the braking quadrant, and then only the section exhibiting deceleration ranges of interest are of direct importance to brake engineers. The optimum curves shown in Fig. 7-5 represent the section of interest relative to deceleration ranges encountered frequently.

The entire optimum braking/acceleration forces diagram, however, is used to develop useful insight and design methods for matching optimum and actual braking forces for brake design purposes. In addition, the methods will also be used in the reconstruction of actual vehicle accidents involving braking and loss of directional stability due to premature rear brake lockup.

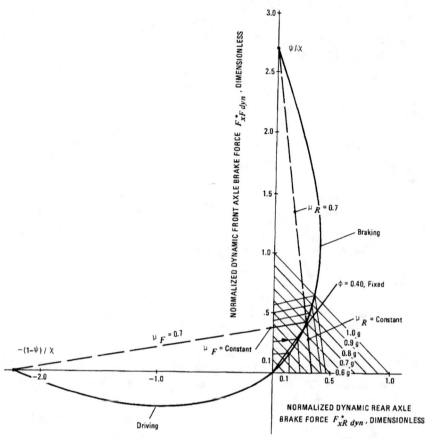

*Figure 7-6. Parabola of normalized dynamic braking and driving forces.*

## 7.3.4 Lines of Constant Friction Coefficient

For increasing deceleration, assuming that the tire-road friction is high enough, the optimum braking of the rear axle begins to decrease and reaches zero where it intercepts the front braking axis. At this point the deceleration of the vehicle is sufficiently high that the rear axle begins to lift off the ground due to excessive load transfer.

Similarly, in the case of increasing acceleration, the front axle begins to lift off the ground when the optimum acceleration curve intercepts the rear braking force axis.

The zero point on the front braking force axis is determined by setting the relative rear braking force equal to zero in Eq. (7-10), and solving for the relative front braking force, resulting in

$$F_{xF} / W = 0 \Rightarrow F_{xR} / W = -(1 - \Psi) / \chi$$

Similarly, setting the relative front braking force equal to zero in Eq. (7-10) yields

$$F_{xR} / W = 0 \Rightarrow F_{xF} / W = \Psi / \chi$$

Any point on the optimum braking forces curve represents the condition under which the front and rear tire-road friction coefficients are equal to each other as well as to the deceleration of the vehicle. Under these conditions all available tire-road friction is utilized for vehicle deceleration. For example, at the 0.6 g point the front and rear tire-road friction coefficients are also equal to 0.6.

At the respective zero points, the tire traction forces, either braking or accelerating, are zero regardless of the level of friction coefficient existing between the tire and the ground due to the normal forces between the tire and the ground being zero.

A straight line connecting the zero point $[-(1 - \Psi)/\chi]$ and a point of the optimum force curve represents a condition of constant coefficient of friction between the front tire and ground. For example, connecting the zero point with the 0.6 g optimum point establishes a line of front tire friction coefficient of 0.6 constant along the entire line. Additional lines of constant friction are obtained by connecting different optimum points.

Similarly, by connecting the rear zero point $(\Psi/\chi)$ with points on the optimum curve, lines of constant rear tire friction coefficient are obtained.

Inspection of Fig. 7-6 reveals that the constant front friction line of 0.7 intercepts the front braking force axis with the rear braking force equal to zero at a deceleration of approximately 0.39 g. In other words, when braking on a road surface having a tire-road friction coefficient of 0.7 with the rear brakes failed or disconnected, the front brakes are at the moment of lockup while the vehicle decelerates at 0.39 g.

On the other hand, when the front brakes are disconnected, the rear brakes lock up at a deceleration of approximately 0.33 g while braking on a road surface with a 0.7 coefficient of friction, as indicated by the interception of the 0.7 constant rear friction line with the rear braking force axis.

The deceleration $a_F$ achievable with the rear axle disconnected is derived from Newton's Second Law and Eq. (7-5a), however, with the traction coefficient equal to the front tire-road friction coefficient since the front brakes are about to lock up:

$$F_{xF} = (1 - \Psi + \chi a_F)\mu_F W = a_F W$$

$$(1 - \Psi)\mu_F = a_F(1 - \chi\mu_F)$$

Solving for deceleration $a_F$ with the rear brakes disconnected, yields

$$a_F = \frac{(1 - \Psi)\mu_F}{1 - \chi\mu_F} \quad , \quad \text{g-units} \qquad (7\text{-}11a)$$

A similar derivation yields the deceleration $a_R$ with the front brakes disconnected:

$$a_R = \frac{\Psi\mu_R}{1 + \chi\mu_R} \quad , \quad \text{g-units} \qquad (7\text{-}11b)$$

where  $\mu_F$ = front tire-road friction coefficient

$\mu_R$ = rear tire-road friction coefficient

It becomes convenient to use Eqs. (7-11a) and (7-11b) along with the optimum points to draw the lines of constant friction coefficient rather than using the zero points.

## 7.3.5 Optimum Braking Forces Parabola Analysis

It is emphasized again that the optimum forces, as well as the lines of constant friction are a function of the vehicle geometrical and weight properties only, and are not a function of the brake system installed. With the optimum curves, lines of constant deceleration, and lines of constant friction drawn into the optimum braking forces diagram, the diagram is complete and ready for comparison or design evaluation of the braking forces actually developed by the brake system installed on the vehicle (see Section 7.4).

In the design of braking systems using limiter or reducer valves, it is helpful to know where the slope of the optimum braking force curve becomes infinite. Taking the derivative of the rear braking force with respect to the front in Eq. (7-10), and evaluating it at zero, yields the maximum optimum rear braking force

$$F_{xR} \ / \ W)_{max} = \Psi^2 \ / \ 4\chi \tag{7-12}$$

The value of the optimum front braking force where the optimum rear braking force is a maximum is given by

$$F_{xF} \ / \ W)_{R=max} = 2\Psi - \chi^2 \ / \ 4\chi \tag{7-13}$$

The slope of the optimum braking force curve at the origin is computed by

$$Slope \, |_0 = \frac{\Psi}{4\chi} \tag{7-14}$$

## 7.4  Actual Braking Forces Developed by Brakes

The braking forces produced by the brake system installed on the vehicle are a function of the individual brake torques generated by the wheel brakes. For a hydraulic brake system, the braking force produced by an axle is computed by Eq. (5-2). A similar expression is used for air brake systems.

The brake force distribution $\Phi$, sometimes called brake balance, is defined by the ratio of rear braking force to the total braking force, or

$$\Phi = F_{xR} \ / \ (F_{xR} + F_{xF}) \tag{7-15}$$

For a vehicle without brake proportioning valves, the brake force distribution is determined from those brake components usually altered between front and rear, resulting in

$$\Phi = \frac{(A_{wc}BFr)_R}{(A_{wc}BFr)_R + (A_{wc}BFr)_F} \tag{7-16}$$

where $A_{wc}$ = wheel cylinder area, cm$^2$ (in.$^2$)

Eq. (7-16) is used to compute the fixed brake force distribution of a braking system. When a brake line pressure-reducing valve is used, then the brake force distribution is fixed up to the knee-point pressure of the proportioning valve, and changes to a different slope for higher pressures. The brake force distribution is computed by use of Eq. (5-2) in Eq. (7-15).

# 7.5 Comparison of Optimum and Actual Braking Forces

When the actual braking forces equal the optimum, then all available tire-road friction is utilized for vehicle deceleration. This optimum condition of locking all brakes simultaneously generally occurs only at one loading and deceleration level for a given vehicle. Of particular interest is the maximum wheels-unlocked deceleration and, hence, minimum stopping distance of a vehicle, and then either with the front or rear brakes approaching lockup first.

It should be remembered that the minimum stopping distance of a vehicle is not only a function of the maximum deceleration achieved but also of the deceleration rise time. Particularly at lower speeds the rise time will have a significant effect on the overall stopping distance (see Section 1.4.3).

For a given vehicle, the maximum wheels-unlocked deceleration can be illustrated and determined most easily from the braking force diagram shown in Figure 7-7. The relative optimum braking force, front and rear, is shown. The lines of constant friction for tire-road friction coefficients of 0.6 and 0.8 are also shown. Two fixed brake force distributions are illustrated in Fig. 7-7 by two straight lines marked stable and unstable. The stable brake force distribution

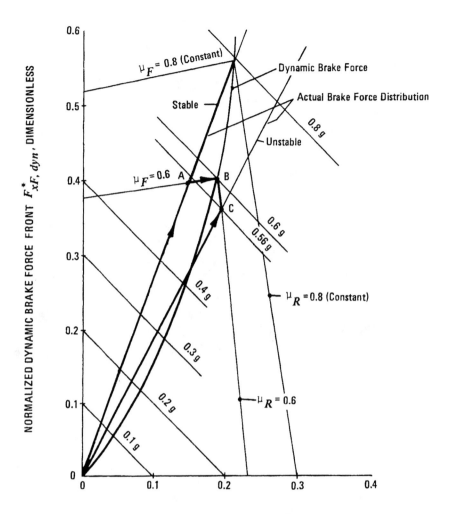

*Figure 7-7. Normalized dynamic and actual brake forces.*

is 27%, the unstable 34% as determined by dropping a vertical line from the respective interception with the 1 g deceleration line to the rear braking force axis.

Relative to Fig. 7-7, consider a vehicle operating on a wet road surface having a tire-road friction coefficient of 0.6. When a pedal force is applied, e.g., for a vehicle having the stable brake force distribution, the actually produced

309

brake forces, front and rear, increase along the stable line to point A. At point A the front brake force crosses the line of constant front friction coefficient equal to 0.6. Since the road friction is 0.6, the front brakes lock at point A. If the driver does not increase pedal force, the vehicle decelerates at approximately 0.55 g with its front brakes locked in a stable and straight direction. If the driver continues to increase pedal force, then the actual braking forces front and rear will move from point A to point B along the line of constant front friction to point B. At point B the rear brakes will lock and the vehicle decelerates with both axles locked at 0.6 g. The slight increase in relative front braking force between points A and B and, hence, deceleration from 0.55 to 0.6 g, occurs due to the increased normal force on the front axle resulting from the additional load transfer caused by the increased rear braking force and associated increase in deceleration.

For a vehicle equipped with the unstable brake force distribution, the braking forces, front and rear, increase to a level indicated by point C, corresponding to a deceleration of 0.56 g. For a maximum tire-road friction coefficient of 0.6, the rear wheels lock at the conditions marked by point C. Further increase in pedal force results in increased deceleration along line CB until all wheels are locked at a deceleration 0.6 g, indicated by point B. This stop is unstable because the rear brakes lock prior to the front brakes.

The unstable brake force distribution line crosses the optimum curve at a deceleration of approximately 0.43 g. For decelerations greater than 0.43 g, the rear brakes will always lock before the front because the actual rear braking forces are greater than the optimum rear braking forces. The deceleration at which the braking process switches from stable to potentially unstable is called *critical deceleration*. For modern passenger cars the critical deceleration should be near 1 g. Even the brake force distribution marked stable in Fig. 7-7 would not be acceptable because its critical deceleration is only 0.8 g. European safety standards require a critical deceleration of greater than 0.82 g, generally considered too low for modern high-speed vehicles and high traction tires. Braking safety standard FMVSS 135 prohibits locking of the rear brakes before the front brakes.

For the safe and efficient design of the front-to-rear brake balance system it is critical that front brakes lock first, hence a safe and stable braking vehicle, and that the actual braking forces are as close as reasonably possible to the optimum braking forces, hence an efficient braking system resulting in minimum stopping distance.

In the late '50s and early '60s in Europe, brake force distribution was designed to optimize (minimize) stopping distance. This practice resulted in premature rear brake lockup when lightly loaded, particularly on dry high-traction roads. For skilled test drivers acceptable results in terms of preventing loss of directional control were seen. For the average driver, however, this was not the case. The number of accidents involving vehicle spinning increased. Brake regulations such as ECE 13 and to a limited extent FMVSS 105 caused the brake force distribution to be shifted toward the front brakes. Many domestic vehicle manufacturers did not front brake bias their vehicles until the early to mid-'80s. Operational and in-use factors still result in loss of directional vehicle control due to premature rear brake lockup. In particular, aftermarket lining manufacturers may inadvertently contribute to rear brake bias by not properly matching their linings to OEM specifications.

## 7.6 Tire-Road Friction Utilization

Several different variations of basically the same physical concept have been used to describe how close the actual braking force is to the optimum.

The tire-road friction utilization relates the maximum wheels-unlocked deceleration to the lowest tire-road friction coefficient with which the deceleration can be achieved. In Section 7.3 the traction coefficient was introduced (Eq. [7-4]). For a two-axle vehicle and a fixed brake force distribution the traction coefficient $\mu_{TR}$ on the rear axle is

$$\mu_{TR} = F_{xR} / F_{zR} = \Phi aW / [(\Psi - \chi a)W] = \Phi a / (\Psi - \chi a) \qquad (7\text{-}17a)$$

where $F_{xR}$ = actual rear brake force, N (lb)

$\quad F_{zR}$ = dynamic rear normal force, N (lb)

Similarly, on the front axle the traction coefficient $\mu_{TF}$ is

$$\mu_{TF} = (1 - \Phi)a / (1 - \Psi + \chi a) \qquad (7\text{-}17b)$$

A graphical representation of Eqs. (7-17a) and (7-17b) is presented in Figure 7-8 for the vehicle geometrical and loading data shown. The friction utilization computed by Eq. (7-17a) is illustrated by the part of the curve labeled "Rear Axle Overbraked," and the part corresponding to Eq. (7-17b)

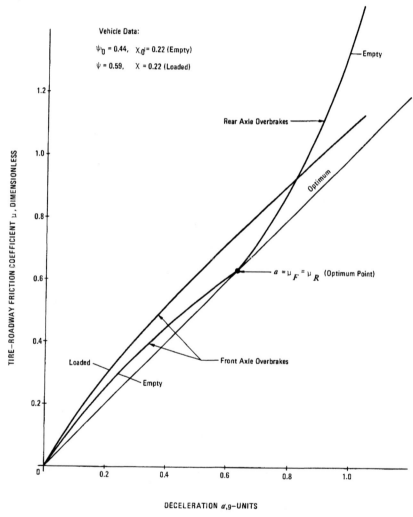

*Figure 7-8. Tire-road friction utilization.*

by "Front Axle Overbraked." The term overbraking means the same as locking up first. The optimum point corresponds to a = 0.62 g, i.e., deceleration and tire-road friction coefficient are equal. For decelerations below 0.62 g, e.g., 0.4 g, a friction coefficient between tires and road of approximately 0.44 is required for wheels-unlocked braking in the empty condition. For decelerations exceeding 0.62 g, e.g., a = 0.9 g, a friction coefficient of about 1.08 is required on the rear wheels.

European brake safety standards for commercial vehicles require that certain tire-road friction utilizations are met. If a special device is used to accomplish this, the device must be automatically acting, such as a load-sensitive brake line pressure reducer valve. In addition, a certain wheel lockup sequence is required such that the tire-road friction utilization curve of the front axle lies above that of the rear axle for all loading conditions. Stated differently, for a given deceleration the friction utilization of the rear wheels is less than that of the front wheels, meaning rear wheels are last to lock. Certain options are provided for specified conditions.

For passenger cars the tire-road friction utilization on the front axle must be greater than that of the rear axle for decelerations between 0.15 and 0.30 g, and between 0.45 and 0.80 g. For decelerations between 0.30 and 0.45 g the utilization of the rear may be greater than that of the front, provided the rear axle utilization curve does not deviate from the optimum by more than 0.05. In addition, vehicle decelerations must be greater than $0.1 + 0.85 (\mu - 0.2)$. The manufacturers must provide theoretical calculations demonstrating compliance with the friction utilization requirements. Because the calculations involve brake system component parameters and, in particular, the brake factor, care must be exercised in obtaining accurate input data.

## 7.7 Braking Efficiency

The concept of tire-road friction utilization may be expanded to be more generally applicable to a braking analysis. Braking efficiency is defined as the ratio of maximum wheels-unlocked vehicle deceleration to tire-road friction coefficient. The braking efficiency expresses the extent to which a given tire-road friction coefficient available to a vehicle is transformed into maximum wheels-unlocked deceleration.

By starting with Eqs. (7-17a) and (7-17b), analytical expressions for the braking efficiency of the rear and front axle may be derived. Eq. (7-17a) can be rewritten as

$$\mu_R \Psi - \mu_R \chi a = \Phi a$$

and collecting terms involving deceleration yields

$$a(\Phi + \mu_R \chi) = \mu_R \Psi$$

The braking efficiency $E_R$ of the rear axle now becomes

$$E_R = (a / \mu)_R = \Psi / (\Phi + \mu_R \chi) \tag{7-18a}$$

Similarly, the braking efficiency $E_F$ of the front axle becomes

$$E_F = (a / \mu)_F = (1 - \Psi) / (1 - \Phi - \mu_F \chi) \tag{7-18b}$$

Only numerical values less than unity are meaningful. If a value greater than unity is computed, e.g., for the front axle, then the corresponding braking efficiency of the rear axle will be less than unity, indicating that it is the limiting axle and will lock up first. A braking efficiency greater than unity indicates also that the particular axle is underbraked, meaning a higher brake force could be used for lockup to occur.

Eqs. (7-18a) and (7-18b) are represented graphically in Figure 7-9, in which the braking efficiency is plotted as a function of tire-road friction coefficient. Inspection of Fig. 7-9 reveals that for $\mu = 0.40$, the efficiency on the front axle is equal to approximately 0.88 for the empty driving condition. A braking efficiency of 88% indicates that 88% of the friction available is used by the braking system of the vehicle for deceleration of $0.88 \times 0.4 = 0.35$ g at the moment the front wheels are about to lock up.

Fig. 7-9 also shows the additional stopping distance over the minimum achievable with optimum braking. The ratio between stopping distance increase $\Delta S$ and minimum stopping distance $S_{min}$ may be derived as

$$\Delta S \,/\, S_{min} = \frac{1 - (a \,/\, \mu)}{(a \,/\, \mu)} \tag{7-19}$$

The braking efficiency diagram can be expanded so that it contains tire-road friction coefficient and deceleration directly, similar to the braking force diagram shown in Fig. 7-7. The lines running under an angle are lines of constant coefficient of friction as illustrated in Figure 7-10 (Ref. 25). Data points falling in the front braking efficiency area indicate front brake lockup before rear, while data points in the rear braking efficiency area indicate premature rear brake lockup. Inspection of Fig. 7-10 reveals that for the lightly laden operating condition or the line closest to the 100% efficiency line, the critical deceleration is approximately 0.86 g, indicating stable braking for all decelerations up to 0.86 g. For the fully laden condition the front brakes will always lock before the rear brakes. Frequently, expanded braking efficiency diagrams are used to compare actual road test data obtained from torque hubs or platform testers to the theoretical optimum, i.e., 100% braking efficiency.

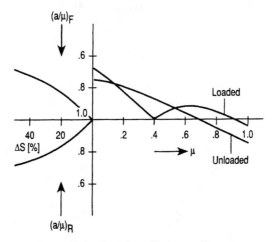

*Figure 7-9. Braking efficiency diagram.*

## 7.8 Fixed Brake Force Distribution Analysis

Before designing the braking system of a motor vehicle, the questions to be answered are: (a) can specific wheels-unlocked decelerations be achieved over a wide range of loading and roadway conditions with a fixed brake force distribution $\Phi$, and (b) if so, what is the required brake force distribution?

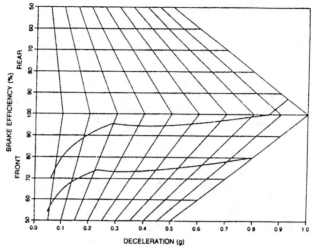

*7.10  Vehicle Stability Analysis*

## 7.8.1  Brake Force Distribution Design Selection

For a two-axle vehicle Eqs. (7-18a) and (7-18b) may be used to develop a limiting relationship on brake force distribution F (Ref. 26)

$$(1 - \mu\chi - (1 - \Psi) / E_{min}) \leq \Phi \leq (\Psi / E_{min} - \mu\chi) \qquad (7\text{-}20)$$

where   $E_{min}$ = minimum braking efficiency to be achieved by vehicle.

Application of the inequality to the limiting conditions corresponding to $0.2 \leq \mu \leq 0.8$ and the laden and empty loading conditions defines an envelope of acceptable values of $\Phi$.  A $\Phi$-value within this envelope may be used for design evaluation.  The $\Phi$-value finally selected for design purposes depends to some extent on the intended vehicle function.  However, regardless of vehicle type, braking stability must be considered by minimizing the potential for rear brakes locking before front brakes.

## 7.8.2  Wheels-Unlocked Deceleration

Wheels-unlocked deceleration achievable with a given brake force distribution may easily be obtained from Eqs. (7-18a) and (7-18b).

316

In the case of locking the rear axle first:

$$a = \mu\Psi / (\Phi + \mu\chi) \quad , \quad \text{g-units} \tag{7-21a}$$

In case of locking the front axle first:

$$a = \mu(1 - \Psi) / (1 - \Phi - \mu\chi) \quad , \quad \text{g-units} \tag{7-21b}$$

Often, it is of interest to know if a two-axle truck can produce certain braking efficiencies and, hence, decelerations and stopping distances, based on its geometrical loading conditions when a fixed brake force distribution is used.

Eqs. (7-21a) and (7-21b) may be used to formulate a requirement on brake force distribution $\Phi$ tailored to the braking limitations of trucks while braking on low-friction ($\mu = 0.2$) and high-friction road surfaces ($\mu = 0.8$) with specified braking efficiencies.

The greatest difficulties exist in preventing premature rear brake lockup when the empty vehicle is braking on dry road surfaces at high deceleration, or preventing front brake lockup when the loaded vehicle is braking at low deceleration on slippery road surfaces. In the first case the static rear axle load is small because the vehicle is empty and large dynamic load transfer off the rear axle occurs due to a large deceleration. In the second case the static front axle load is small and no significant dynamic load transfer to the front axle occurs. For these two operating conditions the braking efficiency generally presents the minimum limit value. For the case of braking the loaded vehicle on a dry road surface or the empty vehicle on a slippery road surface, the braking efficiencies are generally larger than those associated with the limit value.

### 7.8.3 Vehicle Loading-Brake Force Distribution Analysis

Based on these physical constraints, a requirement on the brake force distribution $\Phi$ can be developed.

For the empty vehicle when braking on a high friction road surface ($\mu = 0.8$) with a braking efficiency $E_R = 0.65$, Eq. (7-21a) yields

$$(0.65)(0.8) = 0.52 = 0.8\Psi_o / (\Phi + 0.8\chi_o) \tag{7-22}$$

where $\chi_o$ = relative center-of-gravity height

$\Psi_o$ = static rear axle load divided by total weight

The front axle of the empty vehicle generally operates at a braking efficiency higher than the minimum value when braking on a low-friction road surface. For braking efficiency $E_F = 0.80$ and $\mu = 0.2$, Eq. (7-21b) yields

$$(0.8)(0.2) = 0.16 = \frac{(1 - \Psi_o)0.2}{1 - \Phi - 0.2\chi_o} \tag{7-23}$$

For the loaded vehicle the rear axle generally operates at a braking efficiency higher than the minimum value when braking on a high-friction surface. For $E_R = 0.75$ and $\mu = 0.8$, Eq. (7-21a) yields

$$(0.75)(0.8) = 0.60 = 0.8\Psi / (\Phi + 0.8\chi) \tag{7-24}$$

The front axle of the loaded vehicle operates at the minimum value of braking efficiency when braking on a low-friction road surface. For $E_F = 0.65$ and $\mu = 0.2$, Eq. (7-21b) yields

$$(0.65)(0.2) = 0.13 = \frac{(1 - \Psi)0.2}{1 - \Phi - 0.2\chi} \tag{7-25}$$

The subscript "o" designates the empty operating condition. From Eqs. (7-22) through (7-25) a requirement on the brake force distribution $\Phi = f(\Psi_o, \Psi, \chi_o, \chi)$ as a function of geometric and loading parameters may be formulated. Omitting the algebra, the result from Eqs. (7-22) and (7-23) is

$$\Phi = \frac{\Psi_o(1 + 0.54\chi_o) - 0.65\chi_o}{1 - 0.19(1 - \Psi_o)} \tag{7-26}$$

and from Eqs. (7-24) and (7-25)

$$\Phi = \frac{\Psi(1 + 0.72\chi) - 0.92\chi}{1 + 0.15(1 - \Psi)} \tag{7-27}$$

Application of Eqs. (7-26) and (7-27) generally will result in different values of $\Phi$ for the empty and laden vehicle. But if the values for $\Psi$, $\chi$, $\Psi_o$, and $\chi_o$ are such that the brake force distributions $\Phi$ computed from Eqs. (7-26) and (7-27) are identical, a fixed brake force distribution will be adequate, i.e., the difference in center-of-gravity location between the empty and laden cases are so small that a proportioning braking system is not necessary.

Eqs. (7-26) and (7-27) may be used to eliminate $\Phi$, and it becomes possible to derive a limiting condition on the relative static rear axle load $\Psi_o = f(\Psi, \chi, \chi_o)$ as a function of the remaining vehicle parameters. This condition must be satisfied before a fixed brake force distribution may be considered adequate for the braking process with the minimum braking efficiency specified. Omitting the algebra, the results when plotted for different values of $\Psi$ and $\Delta\chi = \chi - \chi_o$ were found to be described by a functional relationship

$$\Psi - \Psi_o \leq \Delta\chi + 0.09 \tag{7-28}$$

The value of $\Delta\chi$ for trucks is generally small and less than 0.03 and, consequently, an approximate limiting condition on the change in relative static rear axle load is

$$\Psi - \Psi_o \leq 0.12 \tag{7-29}$$

The results indicate that vehicles equipped with fixed brake force distribution systems are capable of achieving decelerations well within the requirements for safe braking performance, provided the vehicle experiences an increase in relative static rear axle load of not more than 12%, i.e., $\Delta\Psi = \Psi - \Psi_o \leq 0.12$. This means, also, that load-sensitive proportioning will yield little or no improvement in braking performance for trucks for which the difference in relative static rear axle load between the empty and laden case is less than 12%. On the other hand, if the difference is greater, proportioning valves may have to be used.

### 7.8.4 Comparison of Theoretical and Road Test Results

The limiting condition on the brake force distribution, i.e., Eq. (7-20), was applied to a variety of commercial vehicles such as light and medium trucks and school buses (Ref. 22). Actual road tests were conducted to determine the maximum braking capabilities of the vehicles. The center of gravity location of the light truck with a GVW of 44,480 N (10,000 lb) remained almost unaffected by the loading as indicated by $\Delta\Psi = \Psi - \Psi_0 = 0.674 - 0.595 = 0.079$ and $\Delta\chi = \chi - \chi_0 = 0.320 - 0.293 = 0.027$. The corresponding values for the medium truck were $\Delta\Psi = 0.29$ and $\Delta\chi = 0.06$, indicating a significant horizontal change in the location of the center of gravity from the laden to the empty case. The location of the center of gravity of the school bus changed little as indicated by $\Delta\Psi = 0.105$ and $\Delta\chi = 0.001$. This result was to be expected due to the long wheelbase of the school bus.

Inspection of the $\Delta\Psi$-values for the light truck and the school bus reveals that no difficulties exist in designing a braking system with a fixed brake force distribution for both vehicles which will yield acceptable braking performance. The $\Delta\Psi$-value for the medium truck is, however, significantly greater than the limit value $\Delta\Psi = 0.12$ and, hence, it becomes impossible to achieve acceptable braking performance with a fixed brake force distribution on the medium truck.

Consider the light truck first. Assume a maximum and minimum value of $\mu$ equal to 0.8 and 0.2, respectively, and a minimum braking efficiency of 0.70. Applying Eq. (7-20) to the light truck results in a theoretical value of $\Phi = 0.51$, as contrasted with the actual brake force distribution of $\Phi = 0.53$. The computed distribution $\Phi = 0.51$ along with the appropriate vehicle data yields a minimum braking efficiency of 77% by use of Eq. (7-18b) for the loaded vehicle operating on a slippery road surface with $\mu = 0.2$. For all other loading and road surface conditions, the theoretical braking efficiencies are higher. For the dry road surface, the braking efficiencies computed by Eqs. (8-18a) for the loaded case and Eq. (7-18b) for the empty case are 87 and 80%, respectively. These braking efficiencies would produce wheels-unlocked decelerations of 6.83 m/s$^2$ (22.4 ft/s$^2$) for the loaded vehicle and 6.28 m/s$^2$ (20.6 ft/s$^2$) for the empty vehicle on a road surface having a tire-road friction coefficient of 0.8. These theoretical values, when compared to the test data measured of 6.28 m/s$^2$ (20 ft/s$^2$) unloaded and 7.01 m/s$^2$ (23 ft/s$^2$) laden, indicate that the braking system of the light truck was operating near optimum condition. Changes in the brake force distribution or even a proportioning braking system would yield no improvement in braking performance. However, as stated earlier, braking efficiencies should be maximized only if braking stability is ensured, i.e., front brakes lock before rear brakes over a wide range of operating conditions.

Application of Eq. (7-20) to the school bus resulted in a brake force distribution $\Phi = 0.50$ to 0.55. The vehicle was equipped with a brake force distribution $\Phi = 0.42$. A brake force distribution of $\Phi = 0.55$ would produce theoretical braking efficiencies of 72 and 93% for the empty and laden vehicle, respectively, on slippery roadways with $\mu = 0.2$; and 92 and 96% for the empty and laden vehicle, respectively, on dry road surfaces with $\mu = 0.8$. For the empty vehicle, a theoretical deceleration of 7.22 m/s² (23.7 ft/s²) may be expected on dry road surfaces. In the case of the school bus, a change in brake force distribution from 0.42 to 0.55 will improve braking performance for the vehicle on slippery road surfaces indicated by an increase from 48 to 72%. Improvements in deceleration capability can be expected from a change in brake force distribution. However, a proportioning braking system will yield only little increase in braking performance indicated by the small change in relative static rear axle loading of $\Delta\Psi = 0.105$.

The design of the braking system for the medium truck is made difficult by a significant change in static rear axle loading indicated by $\Delta\Psi = 0.29$. The brake system of this truck was designed to meet the braking requirements for the loaded operating condition indicated by an actual brake force distribution of $\Phi = 0.74$, i.e., 74% of the total braking effort is concentrated on the rear axle. Although this design will produce desirable results for the loaded vehicle while braking on slippery or dry roads, the braking performance to be expected with the empty vehicle is unacceptable. Application of Eq. (7-20) resulted in the following inequalities for the brake force distribution of the medium-weight truck:

$$0.27 \leq \Phi \leq 0.52 \text{ for } \mu = 0.2, \text{ empty}$$

$$0.12 \leq \Phi \leq 0.36 \text{ for } \mu = 0.8, \text{ empty}$$

$$0.58 \leq \Phi \leq 0.83 \text{ for } \mu = 0.2, \text{ laden}$$

$$0.43 \leq \Phi \leq 0.68 \text{ for } \mu = 0.8, \text{ laden}$$

These results indicate that a brake force distribution $\Phi = 0.74$ will produce acceptable braking performance only for the loaded vehicle on slippery roads. Consider the second and third inequality as a compromise; a brake force distribution of $\Phi = 0.47$ probably produces better braking performance for all road surface and loading conditions than can be expected from $\Phi = 0.74$. The theoretical braking efficiencies with $\Phi = 0.47$ are 67 and 90% for the empty and laden vehicle, respectively, on a road surface with a coefficient of friction of 0.8. For the empty vehicle with $\Phi = 0.47$, a deceleration of approximately 5.24 m/s$^2$ (17.2 ft/s$^2$) may be expected on dry roads; a deceleration of 7.07 m/s$^2$ (23.2 ft/s$^2$) for the loaded vehicle with $\Phi = 0.47$. A further increase in braking capability can be accomplished only by a proportioning brake system. This is also evident from the change in relative static rear axle loading of $\Delta\Psi = 0.29$. A load-sensitive proportioning braking system will increase the maximum wheels-unlocked decelerations to 6.1 to 7.01 m/s$^2$ (20 to 23 ft/s$^2$) for the empty and loaded conditions.

## 7.8.5  Effect of Drive Train on Brake Force Distribution

Engine drag and rotational mass inertias will have an effect on the base brake force distribution that should be used in a vehicle. For manual transmissions the algebraic expressions are straightforward, but they are lengthy and involved. They are a function of gear ratios, rotational inertias, and engine speeds, as well as the basic physical expressions discussed in prior sections.

In general, for rear-wheel-drive vehicles the drive train effects will cause the critical deceleration (above which the rear brakes lock first) to decrease. The decrease will be greater as vehicle speed increases. For example, if the critical deceleration at low speed (or without consideration of speed effects) is 0.9 g, at 50m/s or 180 km/h (112 mph), the high speed critical deceleration may only be 0.6 g. For front wheel drive vehicles the critical deceleration will increase as speeds increase. For example, a low speed critical deceleration of 0.9 g will increase to approximately 0.95 to 1g at speeds of 180 km/h (112 mph).

## 7.8.6  Effect of Aerodynamics on Brake Force Distribution

Aerodynamic forces affect both vehicle handling and limit braking performance. The normal forces on front and rear axle are decreased as a result of the aerodynamic lift. In addition, aerodynamic drag increases vehicle deceleration, and hence, axle load distribution. Eqs. 7-3 will change by a negative aerodynamic lift term in the round bracket. The lift term increases with the speed squared. The decelerations at which either the front or rear brakes lock must be computed for increasing speeds. Eqs. 7-21 are modified by a negative lift term in the numerator. For example, Eq. 7-21b computing front brake lockup will be:

$$a = \frac{\mu(1 - \Psi) - \dfrac{k_{\ell F} V^2}{W}}{1 - \Phi - \mu x} \quad , \quad \text{g-units}$$

where $k_{\ell F}$ = front lift coefficient, $Ns^2/m$ $(lbs^2/ft^2)$

$V$ = vehicle speed, m/s (ft/s)

$W$ = vehicle weight, N (lb)

For standard passenger cars, $k_{\ell F} = 0.125$, $k_{\ell R} = 0.25$; for spoilers front and rear $k_{\ell F} = 0.01$, $k_{\ell R} = 0.38$; for rear spoilers only $k_{\ell F} = 0.125$, $k_{\ell R} = 0.03$.

Brake lockup calculations including aerodynamic lift for a standard passenger car may show front brake lockup for speeds up to 180 km/h (112 mph), while the rear brakes lock first, that is, at a lower deceleration than the front brakes, at a speed of 216 km/h (134 mph).

## 7.9 Variable Brake Force Distribution Analysis

If for a given vehicle the decelerations achieved with a fixed brake force distribution are too low, then a variable brake force distribution must be used. The object of designing variable or proportional braking is to bring braking efficiencies closer to unity over a wide range of loading and road friction conditions, encompassing most winter and summer driving situations. Through a proportioning valve the actual braking forces are brought closer to the optimum. As braking efficiencies are increased, front brakes should lock before the rear brakes.

### 7.9.1 Optimum Brake Line Pressures

In the design of the brake balance front to rear of a brake system, the optimum braking forces or the brake line pressures producing optimum forces may be used. Since proportioning valves affect the brake line pressures reaching front and rear brakes, it is convenient to work with the optimum brake line pressures. The optimum brake line pressures may be computed from the actual braking force expression (Eq. [5-2]), however, replacing the actual braking forces with the optimum (Eqs. [7-7a] and [7-7b]). By equating optimum and actual braking forces and solving for brake line pressure, the optimum brake line pressures $p_{\ell F,opt}$ and $p_{\ell R,opt}$ are

Front axle:

$$p_{\ell F,opt} = \frac{(1 - \Psi + \chi a)aWR}{2(A_{wc}BFr\eta_c)_F} + p_{oF} \quad , \quad N/cm^2 \text{ (psi)} \quad \text{(7-30a)}$$

Rear axle:

$$p_{\ell R,opt} = \frac{(\Psi - \chi a)aWR}{2(A_{wc}BFr\eta_c)_R} + p_{oR} \quad , \quad N/cm^2 \text{ (psi)} \quad \text{(7-30a)}$$

where   $a$ = deceleration, g-units

  $A_{wc}$ = wheel cylinder area, cm$^2$ (in.$^2$)

  $BF$ = brake factor

  $p_{oF}$ = pushout pressure, front brakes, N/cm$^2$ (psi)

  $p_{oR}$ = pushout pressure, rear brakes, N/cm$^2$ (psi)

  $r$ = effective rotor or drum radius, cm (in.)

  $R$ = effective tire radius, cm (in.)

  $W$ = vehicle weight, N (lb)

  $\eta_c$ = wheel cylinder efficiency

The optimum brake line pressure curves are similar to the optimum braking force curves. Lines of constant deceleration are different for the empty and laden conditions.

The optimum brake line pressures for air brake systems are obtained in a similar fashion to those of hydraulic brake systems.

## 7.9.2 Brake Line Pressure Limiter Valve

Pressure limiter valves discussed in Section 5.3.1 use a linearly increasing braking force up to the critical deceleration $a_{crit}$ and a constant rear brake force for deceleration greater than $a_{crit}$. A typical braking force diagram showing optimum and actual braking forces is illustrated in Figure 7-11. The baseline brake force distribution generally is designed so that it would intersect the empty optimum line at a deceleration of 0.5 g. The baseline brake force distribution $\Phi_B$ is computed by equating optimum and actual rear braking forces for a = 0.5 g, resulting in

$$\Phi_B = \Psi - a_{crit}\chi = \Psi - 0.5\chi \tag{7-31}$$

The constant vertical portion of the brake balance is designed so that the actual rear brake force is 90% of the optimum rear brake force given by Eq. (7-12).

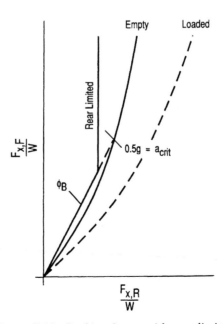

*Figure 7-11. Braking forces with rear limiter.*

$$F_{xR} / W)_{act} \approx 0.9 F_{xR} / W)_{max} = 0.9(\Psi_0^2 / 4\chi_0) \qquad (7\text{-}32)$$

The deceleration $a_{empty}$ of the empty vehicle at the valve shift point is computed by applying Newton's Second Law to the rear brakes and solving for deceleration, or

$$a_{empty} = 1 / \Phi_B(0.9)(\Psi_0^2 / 4\chi_0) \quad , \quad \text{g-units} \qquad (7\text{-}33)$$

The corresponding shift point pressure $p_s$ is given by

$$p_s = \frac{a_{empty} W_0 R}{2[(A_{wc}BF r\eta_c)_F + (A_{wc}BF r\eta_c)_R]} \quad , \quad \text{N/cm}^2 \text{ (psi)} \qquad (7\text{-}34)$$

where $A_{wc}$ = wheel cylinder area, cm$^2$ (in.$^2$)

$\quad$ BF = brake factor

$\quad$ r = effective rotor or drum radius, cm (in.)

$\quad \eta_c$ = wheel cylinder efficiency

The deceleration $a_{s,laden}$ of the laden vehicle existing at the shift pressure is determined by the weight ratio between empty $(W_0)$ and laden $(W_\ell)$ vehicle, or

$$a_{s,laden} = a_{empty}(W_0 / W_\ell) \quad , \quad \text{g-units} \qquad (7\text{-}35)$$

### 7.9.3  Brake Line Pressure Reducer Valve

The operation of reducer valves is discussed in Section 5.3.2. The analysis is similar to that of the limiter valve. The baseline brake force distribution is generally chosen so that it would intersect the "empty" optimum line at approximately 0.4 g, assuming no pressure reducing would occur. The baseline brake force distribution $\Phi_B$ is computed by

$$\Phi_B = \Psi - 0.4\chi_0 \qquad (7\text{-}36)$$

The knee-point pressure $p_k$ of the reducer valve should be chosen so that the pressure switch occurs at 80 to 90% of the baseline critical deceleration for the empty vehicle.

$$p_k = \frac{(0.8 \text{ to } 0.9)a_{crit}W_oR}{2\sum(A_{wc}BFm_c)_{F,R}} \quad , \quad \text{N/cm}^2 \text{ (psi)} \tag{7-37}$$

Reducer valve slopes generally range between 0.3 and 0.5. The shift or knee-point and slope of the reducer valve should be chosen so that for the empty vehicle the reduced brake force distribution intersects the optimum braking force curve at a critical deceleration of 0.9 to 1.0 g.

When a load-sensitive reducer or proportioning valve is used, the critical deceleration for the laden vehicle should not be less than 0.9 to 1.0 g.

A load-sensitive reducer valve braking system design is illustrated in Figure 7-12. The shift point for the fully laden case should be shifted slightly to the left to approximately 0.9 times 524 N/cm² (760 psi) or 472 N/cm² (684 psi) (Eq. [7-37]), i.e., to lower rear brake line pressures to ensure that inadvertent brake factor increases on the rear brakes do not easily cause premature rear brake lockup. The knee-point pressure for the empty case is zero to produce stable braking for all decelerations up to about 1.0 g.

### 7.9.4 Deceleration-Sensitive Pressure Reducer Valve

Deceleration-sensitive reducer valves are discussed in Section 5.3.4. The shift point pressure is a function of the installation angle of the valve and generally ranges from 18 to 20 degrees, corresponding to a shift point deceleration $a = 0.32$ to 0.36 g.

The critical deceleration after the shift point is approximately 1.1 to 1.2 g. The brake force distribution $\Phi_a$ after the shift point is computed by

$$\Phi_a = \Psi_o - a_{crit}\chi_o \tag{7-38}$$

The baseline brake force distribution $\Phi_B$, i.e., the brake force distribution prior to reaching the shift point, is computed by

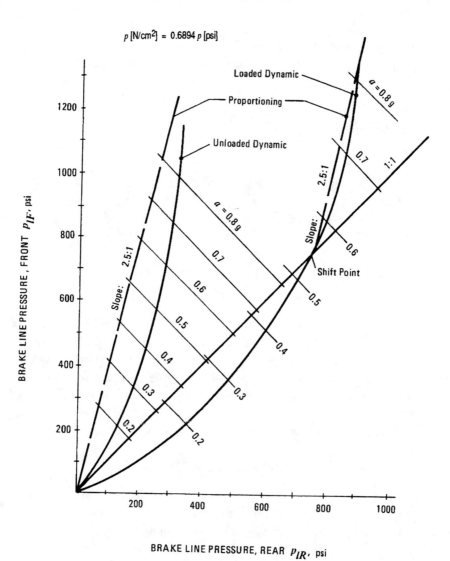

$p$ [N/cm$^2$] = 0.6894 $p$ [psi]

*Figure 7-12. Dynamic and actual brake line pressures.*

$$\Phi_B = \frac{1}{1 + \dfrac{SL}{\Phi_a / (1 - \Phi_a)}} \qquad (7\text{-}39)$$

where   SL = reducer slope (Eq. [5-12])

Front brake lockup occurs at the following conditions:

$$a_{F,lock} = \frac{\mu(1 - \Psi)}{1 - \mu\chi - \Phi_B} \qquad (7\text{-}40)$$

where $a_{F,lock}$ = deceleration at which front brakes lock, g-units

Above shift point but below piston movement:

$$a_{F,lock} = \frac{\mu(1 - \Psi)}{1 - \mu\chi - \Phi_B} \quad , \quad \text{g-units} \qquad (7\text{-}41)$$

$$a_{F,lock} = \frac{\mu(1 - \Psi)}{1 - \mu\chi - \Phi_a} \quad , \quad \text{g-units} \qquad (7\text{-}42)$$

## 7.9.5 Adjustable Step Bore Master Cylinder

Performance details are discussed in Section 5.3.5. Using the component designation shown in Fig. 5-17, the pressure ratio rear to front after the master cylinder has switched to the "laden" condition is determined by

$$(p_R / p_F)_{after} = (p_R / p_F)_{before}(A_{mc,2} / A_{mc,3}) \qquad (7\text{-}43)$$

where $A_{mc,2}$ = cross-sectional area of floating piston, cm$^2$ (in.$^2$)

$A_{mc,3}$ = cross-sectional area of step piston, cm$^2$ (in.$^2$)

$(p_R/p_F)_{after}$ = rear to front pressure ratio after switching to laden

$(p_R/p_F)_{before}$ = rear to front pressure ratio before switching to laden

Example 7-1: Compute the braking efficiencies for a front-wheel-drive passenger car. Use the data that follow.

Vehicle data:

Weight, driver only: 11,564 N (2600 lb)

Laden weight: 15,568 N (3500 lb)

Driver-only rear axle load: 4092 N (920 lb); front axle: 7339 N (1650 lb)

Laden rear axle load: 7784 N (1750 lb); front axle: 7784 N (1750 lb)

Wheelbase: 2616 mm (103 in.)

Tire radius: 29.97 cm (11.8 in.)

Center-of-gravity height empty and laden: 559 mm (22 in.)

## Brake system data:

### Front:

Wheel cylinder diameter: 5.71 cm (2.25 in.)

Rotor radius: 9.9 cm (3.9 in.)

Brake factor: 0.64

Pushout pressure: 3.4 N/cm$^2$ (5 psi)

### Rear:

Wheel cylinder diameter: 1.905 cm (0.75 in.)

Drum diameter: 20.32 cm (8 in.)

Brake factor: 2.9 (LT-shoe brake)

Pushout pressure 55 N/cm$^2$ (80 psi)

Master cylinder diameter: 20.64 mm (0.812 in.)

Pedal ratio: 5.5 to 1 (standard brakes)

Proportioning valve: 276 N/cm$^2$ (400 psi) × 35%

## Calculations (See Table 7-1):

## Relative axle loads:

Rear:

Empty: $\Psi_o$ = 4092 / 11,564 = 0.35

$[\Psi_o = 920 / 2600 = 0.35]$

Loaded: $\Psi_\ell = 7784 / 15{,}508 = 0.5$

$\qquad [\Psi_\ell = 1750 / 3500 = 0.5]$

Relative center-of-gravity height:

Empty: $\chi_o = 559 / 2616 = 0.21$

$\qquad [\chi_o = 22 / 102 = 0.21]$

Loaded: $\chi_o = \chi = 0.21$

The braking forces are computed by Eq. (5-2). For convenience the brake gain is computed individually for the front ($K_F$) and rear ($K_R$) brakes:

Brake gain:

Front: $K_F = 2[A_{wc}BF\eta_c(r / R)]_F$

$\qquad K_F = 2(25.6)(0.64)(0.98)(9.9 / 29.97) = 10.6 \text{ cm}^2$

$\qquad [K_F = 2(3.976)(0.64)(0.98)(3.9 / 11.8) = 1.648 \text{ in.}^2]$

Rear: $K_R = 2[A_{wc}BF\eta_c(r / R)]_R$

$\qquad K_R = 2(2.85)(2.9)(0.96)(10.16 / 29.97) = 5.38 \text{ cm}^2$

$\qquad [K_R = 2(0.44)(2.9)(0.96)(4 / 11.8) = 0.585 \text{ in.}^2]$

### TABLE 7-1

### Example 7-1, Analysis Table

| | | | | | | | |
|---|---|---|---|---|---|---|---|
| ① $p_{mc}$ | N/cm² | 55 | 276 | 552 | 689 | 1034 |
| | psi | 80 | 400 | 800 | 1000 | 1500 |
| ② $p_F$ | N/cm² | 55 | 276 | 552 | 689 | 1034 |
| | psi | 80 | 400 | 800 | 1000 | 1500 |

## TABLE 7-1 *(CONTINUED)*

### Example 7-1, Analysis Table

③ $p_F - p_{oF}$  | N/cm² | 52 | 272 | 548 | 686 | 1031 |
|  | psi | 75 | 395 | 795 | 995 | 1495 |

④ $p_R = 276 + 0.35(p_{mc} - 279)$  N/cm²  55  276  372  421  541
$p_R = 400 + 0.35(p_{mc} - 400)$  psi  80  400  540  610  785

⑤ $p_R - p_{oR}$  | N/cm² | 0 | 221 | 317 | 365 | 486 |
|  | psi | 0 | 320 | 460 | 530 | 705 |

⑥ $F_{xF} = ③ \times K_F$  | N | 552 | 2896 | 5827 | 7295 | 10960 |
|  | lb | 124 | 651 | 1310 | 1640 | 2464 |

⑦ $F_{xR} = ⑤ \times K_R$  | N | 0 | 1183 | 1699 | 1957 | 2602 |
|  | lb | 0 | 187 | 382 | 440 | 585 |

⑧ $F_x = F_{xF} + F_{xR}$  | N | 552 | 4079 | 7526 | 9252 | 13562 |
|  | lb | 124 | 838 | 1692 | 2080 | 3049 |

⑨ $\phi = F_{xR} / F_x$  | | 0 | 0.22 | 0.23 | 0.21 | 0.19 |

⑩ $a_o = F_x / W_o$  | g-units | 0.05 | 0.32 | 0.65 | 0.80 | 1.17 |

⑪ $a_\ell = F_x / W_\ell$  | g-units | 0.04 | 0.24 | 0.48 | 0.59 | 0.87 |

⑫ $F_{zF,dyn}$ (empty)  | N | 7695 | 8367 | 9096 | 9416 | 10359 |
$= (1 - \Psi_o + \chi_a a_o)W_o$  lb  1730  1881  2045  2127  2329
$= (1 - 0.35 + 0.21a_o)11564$

⑬ $F_{zR,dyn}$ (empty)  | N | 3928 | 3198 | 2469 | 2104 | 1205 |
$= (\Psi_o - \chi_a a_o) / W_o$  lb  883  719  555  473  271
$= (0.35 - 0.21a_o)11564$

⑭ $\mu_{TFo} = ⑥ / ⑫$  | | 0.07 | 0.35 | 0.64 | 0.77 | 1.06 |

## TABLE 7-1 *(CONTINUED)*

## Example 7-1, Analysis Table

| | | | | | | |
|---|---|---|---|---|---|---|
| ⑮ $\mu_{TRo}$ = ⑦ / ⑬ | | 0 | 0.37 | 0.69 | 0.93 | 2.16 |
| ⑯ $(a/\mu)_{Fo}$ = ⑩ / ⑭ | | 0.71 | 1.00 | 1.02 | 1.04 | 1.10 |
| ⑰ $(a/\mu)_{Ro}$ = ⑩ / ⑮ | | $\infty$ | 0.95 | 0.94 | 0.86 | 0.54 |
| ⑱ $F_{zF,dyn}$ (laden) | N | 7691 | 8634 | 9354 | 9714 | 10627 |
| | lb | 1729 | 1941 | 2103 | 2184 | 2389 |
| ⑲ $F_{zR,dyn}$ (laden) | N | 7655 | 6934 | 6214 | 5854 | 4942 |
| | lb | 1721 | 1559 | 1397 | 1316 | 1111 |
| ⑳ $\mu_{TF\ell}$= ⑥ / ⑱ | | 0.07 | 0.34 | 0.62 | 0.75 | 1.03 |
| ㉑ $\mu_{TR\ell}$ = ⑦ / ⑲ | | 0 | 0.17 | 0.27 | 0.33 | 0.53 |
| ㉒ $(a / \mu)_{F\ell}$ = ⑪ / ⑳ | | 0.57 | 0.76 | 0.77 | 0.79 | 0.84 |
| ㉓ $(a / \mu)_{R\ell}$ = ⑪ / ㉑ | | $\infty$ | 1.53 | 1.78 | 1.79 | 1.64 |
| ㉔ Pedal Force (Eq. [5-1]) | N | 42 | 209 | 418 | 525 | 787 |
| | lb | 9.4 | 47 | 94 | 118 | 177 |

The braking forces are computed by multiplying brake gain by the brake line pressure, computed by Eq. (5-11). The brake line pressures are shown in the computation table, rows 1 through 5. For example, inspection of column four reveals that for a master cylinder pressure of 689 N/cm$^2$ (1000 psi), the rear brake line pressure is 421 N/cm$^2$ (610 psi) arriving at the rear wheel cylinder, while the effective rear brake line pressure is only 365 N/cm$^2$ (530 psi). Inspection of the results indicates that the braking efficiencies for the empty or driver-only case are near optimum for brake line pressure around 276 N/cm$^2$ (400 psi). For higher brake line pressures the rear brakes will lock up first in the lightly loaded case indicated by braking efficiency values less than unity (row 17).

For the laden case the front brakes always lock up first as indicated by the braking efficiency values less than unity (row 22). Improvement in braking efficiency could be achieved by making the proportioning valve either load- or deceleration-sensitive.

If the pedal forces are too high, a power assist system should be used to bring the pedal force levels within the recommended ranges of 60 to 100 lb per 1 g deceleration (see Section 1.3.1).

# 7.10  Vehicle Stability Analysis

## 7.10.1  General Considerations

The response of a mechanical system to a disturbance force can be stable, unstable, or indifferent. A response is stable when, after the action of a disturbance force, the system returns to its initial stable motion. A response is unstable when a relatively small disturbance force causes greater and greater deviations from the initially stable motion. In the case of an unstable response the energy required for the unstable motion to develop is provided by the kinetic or motion energy of the system, i.e., the speed of the motor vehicle. An indifferent response is when a disturbance force causes a single displacement proportional to the disturbance force.

Accident and vehicle test data, as well as basic engineering analysis, indicate that locking of the rear brakes before the front brakes will result in violent vehicle instability, most frequently causing the vehicle to spin about its vertical axis. The angular spin velocity and corresponding spin angle are a function of vehicle speeds, tire-road friction coefficient, yaw moment of inertia, and vehicle dimensions.

## 7.10.2  Simplified Braking Stability Analysis

The development of vehicle instability due to wheel lockup is illustrated in Figure 7-13. If it is assumed that the front wheels have not yet approached sliding conditions and are still rolling, and that the rear brakes are already locked (Fig. 7-13(A)), any disturbance in the lateral direction due to road grade, side wind, or left-to-right brake imbalance produces a side force $F_y$ acting at the center of gravity of the vehicle. The resultant force $F_r$, stemming from the inertia force $F_x$ induced by braking and the lateral force $F_y$, is now acting under the vehicle slip angle $\alpha_v$. The slip angle $\alpha_v$ is formed by the

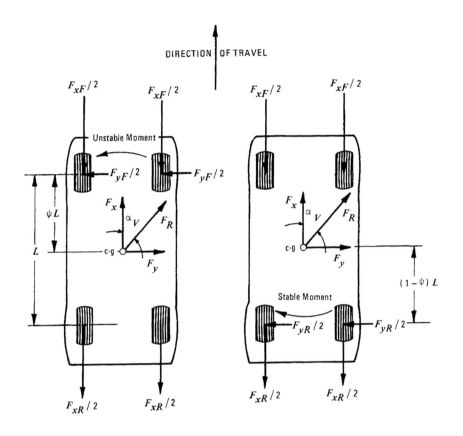

$F_{xF}/2$    $F_{xF}/2$    $F_{xF}/2$    $F_{xF}/2$

DIRECTION OF TRAVEL

Unstable Moment

$-F_{yF}/2$   $F_{yF}/2$

$F_x$   $\alpha_V$   $F_R$   $\psi L$   $L$   c-g   $F_y$

$F_x$   $\alpha_V$   $F_R$   c-g   $F_y$   $(1-\psi)L$

Stable Moment

$-F_{yR}/2$   $F_{yR}/2$

$F_{xR}/2$    $F_{xR}/2$    $F_{xR}/2$    $F_{xR}/2$

(A) Rear Wheels Locked. Unstable Motion    (B) Front Wheels Locked. Stable Motion

*Figure 7-13. Vehicle behavior with rear and front wheels locked.*

longitudinal axis of the vehicle and the direction in which the center of gravity is moving. The lateral side force $F_y$ must be counteracted by side forces produced at the tires. Because the rear wheels are sliding, no tire side forces can be produced at the rear and, consequently, the side forces developed by the still-rotating front wheels produce a yawing moment of magnitude $F_{yF}\Psi L$. This destabilizing moment rotates the vehicle about its vertical axis so that the initial slip angle $\alpha_v$ increases, resulting in vehicle instability.

If the front brakes are locked first, an identical lateral disturbance will be reacted upon by a stabilizing moment $F_{yR}(1-\Psi)L$ produced by the rolling rear wheels. The direction of the moment is such that it rotates the longitudinal axis of the vehicle toward the direction of travel of the center of gravity of the vehicle, thus reducing the initial disturbance slip angle $\alpha_v$ with the vehicle remaining stable. The vehicle will slide straight with its front brakes locked. While the front brakes are locked the vehicle will not respond to steering inputs by the driver. If a collision is unavoidable, a frontal crash typically will result in fewer injuries than a side crash due to better occupant protection.

## 7.10.3 Expanded Braking Stability Analysis

A simplified "bicycle" model of a braking vehicle is shown in Figure 7-14. Although braking, the front and rear wheels are rotating, i.e., they have not yet achieved lockup. Similar to the discussion presented in Section 7.10.1, a disturbance force produces a vehicle slip angle $\alpha_v$. The lateral side force $F_y$ acting at the center of gravity is reacted upon by the tire side forces $F_{yF}$ and $F_{yR}$ on the front and rear axle, respectively. The tire slip angles required to produce side force, front and rear, are assumed to be both equal to each other as well as equal to the vehicle slip angle $\alpha_v$. Because all tire forces are acting on their respective lever arms about the center of gravity, a moment is acting about the center of gravity of the vehicle which is reacted against by the mass moment of inertia of the vehicle.

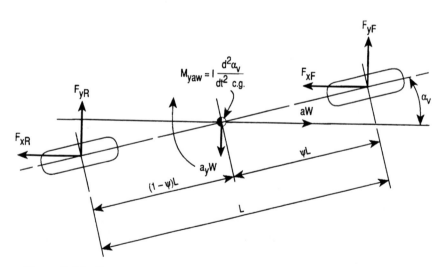

*Figure 7-14. Forces and moments on simplified vehicle model during braking.*

Application of force balance in the x- and y-directions as well as moment balance about the center of gravity yields the governing dynamic equation for angular acceleration in rotation:

$$\varepsilon = d^2\alpha_v \: / \: dt^2 = (L \: / \: I)\left\{\Psi F_{xF} \sin \alpha_v - (1 - \Psi)F_{xR} \sin \alpha_v \right.$$

$$\left. +\Psi F_{yF} \cos \alpha_v - (1 - \Psi)F_{yR} \cos \alpha_v \right\} \quad , \quad 1/s^2 \qquad (7\text{-}44)$$

where $F_{xF}$ = braking force on front axle in x-direction, N (lb)

$\quad\; F_{xR}$ = braking force on rear axle in x-direction, N (lb)

$\quad\; F_{yF}$ = tire side force on front axle, N (lb)

$\quad\; F_{yR}$ = tire side force on rear axle, N (lb)

$\qquad I$ = mass moment of inertia, kgms$^2$ (lb ft s$^2$)

$\qquad L$ = wheelbase, m (ft)

$\qquad t$ = time, s

$\quad\; \alpha_v$ = vehicle slip angle, deg

$\qquad \varepsilon$ = angular acceleration, 1/s$^2$

$\qquad \Psi$ = static rear axle load divided by vehicle weight

Similar to linear acceleration which expresses how quickly a vehicle changes forward speed, angular acceleration expresses how quickly a vehicle changes its rotational speed or angular velocity about its vertical axis. A positive value of angular acceleration in Eq. (7-44) indicates an increase in angular velocity and, hence, vehicle slip angle, i.e., the vehicle is directionally unstable and spins about its vertical axis. On the other hand, a negative value of angular acceleration in Eq. (7-44) indicates a decrease in angular velocity and vehicle slip angle, i.e., the vehicle returns to its initial stable course.

Consider the following cases with respect to Eq. (7-44).

(1) The rear brakes lock with the front wheels still rotating: $F_{yR} = 0$; $\varepsilon$ will be positive resulting in an unstable vehicle.

(2) The front brakes are locked with the rear brakes still rotating: $F_{yF} = 0$; $\varepsilon$ will be negative resulting in a stable vehicle.

Angular acceleration is a measure of the level of directional stability of the vehicle. A negative value is an indication that the vehicle is stable during braking maneuvers involving brake lockup. A positive angular acceleration means there is always a potential for vehicle instability. Typical drivers generally will not be able to control the directional path of a vehicle if the angular acceleration is greater than approximately 0.25 $1/s^2$ (Ref. 2). An angular acceleration of 0.25 $1/s^2$ indicates an angular velocity of approximately 14 deg/s (57.3/4) after one second. The associated yaw or spin angle would be approximately 7 deg one second after the instability began. For vehicles exhibiting rear brake lockup before front brake lockup, relatively small vehicle slip angles of two to five degrees generally are sufficient to generate angular accelerations in excess of the limit recoverable by the driver. If the disturbance slip angle is greater, the unstable angular acceleration and associated spin angle will increase rapidly.

A numerical evaluation of Eq. (7-44) requires computation of the braking and side forces at the front and rear tires. For the braking forces a traction coefficient can be computed by using Eq. (7-4). The braking forces in the numerator of Eq. (7-4) may be expressed as a function of brake line pressure by using Eq. (5-2). When considering that the rear braking forces $F_{xR}$ at the rear tires are related to the braking forces developed by the brake system by the term $\cos\alpha_v$, the following expressions for the traction coefficients $\mu_{TF,x}$ and $\mu_{TR,x}$ may be derived:

Front axle:

$$\mu_{TF,x} = \frac{p_F K_F \cos\alpha_v}{W(1 - \Psi) + (p_F K_F + p_R K_R)\chi \cos\alpha_v} \qquad (7\text{-}45)$$

Rear axle:

$$\mu_{TR,x} = \frac{p_R K_R \cos\alpha_v}{W\Psi - (p_F K_F + p_R K_R)\chi \cos\alpha_v} \qquad (7\text{-}46)$$

where $K_{F,R}$ = axle brake gain, front or rear, cm$^2$ (in.$^2$)

$p_F$ = front brake line pressure, N/cm$^2$ (psi)

$p_R$ = rear brake line pressure, N/cm$^2$ (psi)

W = vehicle weight, N (lb)

$\chi$ = center of gravity height divided by wheelbase

The rear brake line pressure is computed by Eq. (5-11) when a proportioning valve is used.

The front and rear axle brake gains are computed by

$$K_{F,R} = 2[A_{wc}BF\eta_c(r / R)]_{F,R} \quad , \quad cm^2 \text{ (in.}^2) \tag{7-47}$$

where the following terms are evaluated for the front and rear brakes:

$A_{wc}$ = wheel cylinder cross-sectional area, cm$^2$ (in.$^2$)

BF = brake factor

r = effective rotor or drum radius, cm (in.)

R = tire radius, cm (in.)

$\eta_c$ = wheel cylinder efficiency

The traction coefficients computed by Eqs. (7-45) and (7-46) for a given slip angle $\alpha_v$ establish a specific absolute tire slip value $S_a$ by use of basic tire data. Because the total tire traction available has to be shared by braking and side traction, use of the absolute slip $S_a$ and the slip angle $\alpha_v$ establishes values of side traction coefficient $\mu_y$ for a range of brake line pressures.

The side forces front and rear $F_{yF}$ and $F_{yR}$ are computed for the side traction coefficient established above by the following expressions:

Front:

$$F_{yF} = \mu_{yF}\{W(1 - \Psi) + (p_FK_F + p_RK_R)\chi \cos \alpha_v\} \quad , \quad N \text{ (lb)} \tag{7-48}$$

Rear:

$$F_{yR} = \mu_{yR}\{W\Psi - (p_F K_F + p_R K_R)\chi\cos\alpha_v\} \quad , \quad N\ (lb)\ (7\text{-}49)$$

Typical results showing angular acceleration as a function of brake line pressure for the lightly and fully laden operating conditions are illustrated in Figure 7-15. Inspection of the stability curves reveals that the empty vehicle begins to lose lateral stability for brake line pressures above approximately

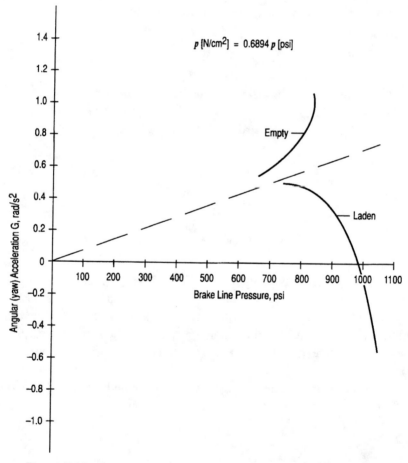

*Figure 7-15. Yaw acceleration as a function of brake line pressure.*

550 N/cm$^2$ (800 psi), while the laden vehicle shows decreasing values of angular acceleration and is absolutely stable for brake line pressures above approximately 690 N/cm$^2$ (1000 psi).

### 7.10.4 Braking on a Split-Coefficient Road Surface

Rotation of a vehicle about its vertical axis during braking may be the result of braking on a split-coefficient road surface. In nearly all cases the angular velocity achieved by the vehicle is considerably less than that caused by premature rear wheel lockup. Vehicles with a negative scrub radius will experience lower rotation than vehicles with positive or zero scrub radius.

The rotation of the vehicle toward the higher traction side may be analyzed by a simplified formulation assuming zero friction under the low traction side. Tire cornering stiffness values, pneumatic trail, and suspension characteristics must be considered.

# 7.11  Braking Dynamics for Three-Axle Vehicle

## 7.11.1  Basic Methodology

The general approach for analyzing friction utilization and braking efficiency is outlined in connection with Example 7-1. The general procedure presented becomes very useful when inter-axle load transfer for a tandem axle truck must be considered, or the braking of combination vehicles is analyzed (see Chapter 8). The approach consists of the following steps: a certain brake line pressure results in certain brake forces on each axle, which produce vehicle deceleration, which causes dynamic axle loads, which may be used to compute the tire-road traction coefficient $\mu_T$ required to prevent wheel lockup.

Because the computations are algebraically involved, computer programs are used to facilitate the analysis. The program consists of five basic steps:

1. *Input.* Vehicle data and specifications, brake data, loading conditions, and tire-road friction coefficient are entered into the program.

2. *Initialization.* Calculations of static axle loads and setting of brake line pressure at initial value.

3. *Calculations.* Necessary calculations including brake line pressure at each axle (if pressure reducer valves are used), brake factor, adjustment and temperature factors, brake force, deceleration, and dynamic axle loads are made. Wheel lockup is indicated if it occurs.

4. *Outputs.* Results from calculations are printed.

5. *Increment/Stop.* If brake line pressure is less than maximum value, it is incremented, and another calculation is made. If calculations have been made for the maximum brake line pressure, or if all wheels are locked for a given brake line pressure, the program is terminated.

## 7.11.2 Three-Axle Truck with Walking Beam Suspension

The forces acting on a decelerating tandem axle truck are illustrated in Figure 7-16. The forces $F_{xF}$, $F_{xRF}$, and $F_{xRR}$ which induce braking deceleration are obtained from the axle brake forces and are considered to be known functions of brake line pressure. In the case of a hydraulic brake system, the brake forces are computed by Eq. (5-2); in the case of an air brake system, by an expression similar to Eq. (6-1), however, with the brake torque divided by tire radius to obtain brake force.

Use the notations from Figs. 7-16 and 7-17(A). The dynamic axle loads during braking on a tandem axle truck with walking beam rear suspension are (Ref. 10)

Truck Front:

$$F_{zF} = W_s - Y_1 \quad , \quad N \text{ (lb)} \tag{7-50}$$

Truck Tandem Forward Axle:

$$F_{zRF} = Y_1 + w_F + w_R - F_{zRR} \quad , \quad N \text{ (lb)} \tag{7-51}$$

Truck Tandem Rear Axle:

$$F_{zRR} = [Y_1 s - X_1 v_1 - (w_F + w_R) a u_1$$
$$+ w_R q] \, 1 / q \quad , \quad N \text{ (lb)} \tag{7-52}$$

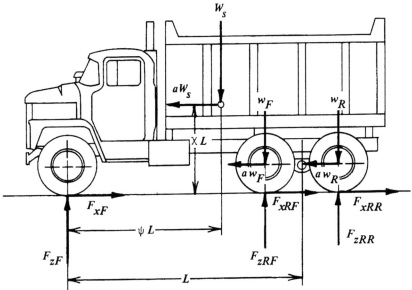

*Figure 7-16. Forces acting on a tandem axle truck.*

The horizontal suspension force $X_1$ is given by

$$X_1 = aW_s - F_{xF} \quad , \quad N \text{ (lb)} \tag{7-53}$$

The vertical suspension force $Y_1$ is given by

$$Y_1 = (W_s \Psi L - aW_s \chi L + X_1 v_1)\, 1 / L \quad , \quad N \text{ (lb)} \tag{7-54}$$

where   a = deceleration, g-units

$F_{xF}$ = actual front axle brake force, N (lb)

$F_{zF}$ = axle load, front axle, N (lb)

$F_{zRF}$ = axle load, second axle, N (lb)

$F_{zRR}$ = axle load, third axle, N (lb)

   L = wheelbase or distance between center of front axle and center of tandem axle, cm (in.)

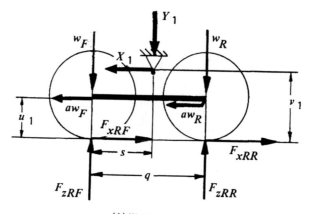

(A) Walking Beam

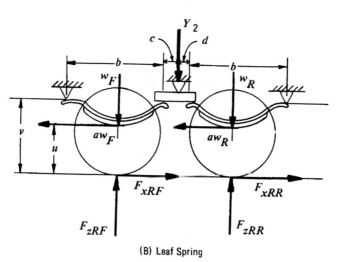

(B) Leaf Spring

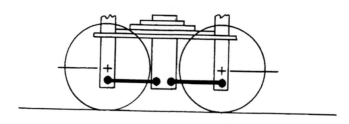

(C) Multiple Leaf-Multiple Rod

*Figure 7-17. Tandem axle suspensions.*

$q$ = dimension, tandem axle, cm (in.)

$s$ = dimension, tandem axle, cm (in.)

$u$ = dimension, tandem axle, cm (in.)

$u_1$ = dimension, tandem axle, cm (in.)

$v$ = dimension, tandem axle, cm (in.)

$v_1$ = dimension, tandem axle, cm (in.)

$W_s$ = weight of truck minus weight of tandem axle, N (lb)

$w_F$ = unsprung weight of tandem forward axle, N (lb)

$w_R$ = unsprung weight of tandem rearward axle, N (lb)

$\chi$ = center of gravity height divided by wheelbase

$\Psi$ = static tandem axle load divided by vehicle weight

The forces $Y_1$ and $X_1$ are the forces transmitted by the suspension to the frame rails of the truck and are important in designing suspension attachments.

The computed dynamic axle loads, using the data shown in Table 7-2, are shown in Figure 7-18. For a deceleration of 0.5 g the axle load $F_{zRR}$ on the tandem rearward axle has decreased to approximately 47% of its static value. The utilization of the tire-road friction coefficient by this suspension design is illustrated in Figure 7-19 where the road friction coefficient required for wheels-unlocked braking is shown as a function of vehicle deceleration. The braking of the tandem forward axle is close to optimum over a wide range of tire-road friction coefficients, whereas the front axle is largely underbraked and the tandem rearward axle overbraked. For example, for $\mu = 0.6$, wheels-unlocked decelerations of not greater than approximately 0.35 g can be expected. Any improvement in braking performance must come by increasing the brake force on the front axle with a corresponding decrease on the rearward axle of the tandem suspension.

## TABLE 7-2

### Tandem Axle Truck Data

| Truck | | | |
|---|---|---|---|
| $W_s$ | = 184,147 N | 41,400 lb | |
| L | = 487.7 cm | 192 in. | |
| $\Psi$ | = 0.74 | | |
| $\chi$ | = 0.40 | | |
| Tire radius R | = 51.4 cm | 20.25 in. | |

| Suspension | | |
|---|---|---|
| $w_F$ | = 10,230 N | 2300 lb |
| $w_R$ | = 10,230 N | 2300 lb |
| q | = 127 cm | 50 in. |
| s | = 61 cm | 24 in. |
| $u_1$ | = 50.8 cm | 20 in. |
| $v_1$ | = 35.6 cm | 14 in. |

| Brakes Type | Front Axle Wedge | | Tandem Forward "S" Cam | | Tandem Rearward "S" Cam | |
|---|---|---|---|---|---|---|
| Drum radius r | 19.05 cm | 7.5 in. | 20.96 cm | 8.25 in. | 20.96 cm | 8.25 in. |
| Brake chamber area $A_C$ | 129 cm² | 20 in.² | 194 cm² | 30 in.² | 194 cm² | 30 in.² |
| Wedge angle $\alpha$ | 12 deg | | — | | — | |
| Slack adjuster length $\ell_s$ | — | | 13.97 cm | 5.5 in. | 13.97 cm | 5.5 in. |
| Brake Factor BF | 4.3 | | 2.7 | | 2.7 | |
| Pushout pressure $p_o$ | 3.4 N/cm² | 5.0 psi | 1.7 N/cm² | 2.5 psi | 1.7 N/cm² | 2.5 psi |
| Cam radius | — | | 1.27 cm | 0.5 in. | 1.27 cm | 0.5 in. |
| Mechanical efficiency $\eta_m$ | 0.88 | | 0.75 | | 0.75 | |

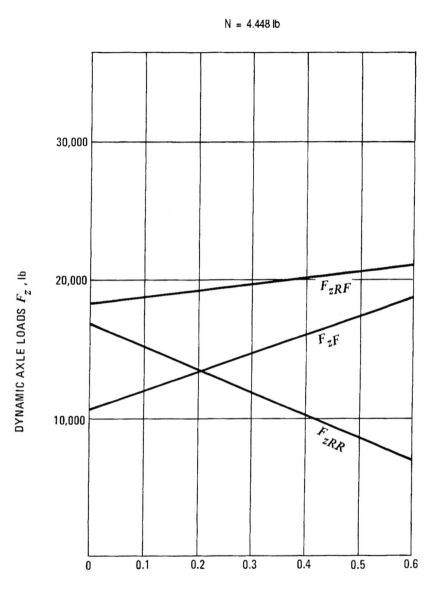

*Figure 7-18. Dynamic axle loads for a truck equipped with walking beam suspension.*

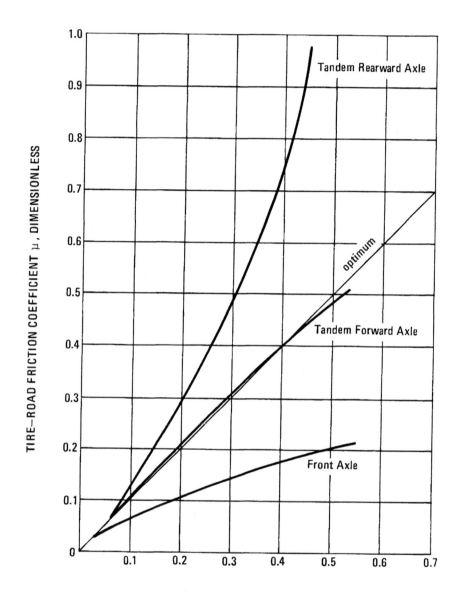

*Figure 7-19. Tire-road friction utilization for a truck equipped with walking beam suspension.*

### 7.11.3 Three-Axle Truck with Two-Elliptic Leaf Spring Suspension

Use the notations from Figs. 7-16 and 7-17(B). The dynamic axle loads during braking on a tandem axle truck with two-elliptic leaf spring suspension are (Ref. 10)

Truck Front:

$$F_{zF} = W_s - Y_2 \left\{ \frac{d[(b/2) - av]}{(c+d)[(b/2) - av]} \right.$$

$$\left. + \frac{c[(b/2) + av]}{(c+d)[(b/2) - av]} + 1 \right\}$$

$$+ \left[ \frac{w_F u}{(b/2) + av} - \frac{w_R u}{(b/2) + av} \right] \quad , \quad N \text{ (lb)} \qquad (7\text{-}55)$$

Truck Tandem Forward:

$$F_{zRF} = \frac{Y_2 bd}{(c+d)[(b/2) + av]} + w_F - \frac{w_F u a}{(b/2) + av} \quad , \quad N \text{ (lb)} \qquad (7\text{-}56)$$

Truck Tandem Rearward:

$$F_{zRR} = \frac{Y_2 bc}{(c+d)[(b/2) - av]} + w_R + \frac{w_R u a}{(b/2) - av} \quad , \quad N \text{ (lb)} \qquad (7\text{-}57)$$

The vertical suspension force $Y_2$ at equalizing bar is given by

$$Y_2 = [W_s L(\Psi - a\chi) + Ga] / H \quad , \quad N \text{ (lb)} \qquad (7\text{-}58)$$

where

$$G = \frac{w_F u}{(b/2) + av}(L - c - b - va)$$

$$- \frac{w_R u}{(b/2) + av}(L + d + b - va) \quad , \quad Ncm \text{ (in-lb)} \qquad (7\text{-}59)$$

349

$$H = \frac{d}{c+d}\left[\frac{(b/2) - av}{(b/2) + av}\right](L - c - b - va) + \frac{c}{c+d}$$

$$\times \left[\frac{(b/2) + av}{(b/2) - av}\right](L + d + b - va) + L - va \quad , \quad \text{cm (in.)} \quad (7\text{-}60)$$

where  a = dimension, tandem axle, cm (in.)

       b = dimension, tandem axle, cm (in.)

       c = dimension, tandem axle, cm (in.)

       d = dimension, tandem axle, cm (in.)

       u = dimension, tandem axle, cm (in.)

       v = dimension, tandem axle, cm (in.)

For the example truck equipped with a two-elliptic leaf spring suspension, the dynamic axle loads are illustrated in Figure 7-20. In addition to the data for the example truck given in Table 7-1, the tandem axle dimensions are b = 86.4 cm (34 in.), c = 17.8 cm (7 in.), d = 17.8 cm (7 in.), m = 53.3 cm (21 in.), v = 67.3 cm (26.5 in.). The dynamic axle load $F_{zRF}$ approaches zero for a deccleration of approximately 0.55 g. The friction utilization diagram, Figure 7-21, indicates that for $\mu = 0.6$ the maximum wheels-unlocked deceleration is only approximately 0.25 g. Consequently, braking skid marks may be produced at decelerations as low as 0.25 g for a road surface having a friction coefficient of 0.6. In the reconstruction of accidents, significant differences in speed prediction may result when the improper deceleration values are used.

## 7.12  Braking Dynamics While Turning

### 7.12.1  Basic Considerations

When braking in a turn, tires are required to produce longitudinal or braking forces and side forces. Vehicles that have their brake force distribution optimized for straight-line braking do not provide optimal braking while turning. Wheels-unlocked decelerations decrease as turning severity increases. The

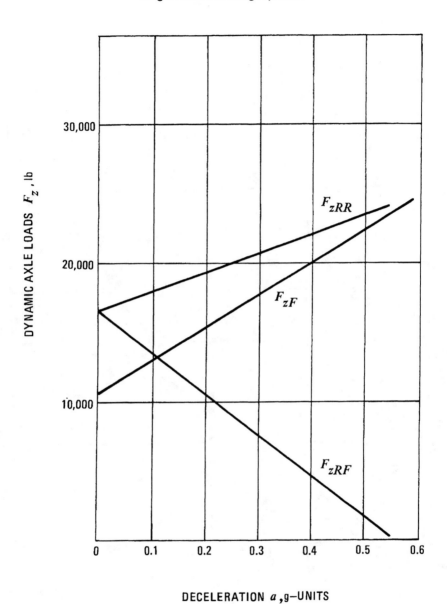

*Figure 7-20. Dynamic axle loads for a truck equipped with two-leaf suspension.*

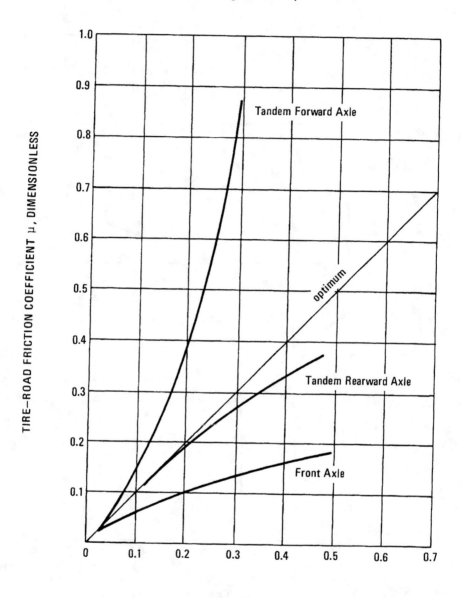

Figure 7-21. *Tire-road friction utilization for a truck equipped with two-leaf suspension.*

rear braking efficiency as a function of speed and path radius or road curvature is illustrated in Figure 7-22 for a tire-road friction coefficient of 0.6. Inspection of Fig. 7-22 indicates that for straight-line braking, i.e., the radius of the curve is infinite, the rear axle braking efficiency is approximately 0.86, resulting in a maximum wheels-unlocked deceleration of 0.52 g. For braking on a 300-foot radius curve from a speed of 72.4 km/h (45 mph) the braking efficiency reduces to approximately 0.45, yielding a deceleration of only 0.27 g when the inner rear wheel is near lockup. The braking dynamics used in developing Fig. 7-22 were based on a bicycle vehicle model, i.e., lateral load transfer was excluded.

## 7.12.2 Detailed Analysis for Braking in a Turn

The tire normal forces change due to the longitudinal deceleration of braking and the lateral acceleration or cornering force as illustrated in Figure 7-23. Additional longitudinal load transfer occurs due to the centripetal force component $F_{cx}$ acting in the direction of the longitudinal vehicle axis. The development of a combined braking and turning braking efficiency analysis requires the determination of the tire normal forces, tire side forces, and braking forces.

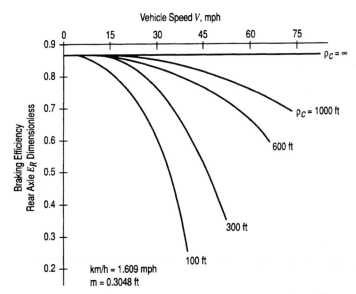

*Figure 7-22. Rear braking efficiency as function of speed and road curvature for a tire-road friction coefficient of 0.6.*

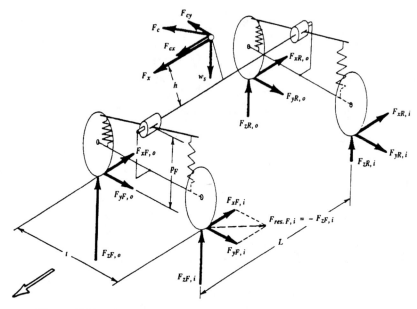

*Figure 7-23. Forces acting on a braking and turning vehicle.*

For a four-wheel vehicle with its center of gravity located midway between the left and right wheels, the tire normal forces on the outer and inner front, and outer and inner rear wheels are, respectively (Ref. 1):

$$F_{zF,o} = (1 / 2)F_{zF,static} + \Delta F_{zF} \quad , \quad N \ (lb)$$

$$F_{zF,i} = (1 / 2)F_{zF,static} - \Delta F_{zF} \quad , \quad N \ (lb)$$

$$F_{zR,o} = (1 / 2)F_{zR,static} + \Delta F_{zR} \quad , \quad N \ (lb)$$

$$F_{zR,i} = (1 / 2)F_{zR,static} - \Delta F_{zR} \quad , \quad N \ (lb) \tag{7-61}$$

where  $F_{zF,i}$ = normal force of inner front wheel, N (lb)

$F_{zF,o}$ = normal force of outer front wheel, N (lb)

$F_{zF,static}$ = static normal force of front axle, N (lb)

$F_{zR,i}$ = normal force of inner rear wheel, N (lb)

$F_{zR,o}$ = normal force of outer rear wheel, N (lb)

$F_{zR,static}$ = static normal force of rear axle, N (lb)

$\Delta F_{zF}$ = load transfer of one front wheel due to turning, N (lb)

$\Delta F_{zR}$ = load transfer of one rear wheel due to turning, N (lb)

Including load transfer due to the distance between roll centers and the road surface, the centripetal forces associated with the unsprung axles, the suspension moment, and the longitudinal and lateral components of the centripetal force, the tire normal forces are

Front, inner wheel:

$$F_{zF,i} = \left\{ 1 - \Psi + a_x \chi + \frac{\chi[L(1-\Psi)-\chi]\,a_y}{\rho_c} \right\}$$

$$\times\, W/2 - a_y S_F W_s \quad , \quad \text{N (lb)} \tag{7-62}$$

Front, outer wheel:

$$F_{zF,o} = \left\{ 1 - \Psi + a_x \chi + \frac{\chi[L(1-\Psi)-\chi]\,a_y}{\rho_c} \right\}$$

$$\times\, W/2 + a_y S_F W_s \quad , \quad \text{N (lb)} \tag{7-63}$$

Rear, inner wheel:

$$F_{zR,i} = \left\{ \Psi - a_x \chi - \frac{\chi[L(1-\Psi)-\chi]\,a_y}{\rho_c} \right\}$$

$$\times\, W/2 - a_y S_R W_s \quad , \quad \text{N (lb)} \tag{7-64}$$

Rear, outer wheel:

$$F_{zR,o} = \left\{ \Psi - a_x \chi - \frac{\chi[L(1-\Psi)-\chi]\,a_y}{\rho_c} \right\}$$

$$\times W/2 + a_y S_R W_s \quad , \quad N \text{ (lb)} \tag{7-65}$$

where  $a_x$ = longitudinal acceleration, g-units

$a_y$ = lateral acceleration, g-units

$L$ = wheelbase, m (ft)

$S_F$ = front normalized roll stiffness

$S_R$ = rear normalized roll stiffness

$W$ = total vehicle weight, N (lb)

$W_s$ = sprung weight, N (lb)

$\rho_c$ = radius of curvature of turn, m (ft)

The normalized roll stiffness $S_F$ on the front suspension is (Ref. 8)

$$S_F = (L_R / L)(p_F / t_F) + [K_F / (K_F + K_R - W_s h_r)](h_r / t_F)$$

$$+ (w_F / W_s)(h_F / t_F) \tag{7-66}$$

where  $h_F$ = center of gravity height of front unsprung mass, m (ft)

$h_r$ = perpendicular distance between center of gravity and roll axis, m (ft)

$K_F$ = front roll stiffness, Nm/rad (ft-lb/rad)

$K_R$ = rear roll stiffness, Nm/rad (ft-lb/rad)

$L_R$ = horizontal distance between center of gravity and rear axle, m (ft)

$p_F$ = front roll center-to-ground distance, m (ft)

$t_F$ = front track width, m (ft)

$w_F$ = front suspension unsprung weight, N (lb)

The rear normalized roll stiffness is obtained from Eq. (7-66) by replacing subscript F by R and using the appropriate vehicle data.

The calculation of the general braking efficiency consists of the following eight steps:

1. Obtain brakeline pressures from proportioning characteristics—front and rear.

2. Compute braking forces for individual wheels.

3. Compute total brake force $F_{x,total} = \Sigma F_x$

4. Compute vehicle deceleration $a_x$ by

$$a_x = F_{x,total} / W \quad , \quad \text{g-units} \tag{7-67}$$

5. For a given lateral acceleration $a_y$ compute the individual tire normal forces by Eqs. (7-62) through (7-65).

6. Compute tire side forces for each wheel, $F_{yF,i}$, $F_{yF,o}$, $F_{yR,i}$, and $F_{yR,o}$, by the equations that follow:

Front, inner wheel:

$$F_{yF,i} = (1 - \Psi)W(a_y / \rho_c)(F_{zF,i} / F_{zF})$$

$$\times \left\{ \rho_c^2 - [L(1 - \Psi) - \chi]^2 \right\}^{1/2} \quad , \quad \text{N (lb)} \tag{7-68}$$

Front, outer wheel:

$$F_{yF,o} = (1 - \Psi)W(a_y / \rho_c)(F_{zF,o} / F_{zF})$$

$$\times \left\{ \rho_c^2 - [L(1 - \Psi) - \chi]^2 \right\}^{1/2} \quad , \quad \text{N (lb)} \tag{7-69}$$

Rear, inner wheel:

$$F_{yR,i} = \Psi W(a_y / \rho_c)(F_{zR,i} / F_{zR})$$

$$\times \left\{ \rho_c^2 - [L(1-\Psi) - \chi]^2 \right\}^{1/2} \quad , \quad N \text{ (lb)} \qquad (7\text{-}70)$$

Rear, outer wheel:

$$F_{yR,o} = \Psi W(a_y / \rho_c)(F_{zR,o} / F_{zR})$$

$$\times \left\{ \rho_c^2 - [L(1-\Psi) - \chi]^2 \right\}^{1/2} \quad , \quad N \text{ (lb)} \qquad (7\text{-}71)$$

where $F_{yF,i}$ = side force of inner front wheel, N (lb)

$F_{yF,o}$ = side force of outer front wheel, N (lb)

$F_{yR,i}$ = side force of inner rear wheel, N (lb)

$F_{yR,o}$ = side force of outer rear wheel, N (lb)

7. Compute the tire-road friction coefficient $\mu_{F,i,req}$ to prevent wheel lockup on the inner front wheel by

$$\mu_{F,i,req} = \left[ \frac{(F_{xF,i})^2 + (m^2 F_{yF,i})^2}{(F_{zF,i})^2} \right]^{1/2} \qquad (7\text{-}72)$$

where $F_{xF,i}$ = braking force of inner front wheel, N (lb)

$m$ = tire factor

The tire factor accounts for the difference in brake and side force produced by the tires. For m = 1 the so-called "friction circle" exists. The friction circle assumes that both longitudinal and lateral tire forces are related by the equation of a circle. A roadway having a friction coefficient of $\mu$ = 1.0 is capable of producing a maximum vehicle deceleration of 1.0 g. The maximum lateral acceleration capability of the same vehicle is usually less than 1.0 g. This is caused by differences in the mechanisms involved in producing braking and

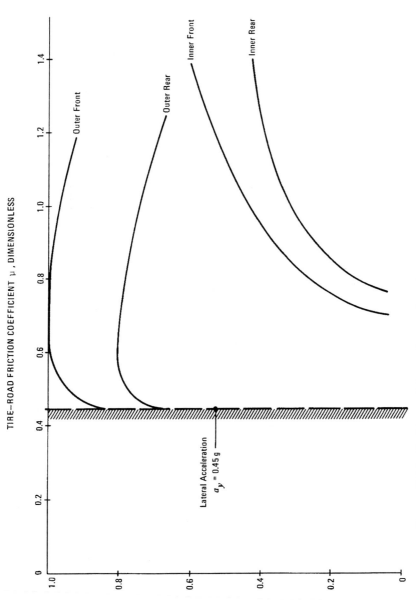

*Figure 7-24. Braking efficiency for combined braking and turning (Fiat 124).*

side forces. A tire-road surface having a braking friction coefficient of 0.9 and thus $a_{x(max)} = 0.9$ g tends to produce only about 0.7 g lateral acceleration. The friction circle concept does not describe accurately the relationship between limit braking and turning performance for a given tire-road surface condition. To describe this behavior more accurately, Eq. (7-72) is proposed. The coefficient m is the ratio of $a_{xo}$ to $a_{yo}$, where $a_{xo}$ designates the maximum braking deceleration in the absence of any lateral acceleration and thus is equal to the conventional tire-road friction coefficient $\mu$, and $a_{yo}$ designates the maximum lateral acceleration in the absence of any braking. The value of m for most tires and dry road surfaces ranges from 1.1 to 1.2.

8. Compute the braking efficiency $E_{Fi}$ by dividing deceleration $a_x$ (step 4) by the required friction coefficient $\mu_{F,i,req}$, e.g., for the inner front wheel

$$E_{Fi} = (a_x / \mu_{F,i,req}) \tag{7-73}$$

The computations were carried out using the geometrical and brake system data of a Fiat 124. Inspection of the results presented in Figure 7-24 indicates that the inner rear wheel locks up first, the inner front second, the outer rear third, and the outer front last. A stable braking efficiency, limited by the outer rear, of 0.78 can be achieved with the tire-road friction coefficient of 0.8. This value corresponds to a deceleration of about 6.1 m/s² (20 ft/s²) at a lateral acceleration of 0.45 g. For a tire-road friction coefficient of 0.45, no braking is possible because all friction available is used for the turning maneuver.

# CHAPTER 8

# Braking Dynamics of Combination Vehicles

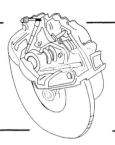

*In this chapter the braking dynamics of car-trailer and commercial tractor-trailer combinations is presented. The influence of brake line pressure reducing valves on braking efficiency is discussed.*

## 8.1 Car-Trailer—No Brakes on Trailer

The braking analysis for a combination vehicle is more complicated than for a two-axle vehicle because the summation of the axle loads of the tow vehicle is not equal to the weight of the tow vehicle. The tow vehicle axle loads are a function of the loading and braking of the trailer.

### 8.1.1 Simplified Analysis

In the case of a trailer without brakes, the entire braking force is produced by the car. The deceleration achievable is reduced by the additional weight of the trailer. An approximate expression can be derived from Newton's Second Law with the resultant external force equal to the braking force of the tow vehicle.

For the case where the front or all tires of the tow vehicle are skidding, the approximate deceleration $a_c$ is

$$a_c = W_c \mu / (W_c + W_T) \quad , \quad \text{g-units} \tag{8-1}$$

where $W_c$ = weight of car, lb

$W_T$ = weight of trailer, lb

$\mu$ = tire-road friction coefficient of tow vehicle

Eq. (8-1) is approximate because the effects of tongue force on the weight distribution of the tow vehicle, the brake force distribution of the tow vehicle, and all wheel lockup are neglected.

Braking tests showed the following results for a roadway friction coefficient of about 0.76 and a car weight of 2945 lb (Refs. 27, 28).

| Combination Weights | Deceleration measured, g-units | Deceleration, Eq. (8-1) |
|---|---|---|
| 15,399 N (3462 lb) | 0.63 | 0.65 |
| 17,401 N (3912 lb) | 0.55 | 0.57 |
| 19,100 N (4294 lb) | 0.50 | 0.50 |

## 8.1.2 Detailed Analysis

The forces acting on a braking car-trailer combination are illustrated in Figure 8-1. For the unbraked trailer the trailer braking force $F_{xT}$ is equal to zero.

The traction coefficients or friction utilization determined by the ratio of braking to dynamic normal force (Eq. [7-4]) may be derived from force and moment balance equations.

For the front axle the traction coefficient $\mu_{TF}$ is

$$\mu_{TF} = \frac{a(1 - \Phi)\left(1 + \dfrac{W_c}{W_c + W_T}\right)}{(1 - \Psi + a\chi) - \left(\dfrac{W_c}{W_c + W_T}\right)\left\{(1 - \Psi_T)\dfrac{\ell_B}{L} + a\left[\left(\dfrac{h_T}{L_T} - \dfrac{h_B}{L_T}\right)\dfrac{\ell_B}{L} - \dfrac{h_B}{L}\right]\right\}}$$

$$(8\text{-}2)$$

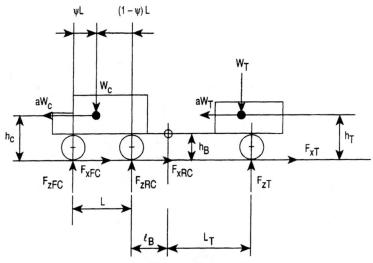

*Figure 8-1. Car-trailer combination.*

For the rear axle the traction coefficient $\mu_{TR}$ is

$$\mu_{TR} = \frac{a\Phi\left(1 + \dfrac{W_c}{W_c + W_T}\right)}{(\Psi - a\chi) + \left(\dfrac{W_c}{W_c + W_T}\right)\left\{(1 - \Psi_T)\left(1 + \dfrac{\ell_B}{L}\right) + a\left[\left(\dfrac{h_T}{L_T} - \dfrac{h_B}{L_T}\right)\left(1 - \dfrac{\ell_B}{L}\right) - \dfrac{h_B}{L}\right]\right\}}$$

(8-3)

where   a   = deceleration of combination, g-units

$h_B$ = vertical distance from ground to hitch ball, cm (in.)

$h_T$ = vertical distance from ground to center of gravity of trailer, cm (in.)

L   = wheelbase of car, cm (in.)

$\ell_B$ = horizontal distance from rear axle of car to hitch ball, cm (in.)

$L_T$ = horizontal distance from hitch ball to trailer axle, cm (in.)

363

$W_c$ = weight of car, N (lb)

$W_T$ = weight of trailer, N (lb)

$\Phi$ = rear brake force of car divided by total brake force of car

$\chi$ = center-of-gravity height divided by wheelbase

$\Psi$ = car rear axle load (without trailer) divided by car weight

$\Psi_T$ = trailer axle load divided by trailer weight

The horizontal hitch force $F_{BxT}$ between trailer and car is computed by

$$F_{BxT} = aW_T \quad , \quad N \text{ (lb)} \tag{8-4}$$

The vertical hitch force $F_{BzT}$ may be computed by

$$F_{BzT} = W_T[1 - \Psi_T + a(h_T / L_T - h_B / L_T)] \quad , \quad N \text{ (lb)} \tag{8-5}$$

Eqs. (8-2) and (8-3) may be solved for the case of traction coefficient equal to the tire-road friction coefficient, i.e., the maximum deceleration achievable for a given road surface.

## 8.2 Braking Dynamics for Trailer with Surge Brakes

### 8.2.1 Simplified Analysis

For surge brakes, the brakes of the tow vehicle generally decelerate between 15 and 20% of the weight of the trailer. The approximate deceleration of the car-trailer combination $a_{CT}$ is determined by

$$a_{CT} = \mu W_c / [W_c + (0.15 \text{ to } 0.20)W_T] \quad , \quad \text{g-units} \tag{8-6}$$

When no braking skid marks are found at the scene, the tire-road friction coefficient $\mu$ is replaced by the deceleration the tow vehicle would have achieved alone.

## 8.2.2 Detailed Braking Analysis of Trailer with Surge Brake

The trailer braking force produced between the trailer tires and the ground is a function of the horizontal hitch ball force actuating the trailer master cylinder. Due to the friction in the sliding or roller mechanism of the hitch assembly the usable actuation force is reduced.

The trailer brake gain $\rho_T$ is expressed by the ratio of trailer braking force to actuation force between tow vehicle and trailer:

$$\rho_T = F_{xT} / F_{BxT} \tag{8-7}$$

where $F_{BxT}$ = horizontal hitch ball force, N (lb)

$F_{xT}$ = trailer braking force, N (lb)

The friction in the tongue reduces the actuation force against the master cylinder and opposes the release force. Consequently, the trailer brake gain actually usable, $\rho_{T,act}$, is given by

$$\rho_{T,act} = \rho_T (F_{BxT} \pm F_{BF}) / F_{BxT} \tag{8-8}$$

where $F_{BF}$ = friction force in hitch assembly, N (lb)

The minus sign is used for application, the plus sign for release of the brakes. The ratio of friction force to hitch ball force—an indication of the quality of a surge brake hitch—should not exceed 0.2 for roller-guided hitches. The horizontal hitch ball force $F_{BxT}$ is given by

$$F_{BxT} = aW_T / (1 + \rho_{T,act}) \quad , \quad N \text{ (lb)} \tag{8-9}$$

The vertical hitch ball force $F_{BzT}$ is determined by

$$F_{BzT} = W_T \left[ 1 - \Psi_T + a\left( \frac{h_T}{L_T} - \frac{h_B / L_T}{1 + \rho_{T,act}} \right) \right] \quad , \quad N \text{ (lb)} \tag{8-10}$$

with the terms defined in Section 8.1.2.

The tire-road traction coefficients are determined from the ratios of axle braking force divided by axle normal force. The normal forces are obtained from an application of force and moment balance on the trailer and car.

For the front axle of the tow vehicle the traction coefficient $\mu_{TF}$ is given by

$$\mu_{TF} = \frac{a(1-\Phi)\left(1+\dfrac{W_c}{(W_c+W_T)(1+\rho_{T,act})}\right)}{(1-\Psi+a\chi)-\left(\dfrac{W_c}{W_c+W_T}\right)\left[(1-\Psi_T)\dfrac{\ell_B}{L}+a\left(\dfrac{h_T\ell_B}{L_T L}-\dfrac{\dfrac{h_B\ell_B}{L_T L}+\dfrac{h_B}{L}}{1+\rho_{T,act}}\right)\right]} \tag{8-11}$$

For the rear axle of the tow vehicle the traction coefficient $\mu_{TR}$ is given by

$$\mu_{TR} = \frac{a\Phi\left(1+\dfrac{W_c}{(W_c+W_T)(1+\rho_{T,act})}\right)}{(\Psi-a\chi)+\left(\dfrac{W_c}{W_c+W_T}\right)\left\{(1-\Psi_T)\left(1+\dfrac{\ell_B}{L}\right)+a\left[\dfrac{h_T}{L_T}\left(1+\dfrac{\ell_B}{L}\right)-\dfrac{\dfrac{h_B}{L_T}\left(1+\dfrac{\ell_B}{L}\right)+\dfrac{h_B}{L}}{1+\rho_{T,act}}\right]\right\}} \tag{8-12}$$

For the trailer axle the traction coefficient $\mu_{TT}$ is

$$\mu_{TT} = \frac{a\dfrac{\rho_{T,act}}{1+\rho_{T,act}}}{\Psi_T-a\left(\dfrac{h_T}{L_T}-\dfrac{h_B/L_T}{1+\rho_{T,act}}\right)} \tag{8-13}$$

For tow vehicles equipped with four-wheel anti-lock brakes (ABS), the tire-road friction is nearly always fully utilized, assuming the pedal forces are sufficiently high for the ABS to operate. Under these conditions, approximately 90 to 95% of the friction available is actually used for braking. The deceleration a achievable with an ABS-equipped tow vehicle may be determined by

$$a = \frac{\mu_{max}\eta_{ABS}\left[1+\left(\dfrac{W_c}{W_c+W_T}\right)(1-\Psi_T)\right]}{1+\dfrac{W_c}{(W_c+W_T)(1+\rho_{T,act})}-\mu_{max}\eta_{ABS}\left(\dfrac{W_c}{W_c+W_T}\right)\left(\dfrac{h_T}{L_T}-\dfrac{h_B/L_T}{1+\rho_{T,act}}\right)} \quad , \text{g-units}$$

(8-14)

where $\eta_{ABS}$ = efficiency of ABS (90 to 95%)

$\mu_{max}$ = maximum tire-road friction coefficient

For tow vehicles using rear ABS only, the increased rear axle normal load due to the trailer tongue force may be sufficiently large to prevent wheel lockup, resulting in front axle lockup first.

# 8.3  Braking of Three-Axle Tractor-Trailer Combination (2-S1)

For the correct design and brake balance analysis of a vehicle combination, it is essential that the optimum braking forces for each axle are known for the empty and laden cases. For a two-axle tractor towing a single-axle trailer (2-S1 combination), the equations are still algebraically manageable. For tandem axle tractors or trailer a computerized procedure is used.

## 8.3.1  Optimum Braking Forces

A detailed discussion of optimum braking forces is presented in Section 7.2. In the optimum condition, i.e., $a = \mu$, all road friction available is used, and the braking forces are directly related to the dynamic axle loads. Use the terminology shown in Figure 8-2. The equations for force and moment equilibrium yield as the normalized optimum braking forces on each axle (Ref. 10):

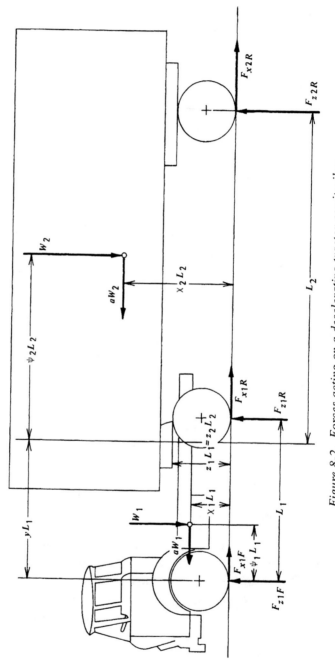

Figure 8-2. Forces acting on a decelerating tractor-semitrailer.

Tractor rear axle:

$$F_{x1R,dyn} / W_1 = a(\Psi_1 - a\chi_1)$$
$$+a(W_2 / W_1)(y - az_1)[(1 - \Psi_2 + a\chi_2) / (1 + az_2)] \quad (8\text{-}15)$$

Tractor front axle:

$$F_{x1F,dyn} / W_1 = a(1 - \Psi_1 + a\chi_1)$$
$$+a(W_2 / W_1)(1 - y + az_1)[(1 - \Psi_2 + a\chi_2) / (1 + az_2)]$$
$$(8\text{-}16)$$

Trailer axle:

$$F_{x2R,dyn} / W_2 = a\left[\frac{\Psi_2 + a(z_2 - \chi_2)}{1 + az_2}\right] \quad (8\text{-}17)$$

where   $a$ = deceleration, g-units

    $W_1$ = tractor weight, N (lb)

    $W_2$ = semitrailer weight N (lb)

    $y$ = horizontal distance between front wheels and fifth wheel divided by tractor wheelbase $L_1$

    $z_1$ = fifth wheel height divided by tractor wheelbase $L_1$

    $z_2$ = fifth wheel height divided by semitrailer base $L_2$

    $\chi_1$ = tractor center-of-gravity height divided by tractor wheelbase $L_2$

    $\chi_2$ = semitrailer center-of-gravity height divided by semitrailer base $L_2$

    $\Psi_1$ = empty tractor rear axle load (without semitrailer) divided by tractor weight

    $\Psi_2$ = static semitrailer axle load divided by semitrailer weight

If $W_2 = 0$, i.e., no trailer is attached to the tractor, Eqs. (8-15) and (8-16) may be rearranged to yield the equations applicable to a straight truck.

The optimum braking forces normalized by dividing by the total combination weight ($W_1 + W_2$) are shown in Figure 8-3. The nature of the curves shows that it will be difficult to design a fixed brake force distribution braking system that will produce actual braking forces on each axle that come close to the optimum braking forces for the laden and empty vehicle on both slippery and dry road surfaces (high and low deceleration).

Since it will be impossible to lock up all three axles simultaneously for all loading and road friction conditions, the design engineer must decide which axles to lock first.

Vehicle stability, much like for cars, requires that the tractor front axle locks first, followed by the trailer axle, with the tractor rear axle locking up last. European design practices are directed toward locking the tractor front axle first. Demonstration tests conducted by the National Highway Traffic Safety Administration with five different commercial vehicles and several drivers clearly showed the safety benefits of locking front brakes first in terms of shorter stopping distances, improved stability, even when braking on a slippery roadway while turning.

In terms of apply sequence, all brakes should ideally be applied at the same moment to ensure a stretched combination, which is of particular importance when braking empty on slippery road surfaces. Eqs. (8-15), (8-16), and (8-17) may be rewritten to yield the optimum tractor braking forces as a function of the trailer loading condition and trailer brake force:

$$F_{x1F,opt} = aW_1(1 - \Psi_1 + a\chi_1)$$

$$+(aW_2 + F_{x2R})(1 - y + az_1) \quad , \quad N \text{ (lb)} \tag{8-18}$$

$$F_{x1R,opt} = aW_1(\Psi_1 - a\chi_1)$$

$$+(aW_2 - F_{x2R})(y - az_1) \quad , \quad N \text{ (lb)} \tag{8-19}$$

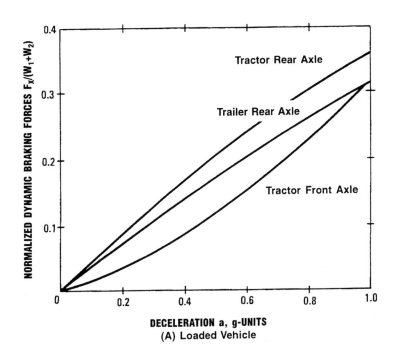

**(A) Loaded Vehicle**

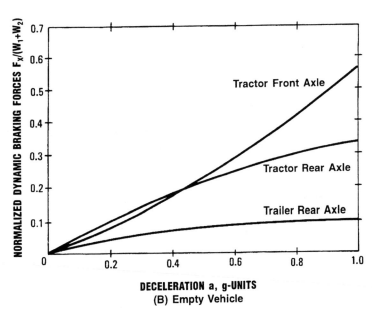

**(B) Empty Vehicle**

*Figure 8-3. Normalized dynamic braking forces of a tractor-semitrailer.*

where   a = deceleration, g-units

     $F_{x2R}$ = actual brake force of semitrailer, N (lb)

     $W_1$ = tractor weight, N (lb)

     $W_2$ = trailer weight, N (lb)

     y = horizontal distance between front wheels and fifth wheel divided by tractor wheelbase $L_1$

     $z_1$ = fifth wheel height divided by semitrailer base $L_2$

The last term in each of the Eqs. (8-18) and (8-19) represents the influence of the trailer on the tractor. A graphical representation of Eqs. (8-18) and (8-19) is shown in Figure 8-4 for a typical vehicle and several loading conditions. Inspection of the three curved lines for the empty, half laden, and laden operating conditions reveals that a fixed brake force distribution on the tractor will not result in optimum braking for most loading conditions and braking forces or deceleration levels.

The actual brake forces generated at each axle are determined by the brake line pressures supplied to the brakes, the brake geometry, lining friction coefficient, and tire radius. For air brakes, the brake forces may be computed from Eq. (6-1), however, multiplied by tire radius, and for hydraulic brakes from Eq. (5-2).

A fixed brake force distribution on the tractor is presented in Fig. 8-4 as a straight line. The location or closeness of this straight line relative to the optimum braking forces determines the utilization of the given road friction by the brake system and, hence, the overall braking performance of the system combination.

## 8.3.2. Optimum Brake Force Distribution

The optimum brake force distribution is that fixed ratio of brake force distribution among the axles which will result in maximum wheels-unlocked decelerations on dry and slippery roadways for both the empty and laden conditions. In this section an approach is outlined, similar to that developed for the two-axle vehicle (Eq. [7-20]), which allows the determination of the approximate optimum tractor brake force distribution for a given trailer brake system.

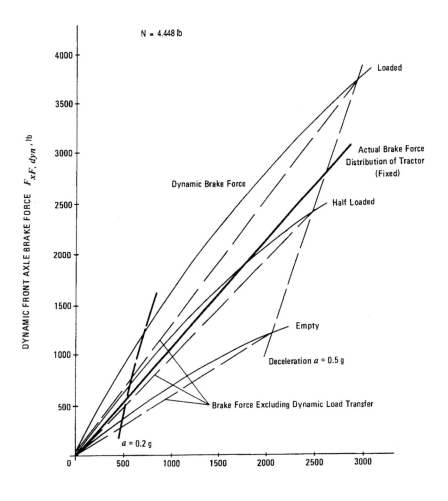

Figure 8-4. *Dynamic braking forces of the tractor of a tractor-semitrailer combination.*

The tire-road traction coefficient required to prevent wheel lockup on a particular axle during braking can be computed by dividing the axle brake force by the instantaneous dynamic axle load (Eq. [7-4]). To simplify the analysis, the influence of dynamic weight transfer during braking is neglected; this is equivalent to replacing the curves of the optimum forces in Fig. 8-4 by straight lines, indicated by the broken lines. This procedure will result in a small error; however, the influence of the braking forces through the fifth wheel connection is retained. It should be noted that this simplification is introduced only to arrive at simple expressions considered important in designing a brake system for a combination vehicle. The design parameters hereby obtained subsequently should be checked and evaluated by means of the braking performance calculation program discussed later in this chapter.

The tire-road friction coefficient $\mu_{1F}$ required to prevent lockup on the tractor front axle is

$$\mu_{1F} = F_{x1F} / F_{z1F} = \Phi_{1F} a W / F_{z1F} \tag{8-20}$$

where   $F_{x1F}$ = actual brake force of tractor front axle, N (lb)

$F_{x1R}$ = actual brake force of tractor rear axle, N (lb)

$F_{x2R}$ = actual brake force of semitrailer axle, N (lb)

$F_{z1F}$ = front axle normal force of tractor, N (lb)

$W$ = total combination weight, N (lb)

The approximate normalized dynamic axle loads of the combination, $F_{z1F,ap}$, $F_{z1R,ap}$, and $F_{z2R,ap}$, can be computed and are

Tractor front axle:

$$F_{z1F,ap} / W_1 = 1 - \Psi_1 + (W_2 / W_1)(1 - \Psi_2)(1 - y) + [(F_{x1F} + F_{x1R}) / W]$$
$$\times [\chi_1 + (W_2 / W_1)z_1 + (W_2 / W_1)(\chi_2 - z_2)(1 - y)]$$
$$+ (F_{x2R} / W)[z_2(1 - y) - (z_1 - \chi_1) + (W_2 / W_1)\chi_2(1 - y)]$$

$$\tag{8-21}$$

Tractor rear axle:

$$F_{z1R,ap} / W_1 = \Psi_1 + (W_2 / W_1)(1 - \Psi_2)y - [(F_{x1F} + F_{x1R}) / W]$$

$$\times [\chi_1 + (W_2 / W_1)z_1 + (W_2 / W_1)(\chi_2 - z_2)y]$$

$$+(F_{x2R} / W[z_2y + (z_1 - \chi_1) + (W_2 / W_1)\chi_2y] \qquad (8\text{-}22)$$

Trailer axle:

$$F_{z2R,ap} / W_2 = \Psi_2 - [(F_{x1F} + F_{x1R}) / W](\chi_2 - z_2)$$

$$-(F_{x2R} / W)[\chi_2 + (W_1 / W_2)z_2] \qquad (8\text{-}23)$$

where $F_{x1F}$ = actual brake force of tractor front axle N (lb)

$\qquad F_{x1R}$ = actual brake force of tractor rear axle, N (lb)

$\qquad F_{x2R}$ = actual brake force of semitrailer axle, N (lb)

Eqs. (8-21), (8-22), and (8-23) may be rewritten to be functions of the trailer axle brake force $F_{x2R}$ only by using the relationship $F_{x1F} + F_{x1R} + F_{x2R} = aW$. The following expressions are obtained for the approximate axle loads on the tractor-semitrailer combination.

Tractor front axle:

$$F_{z1F,ap} = W_1(1 - \Psi_1) + W_2(1 - \Psi_2)(1 - y) + F_{x2R}(z_2 - z_2y - z_1)$$

$$+a[W_1\chi_1 + W_2z_1 - W_2(\chi_2 - z_2)(\chi_1 - y)] \quad , \quad N \text{ (lb)} \quad (8\text{-}24)$$

Tractor rear axle:

$$F_{z1R,ap} = W_1\Psi_1 + W_2(1 - \Psi_2)y + F_{x2R}(z_2y + z_1)$$

$$-a[W_1\chi_1 + W_2z_1 - W_2(\chi_2 - z_2)y] \quad , \quad N \text{ (lb)} \qquad (8\text{-}25)$$

Trailer axle:

$$F_{z2R,ap} = W_2\Psi_2 - F_{x2R}z_2 - aW_2(\chi_2 - z_2) \quad , \quad N \text{ (lb)} \qquad (8\text{-}26)$$

With the expressions for the approximate dynamic axle loads substituted into the friction relationship, Eq. (8-20), the decelerations $a_{1F}$, $a_{1R}$, and $a_{2R}$ achievable on an axle prior to wheel lockup for a particular tire-road friction coefficient, vehicle geometry, and trailer brake force may now be computed by the following expressions.

Tractor front axle:

$$a_{1F} = \frac{\mu_{1F}[\lambda(1 - \Psi_1) + (1 - \lambda)(1 - \Psi_2)(1 - y) + \rho(z_2 - z_2 y - z_1)]}{\Phi_{1F} - \mu_{1F}[\lambda\chi_1 + (1 - \lambda)z_1 + (1 - \lambda)(\chi_2 - z_2)(1 - y)]} \quad (8\text{-}27)$$

Tractor rear axle:

$$a_{1R} = \frac{\mu_{1R}[\lambda\Psi_1 + (1 - \lambda)(1 - \Psi_2)y + \rho(z_2 y + z_1)]}{\Phi_{1R} + \mu_{1R}[\lambda\chi_1 + (1 - \lambda)z_1 - (1 - \lambda)(\chi_2 + z_2)y]} \quad (8\text{-}28)$$

Trailer axle:

$$a_{2R} = \frac{\mu_{2R}[(1 - \lambda)\Psi_2 - \rho z_2]}{\Phi_{2R} + \mu_{2R}(1 - \lambda)(\chi_2 - z_2)} \quad (8\text{-}29)$$

where $F_{x, total}$ = total braking force of combination, N (lb)

$\lambda = W_1/W$

$\mu_{1F}$ = tire-road friction coefficient on tractor front wheels

$\mu_{1R}$ = tire-road friction coefficient on tractor rear wheels

$\mu_{2R}$ = tire-road friction coefficient of trailer wheels

$\rho = F_{x2R}/W$

$\Phi_{1F} = F_{x1F}/F_{x,total}$

$\Phi_{1R} = F_{x1R}/F_{x,total}$

$\Phi_{2R} = F_{x2R}/F_{x,total}$

For $\rho = 0$ and $\lambda = 1$, i.e., no trailer is connected to the tractor, Eqs. (8-27) and (8-28) reduce to those of a two-axle truck.

A graphical representation of Eqs. (8-27), (8-28), and (8-29) is shown in Figures 8-5 and 8-6 for the laden and empty driving condition, respectively. In this example the brake forces were distributed to match the optimum braking forces existing during the loaded driving condition. Consequently, for the empty vehicle the trailer axle is always overbraked, i.e., tends to lock up first, compared to the two other axles. For example (Fig. 8-6), a deceleration of 0.4 g requires a road coefficient of friction $\mu = 0.6$ on the trailer tires to prevent wheel lockup. The required friction coefficient for the loaded condition shown in Fig. 8-5 indicates an acceptable brake balance design of the

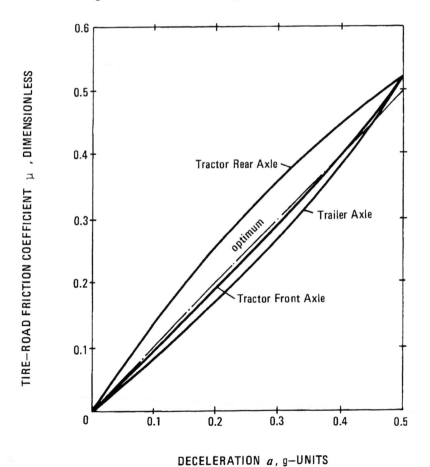

DECELERATION $a$, g–UNITS

*Figure 8-5. Tire-road friction utilization for a loaded tractor-semitrailer.*

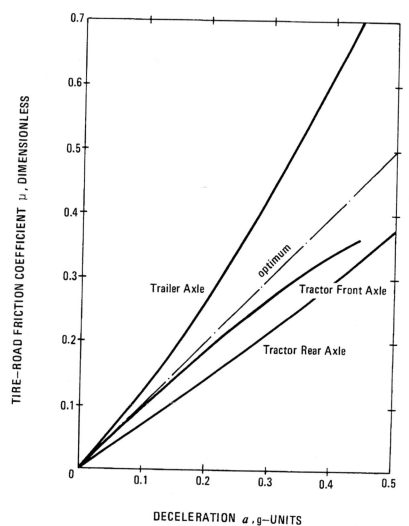

*Figure 8-6. Tire-road friction utilization for an empty tractor-semitrailer.*

vehicle with the rear axle of the tractor slightly overbraked. The optimum solution would be such that the tractor rear axle would be near the optimum line; however, it would lockup last.

Eqs. (8-27) and (8-28) may be used to develop limiting relationships on the relative rear axle brake force of the tractor similar to Eq. (7-20). With the braking efficiency $E = a/\mu$ this results in the expression

$$\Phi_{1R} = A \: / \: E_{min} - \mu_{1R}B \qquad\qquad (8\text{-}30)$$

where $A = \lambda\Psi_1 + (1 - \lambda)(1 - \Psi_2)y + \rho(z_2 y + z_1)$

$B = \lambda\chi_1 + (1 - \lambda)z_1 - (1 - \lambda)(\chi_2 - z_2)y$

$E_{min}$ = minimum braking efficiency to be achieved by vehicle

The relative brake force $\Phi_{2R}$ on the trailer axle can be computed from $\Phi_{2R} = \rho/a$, where a is equal to the deceleration of the combination in g-units. The relative front axle brake force is determined from $1 = \Phi_{1F} + \Phi_{1R} + \Phi_{2R}$. For $\rho = 0$ and $A = 1$, i.e., no trailer is connected to the tractor, Eq. (8-30) reduces to that for a two-axle vehicle (Eq. [7-20]).

The relative tractor rear axle brake force $\Phi_{1R}$ for most cases should not exceed 0.50. A relatively small value of $\Phi_{1R}$ and, hence, moderate utilization of the friction in longitudinal direction on the tractor rear axle, means that a considerable lateral tire force is still available for directional stability. This is of importance since the danger of jackknifing is directly related to the lateral forces available at the tractor rear axle.

The decelerations achievable with a fixed brake force distribution on slippery ($\mu = 0.2$) and dry ($\mu = 0.8$) road surfaces must be checked with Eqs. (8-27), (8-28), and (8-29) for the empty and laden operating condition. If the results indicate too low a performance for one loading condition, the distribution must be altered until an acceptable value between empty and laden brake force distribution has been found. If a fixed brake force distribution does not yield acceptable results, a variable brake force distribution by use of brake line pressure modulating valves must be employed.

For a typical tractor-semitrailer combination geometry with the vehicle data given in Table 8-1, the optimum brake force distribution was determined for $\rho = 0.23$, i.e., a brake force of 23% of the total combination weight is acting on the trailer axle at maximum reservoir pressure, to be equal to $\Phi_{1F} = 0.17$, $\Phi_{1R} = 0.47$, and $\Phi_{2R} = 0.36$ for a braking efficiency of 75%. When checked with Eqs. (8-27), (8-28), and (8-29), this brake force distribution yielded braking efficiencies of 75% for the empty and laden vehicle on both slippery ($\mu = 0.2$) and dry ($\mu = 0.8$) road surfaces. This theoretical result is supported

**TABLE 8-1**

### Vehicle Data for Tractor-Semitrailer Calculations

|  | Empty | Loaded |
|---|---|---|
| $\Psi_1$ | 0.52 | 0.52 |
| $\Psi_2$ | 0.738 | 0.51 |
| $\chi_1$ | 0.22 | 0.22 |
| $\chi_2$ | 0.12 | 0.26 |
| $\lambda$ | 0.6 | 0.21 |
| $\rho$ | 0.23 | 0.23 |
| $(z_1 = 0.193; z_2 = 0.12; y = 0.93)$ | | |

by actual road tests of combination vehicles (Ref. 18). The brake force distribution originally used was equal to 0.12, 0.44, 0.44, front to rear, resulting in trailer axle lockup for the empty case at approximately 0.53 g, indicating a braking efficiency of about 60%. No tractor axle lockup was observed for the laden case. Changing the basic brake force distribution by means of Eq. (8-30) to 0.17, 0.44, and 0.39, front to rear, increased the braking efficiency to approximately 70%.

## 8.3.3 Braking Analysis for 2-S1 Combination Using Brake Line Pressure Reducing Valves

The normalized optimum braking forces of a typical tractor-semitrailer vehicle for the empty and loaded operating conditions are shown in Fig. 8-3. These curves show that the optimum braking forces are heavily influenced by the loading condition of the trailer. If the brake system is designed to near optimum for the laden vehicle, it will perform poorly for the empty case unless a proportioning brake system is provided that will vary the brake force distribution according to the loading conditions of the vehicle combination.

An additional difficulty arises from the fact that a particular tractor may be used with different trailers, each having a variety of loading configurations and brake force levels.

The optimum braking forces of a typical combination vehicle are shown in Fig. 8-3 and are redrawn for convenience in Figure 8-7. The actual brake forces for the empty and laden cases which would best approximate the optimum braking forces are illustrated by the broken lines in Fig. 8-7. The optimum curves indicate that for a deceleration of 0.8 g the actual brake forces, front to rear, are approximately equal to 0.22 W, 0.30 W, 0.24 W for the laden vehicle, and 0.40 W, 0.30 W, and 0.10 W for the empty vehicle. With a laden and empty combination weight of W = 195,712 N (44,000 lb) and W = 88,960 N (20,000 lb), respectively, the brake forces for optimum braking at a = 0.8 g must be proportioned between 35,584 N to 43,146 N (8000 lb to 9700 lb) on the tractor front axle 26,688 N to 58,714 N (6000 lb to 13,200 lb) on the tractor rear axle, and 8896 N to 46,704 N (2000 lb to 10,500 lb) on the trailer axle to best adjust to the empty and laden conditions. The numbers indicate that, whereas the optimum brake force on the front axle varies little with change in vehicle loading, the optimum brake forces on the rear axle of the tractor and on the trailer axle are heavily influenced by the loading condition.

In European applications it has been found convenient to implement a variable brake force distribution in articulated vehicles in the manner described next. The front axle brake force of the tractor is designed to be proportional to the application valve exit pressure. The brake force at the rear axle of the tractor is determined by the load or suspension height-sensitive pressure reducing valve. Depending on the design of the pressure reducing valve of the tractor, the brake torque on the tractor rear axle may vary, e.g., from 60 to 140% of the tractor front axle brake torque. The brake torque on the trailer axle is determined by either a pressure reducing or limiting valve. Depending on the design of the trailer proportioning valve, the brake torque on the trailer may vary from 20 to 100% (or more) of the front brake torque. It may be sufficient to control the brake force of the trailer axle by a manual or automatically positioned limiting valve which has settings for the empty, half-loaded, and loaded conditions resulting in different limiting brake torque/line pressure relationships on the trailer axle.

The results of the friction utilization calculations carried out for several loading and proportioning valve settings are presented for a 2-S1 tractor-semitrailer combination, i.e., the tractor has two axles, the trailer one axle. The vehicle combination having the basic friction utilizations shown in Figs. 8-5 and 8-6 was analyzed relative to the effects of different brake line pressure reducing valve characteristics.

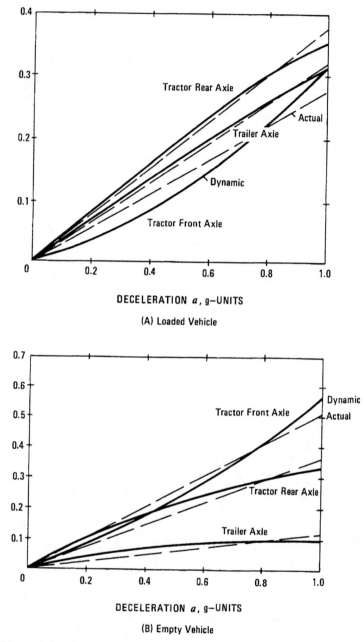

*Figure 8-7. Normalized dynamic braking force distribution.*

1. Case 1. The vehicle combination was loaded to GVW capacity, and the proportioning valve setting on the tractor axle and the limiting valve setting on the trailer axle are as shown in Figure 8-8. The tire-road friction utilization diagram shown in Figure 8-9 illustrates a near optimum braking of the vehicle. For all deceleration levels up to approximately 0.5 g, the tractor front axle will lock first, while the tractor rear and trailer axle are close to optimum braking conditions and slightly underbraked.

2. Case 2. The conditions are identical to Case 1, except the trailer setting is incorrect as indicated in Fig. 8-8 by the broken lines. Although the trailer is fully loaded, the trailer brake force is set at lower value corresponding to the half-laden case. As an inspection of Fig. 8-9 reveals, the same tire-road friction utilization exists as in Case 1 up to a deceleration of about 0.36 g. For decelerations greater than approximately 0.42 g, the danger of locking the tractor rear axle first exists, resulting in a possible instability of the combination, most likely seen in the form of jackknifing of the tractor.

3. Case 3. The trailer is loaded to half of GVW with the proportioning valve setting as indicated in Figure 8-10. The trailer valve setting is for the half-laden case also. The danger of first locking the tractor rear axle exists for decelerations greater than 0.49 g, below which the front axle locks up first as illustrated in Figure 8-11. The trailer brakes tend to lock up for decelerations greater than about 0.32 g.

4. Case 4. The loading conditions are identical to Case 3, except the trailer valve is mistakenly set to the empty condition as illustrated in Fig. 8-10. As noted from the friction utilization diagram (Fig. 8-11), now the tractor rear axle tends to overbrake at decelerations of 0.36 g and greater, requiring relatively high coefficients of friction between tire and road.

5. Case 5. For the empty vehicle combination the valve settings are indicated in Figure 8-12. The tire-road friction utilization is illustrated in Figure 8-13. The trailer axle tends to overbrake compared to the two other axles for decelerations below 0.53 g. For decelerations above 0.53 g the tractor rear axle begins to lock up.

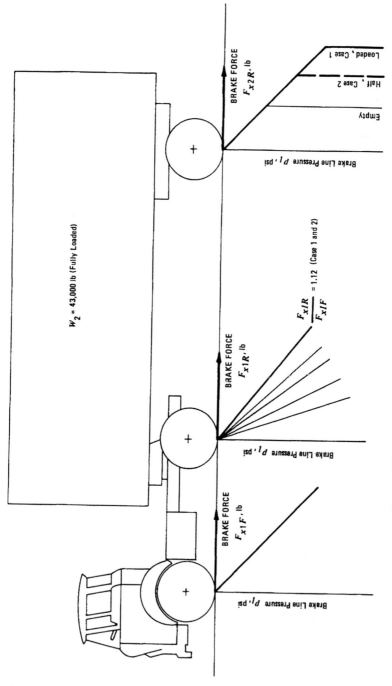

*Figure 8-8. Schematic brake force distribution, Case 1 and 2.*

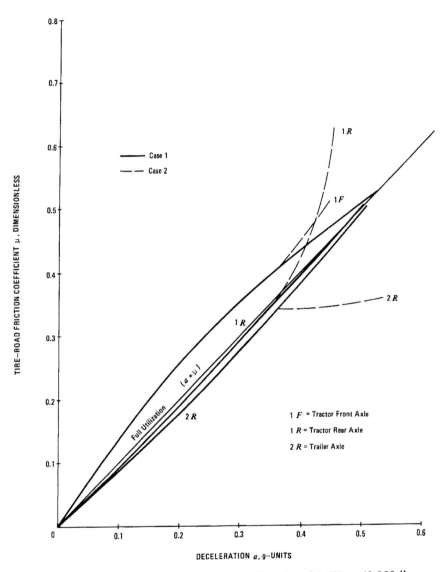

*Figure 8-9. Tire-road friction utilization, Case 1 and 2, $W_2 = 43,000$ lb.*

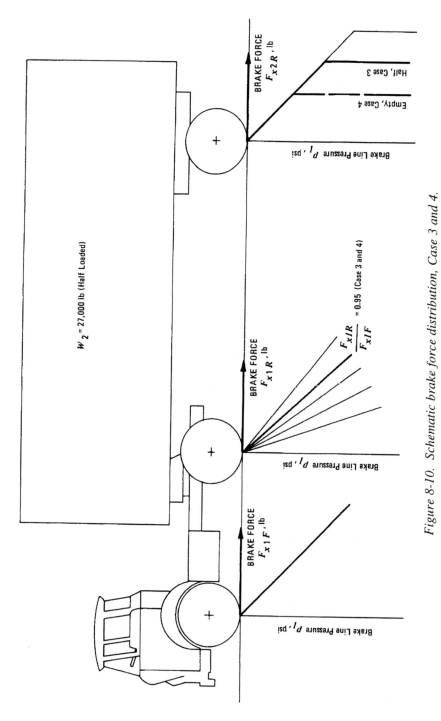

*Figure 8-10. Schematic brake force distribution, Case 3 and 4.*

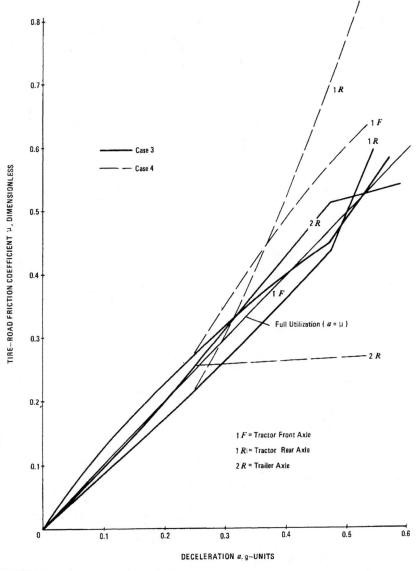

*Figure 8-11. Tire-road friction utilization, Case 3 and 4, $W_2 = 27,000$ lb.*

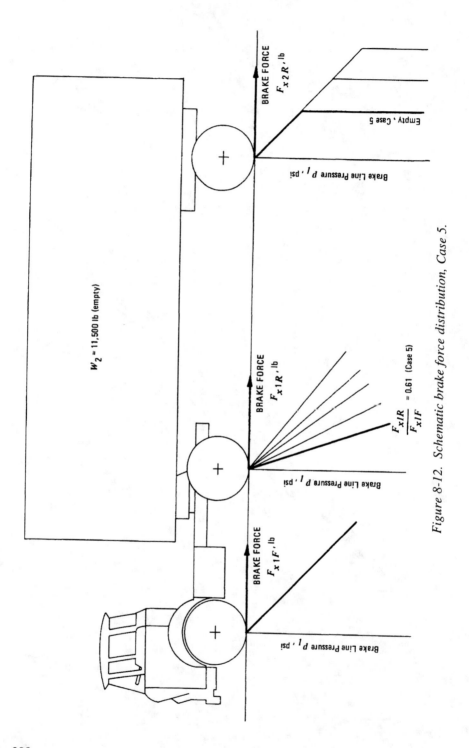

*Figure 8-12. Schematic brake force distribution, Case 5.*

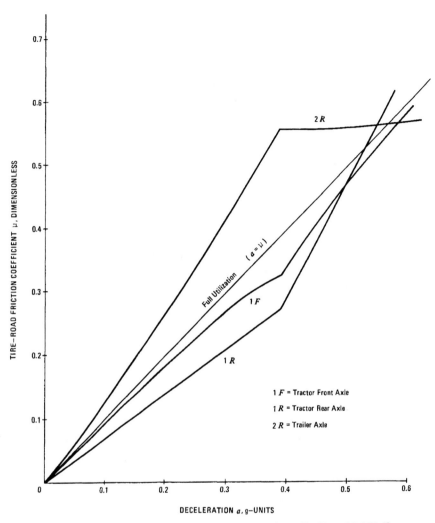

*Figure 8-13. Tire-road friction utilization, Case 5, $W_2$ = 11,000 lb.*

6. <u>Case 6</u>. The empty vehicle combination is braked with the valve
   settings as indicated in Figure 8-14. The tractor rear axle brake
   force is set for the laden condition. Since domestic tractors in
   current use generally do not have brake line pressure reducing
   valves for the rear axle, Case 6 represents operation of a tractor
   pulling an empty trailer that may be found on public highways.

The trailer brake force is set for the empty condition. Examination of the friction utilization diagram shown in Figure 8-15 reveals that a deceleration of approximately 0.4 g tends to be critical with respect to jackknifing because the tractor rear axle is always locking up first. This case illustrates the importance of automatic load-dependent and driver-independent brake torque variation of the tractor rear axle.

7. <u>Case 7</u>. The empty combination vehicle is equipped with the tractor proportioning valve automatically set to the empty position as indicated in Figure 8-16. The trailer brake force is no longer limited as in the previous cases. The trailer brake force is not controlled by a valve. The results of the friction utilization calculations shown in Figure 8-17 demonstrate an almost optimum braking indicated by the fact that all three curves are close to the optimum or full utilization line.

8. <u>Case 8</u>. The laden vehicle is braked with the tractor rear axle valve set at the laden condition. The trailer axle is not controlled and produces the brake force illustrated in Figure 8-18, i.e., the same brake force as in the previous case. As illustrated in Figure 8-19, the front axle tends to overbrake for decelerations below 0.43 g. Although locking of the front axle will not allow any steering wheel corrections by the driver, the combination remains stable traveling straight ahead. Above decelerations of 0.43 g there is a danger of overbraking of the tractor rear axle.

Many tractors are equipped with a front axle brake line pressure automatic limiting valve (see Section 6.3.10). Improvements in braking performance can be achieved by means of modulating the brake force of the tractor front axle if the static laden-to-empty axle ratio of the tractor front axle is greater than approximately 1.4. If the laden-to-empty ratio is less than 1.4, then only the tractor rear and trailer axle must be modulated.

Road tests have shown that a proper brake force distribution among axles of a tractor-semitrailer combination has been achieved when no wheels lock below decelerations of 0.5 g with the laden combination braking on dry road surfaces. This brake force distribution generally yields acceptable braking

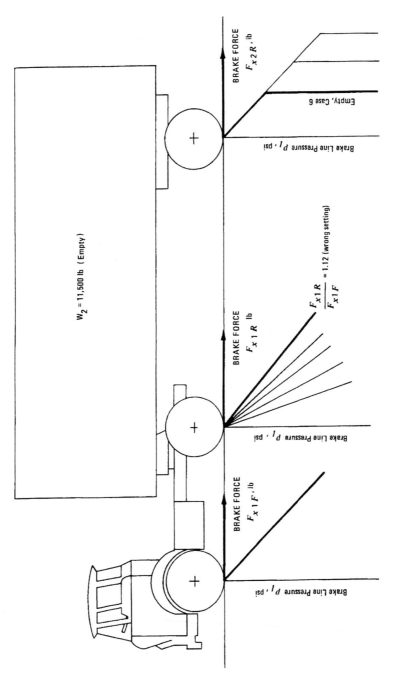

*Figure 8-14. Schematic brake force distribution, Case 6.*

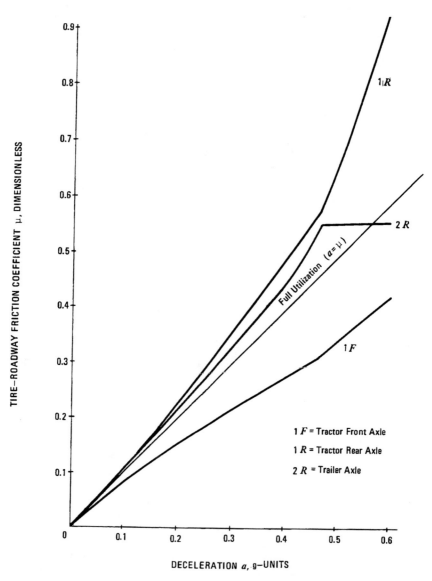

*Figure 8-15. Tire-road friction utilization, Case 6, $W_2 = 11,500$ lb.*

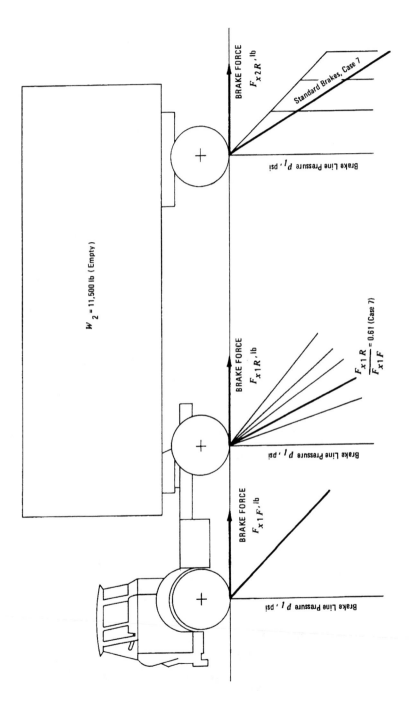

*Figure 8-16. Schematic brake force distribution with proportioning valves on tractor rear and standard brakes on trailer axle.*

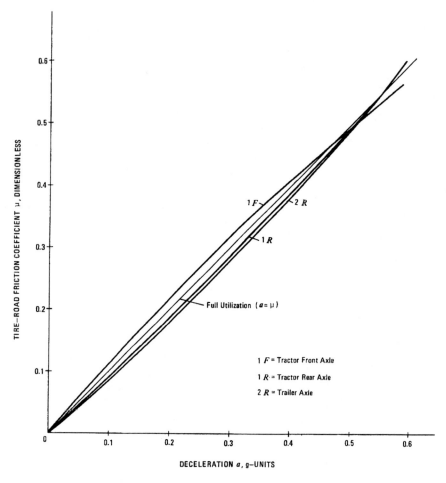

*Figure 8-17. Tire-road friction utilization, Case 7, $W_2 = 11,500$ lb.*

performance with the empty combination. However, if some axle(s) lock below 0.5 g, the advantages of load-dependent brake force distribution are not fully utilized.

Since manually operated or automatic load-dependent proportioning valves still allow wheel lockup to occur, ABS brake systems should be used to achieve directional control and stability of combinations during braking involving wheel lockup. FMVSS 121 requires ABS brakes on all commercial vehicles.

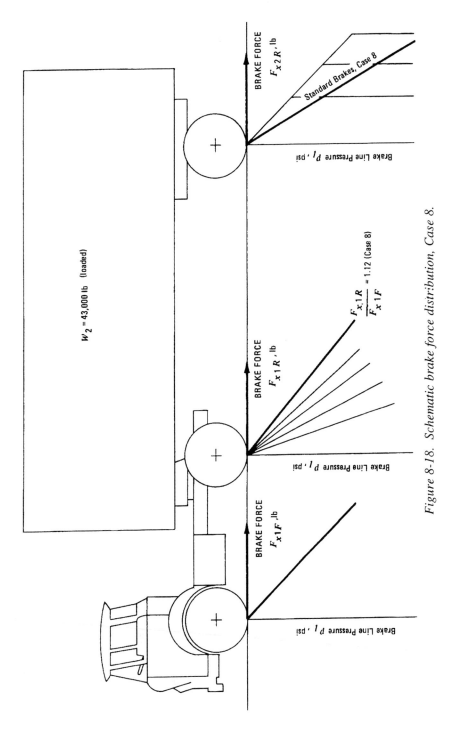

*Figure 8-18. Schematic brake force distribution, Case 8.*

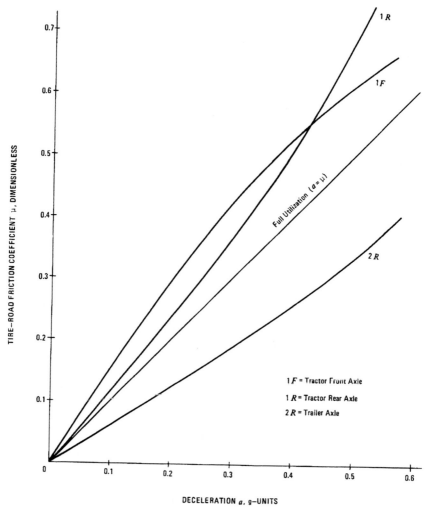

*Figure 8-19. Tire-road friction utilization, Case 8.*

## 8.3.4 Optimum Brake Line Pressures

The use of optimum brake line pressures for the design of brake balance, front to rear, for a two-axle vehicle is discussed in Section 7.9.1. For a particular vehicle combination, the optimum brake line pressures may be computed from the optimum braking forces given by Eqs. (8-15) through (8-17), and Eq. (5-2) for hydraulic brakes, or Eq. (6-1) for air brakes.

The optimum brake line pressures $p_{\ell 1F,opt}$, $p_{\ell 1R,opt}$, and $p_{\ell 2R,opt}$ for a 2-S1 vehicle combination equipped with air foundation brakes are computed by

Tractor front axle:

$$p_{\ell 1F,opt} = \frac{a\,[(1 - \Psi_1 + a\chi_1)W_1 - (1 - y + az_1)W_2 N]}{[2A_c BF\eta_m (r / R)\rho k_A k_T]_{1F}}$$

$$+p_{o1F} \quad , \quad N/cm^2 \text{ (psi)} \tag{8-33}$$

Tractor rear axle:

$$p_{\ell 1R,opt} = \frac{a\,[(\Psi_1 - a\chi_1)W_1 + (y - az_1)W_2 N]}{[2A_c BF\eta_m (r / R)\rho k_A k_T]_{1R}}$$

$$+p_{o1R} \quad , \quad N/cm^2 \text{ (psi)} \tag{8-34}$$

Trailer axle:

$$p_{\ell 2R,opt} = \frac{a(1 - N)W_2}{[2A_c BF\eta_m (r / R)\rho k_A k_T]_{2R}}$$

$$+p_{o2R} \quad , \quad N / cm^2 \text{ (psi)} \tag{8-35}$$

where $A_c$ = brake chamber area, $cm^2$ ($in.^2$)

$BF$ = brake factor

$k_A$ = adjustment factor

$k_T$ = temperature factor

$L_1$ = tractor wheelbase, cm (in.)

$L_2$ = distance between fifth wheel and trailer axle or trailer base, cm (in.)

$N = (1 - \Psi_2 + a\chi_2) / (1 + az_2)$

$p_{o1F}$ = pushout pressure, tractor front brakes $N/cm^2$ (psi)

$p_{o1R}$ = pushout pressure, tractor rear brakes $N/cm^2$ (psi)

397

$P_{o2R}$ = pushout pressure, trailer brakes $N/cm^2$ (psi)

$r$ = drum or effective rotor radius, mm (in.)

$R$ = tire radius, mm (in.)

$W_1$ = tractor weight, N (lb)

$W_2$ = trailer weight, N (lb)

$y$ = horizontal distance between tractor front wheels and fifth wheel divided by tractor wheelbase $L_1$

$z_1$ = fifth wheel height divided by tractor wheelbase $L_1$

$z_2$ = fifth wheel height divided by trailer base $L_2$

$\eta_m$ = mechanical efficiency between brake chamber and brake shoe

$\rho$ = application gain (lever ratio) between brake chamber and brake shoe

$\chi_1$ = tractor center-of-gravity height divided by tractor wheelbase $L_1$

$\chi_2$ = trailer center-of-gravity height divided by trailer base $L_2$

$\Psi_1$ = static tractor rear axle load divided by tractor weight (without trailer)

$\Psi_2$ = static trailer axle load divided by trailer weight

Eqs. (8-33) through (8-35) may be presented in terms of either optimum brake line pressure versus deceleration, or in terms of individual optimum brake line pressures versus application valve exit pressure. The latter probably is more suited for brake design purposes. The graphical relationship representing the actual brake line pressures delivered to the brake chambers of different axles is an effective means for obtaining the desired proportioning valve control range on each axle.

If a proportioning device is to be installed into a tractor brake system, and if the loading and brake force levels of the trailer are specified, i.e., a given trailer is to be connected to a tractor, the optimum brake line pressures on the tractor may be obtained by means of Eqs. (6-1), (8-18), and (8-19), yielding the following expressions:

Tractor front axle:

$$p_{\ell 1F,opt} = [aW_1(1 - \Psi_1 + a\chi_1) + (aW_2 - F_{x2R})(1 - y + az_1)]$$

$$\times \left\{ 2[A_c\eta_m(BF)(r/R)\rho]_{1F} \right\}^{-1} + p_{o1F} \quad , \quad N/cm^2 \text{ (psi)} \quad (8\text{-}36)$$

Tractor rear axle:

$$p_{\ell 1R,opt} = [aW_1(\Psi_1 - a\chi_1) + (aW_2 - F_{x2R})(y - az_1)]$$

$$\times \left\{ 2[A_c\eta_m(BF)(r/R)\rho]_{1R} \right\}^{-1} + p_{o1R} \quad , \quad N/cm^2 \text{ (psi)} \quad (8\text{-}37)$$

where $F_{x2R}$ = trailer rear axle brake force, lb

The graphical representation of Eqs. (8-36) and (8-37) for typical tractor and trailer data is shown in Figure 8-20 for several loading conditions. The brake line pressure curves presented in Fig. 8-20 may be used to design the variable brake force distribution of the tractor. When a proportional ratio between front and rear has been selected, the braking performance obtained in terms of friction utilization or braking efficiency must be calculated.

The load-sensitive variation of the ratio of the brake line pressures—tractor rear axle to tractor front axle—i.e., $p_{\ell 1R} / p_{\ell 1F}$ or the variation of the relative tractor rear axle brake force $\Phi_{1R}$ with respect to the total brake force of the combination, may be obtained directly as a function of the tractor rear axle suspension deflection which is a direct measure of the tractor axle loads. The values for the variation of $p_{\ell 1R} / p_{\ell 1F}$ and the relative tractor rear axle brake force $\Phi_{1R}$ are shown in Figure 8-21 as a function of the rear suspension deflection illustrated for a typical standard and heavy-duty suspension.

## 8.4 Braking Dynamics of Combination Vehicle Equipped with Tandem Axles

The basic methodology used in analyzing the braking dynamics of trucks equipped with tandem axles is discussed in Section 7.11.

First, only the trailer will be equipped with a tandem axle suspension. In later sections, both the tractor and trailer will be equipped with tandem axle suspensions.

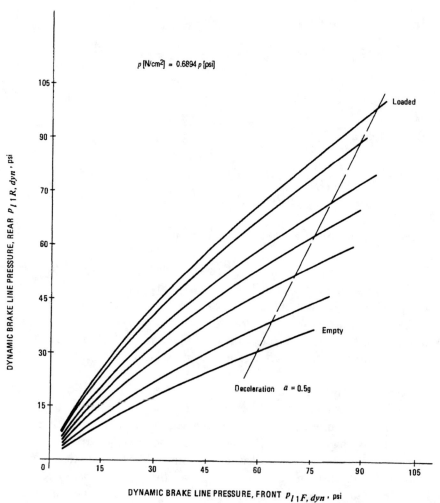

*Figure 8-20. Dynamic brake line pressures for the tractor of a tractor-semitrailer combination.*

## 8.4.1 2-S2 Combination—Trailer with Two-Elliptic Leaf Springs Suspension

Since not all different suspension designs can be analyzed in this book, the basic approach is outlined in detail for the tandem axle design considered here. Similar methods can be used to derive expressions for air suspensions, and suspensions utilizing special kinematic components to optimize axle loads during braking.

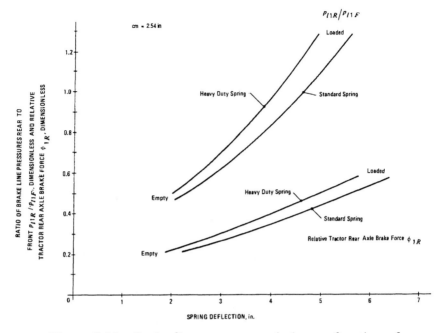

*Figure 8-21. Brake line pressure variation as function of spring deflection on trailer axle.*

We need to derive expressions for the individual dynamic normal axle loads on the tractor and the trailer. The forces acting on a decelerating tractor-semitrailer are shown in Figure 8-22. Use the terminology of Fig. 8-22. The force and moment balance equations applied to the free body of the tractor, sprung trailer, forward trailer axle, and rearward trailer axle yield a set of equations which may be solved for the individual dynamic normal axle loads. When aerodynamic drag, rotational energies, and rolling resistance are neglected, the equilibrium equations are (Ref. 10)

Tractor:

$$Xz_1L_1 + Y(1 - y)L_1 - F_{z1F}L_1 + W_1(1 - \Psi_1)L_1$$

$$+W_1a\chi_1L_1 = 0 \quad , \quad \text{Ncm (lb-in.)} \tag{8-38}$$

$$F_{z1F} + F_{z1R} - W_1 - Y = 0 \quad , \quad \text{N (lb)} \tag{8-39}$$

401

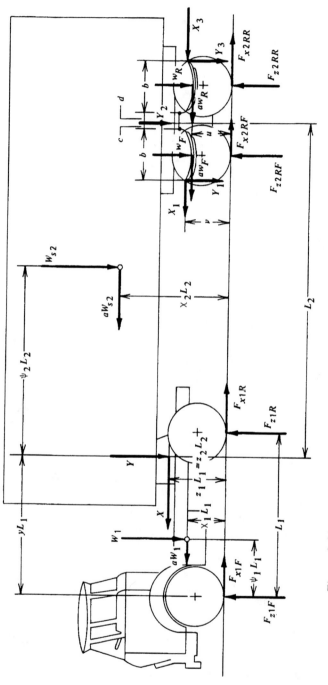

*Figure 8-22. Forces acting on a tractor-semitrailer equipped with two-leaf suspension.*

Sprung trailer:

$$W_{s2}a - X - X_1 - X_3 = 0 \quad , \quad \text{N (lb)} \tag{8-40}$$

$$Y + Y_1 + Y_2 + Y_3 - W_{s2} = 0 \quad , \quad \text{N (lb)} \tag{8-41}$$

$$W_{s2}a(\chi_2 - z_2)L_2 + (X_1 + X_3)(z_1L_1 - v) + Y_1(L_2 - c - b)$$

$$+Y_2L_2 + Y_3(L_2 + d + b) - W_{s2}\Psi_2L_2 = 0 \quad , \quad \text{Ncm (lb-in.)} \tag{8-42}$$

Trailer forward axle:

$$X_1 - F_{x2RF} + w_Fa = 0 \quad , \quad \text{N (lb)} \tag{8-43}$$

$$F_{z2RF} - Y_1 - Y_2d / (c + d) - w_F = 0 \quad , \quad \text{N (lb)} \tag{8-44}$$

$$X_1v + [Y_1 - Y_2d / (c + d)](b / 2) + aw_Fu = 0 \quad , \quad \text{Ncm (lb-in.)} \tag{8-45}$$

Trailer rearward axle:

$$X_3 - F_{x2RR} + aw_R = 0 \quad , \quad \text{N (lb)} \tag{8-46}$$

$$F_{z2RR} - Y_3 - Y_2c / (c + d) - w_R = 0 \quad , \quad \text{N (lb)} \tag{8-47}$$

$$X_3v + [Y_2c / (c + d) - Y_3](b / 2) + aw_Ru = 0 \quad , \quad \text{N (lb)} \tag{8-48}$$

where  $a$ = deceleration, g-units

$b$ = dimension, tandem axle, cm (in.)

$c$ = dimension, tandem axle, cm (in.)

$d$ = dimension, tandem axle, cm (in.)

$F_{x2RF}$ = actual brake force of trailer tandem forward axle, N (lb)

$F_{x2RR}$ = actual brake force of trailer tandem rearward axle, N (lb)

403

$F_{z1F}$ = normal force of tractor front axle, N (lb)

$F_{z1R}$ = normal force of tractor rear axle, N (lb)

$F_{z2RF}$ = normal force of trailer tandem forward axle, N (lb)

$F_{z2RR}$ = normal force of trailer tandem rearward axle, N (lb)

u = dimension, tandem axle, cm (in.)

v = dimension, tandem axle, cm (in.)

$W_{s2}$ = semitrailer weight minus weight of tandem axle, N (lb)

X = horizontal fifth wheel force, N (lb)

$X_1$ = horizontal suspension force, forward axle, N (lb)

$X_3$ = horizontal suspension force, rearward axle, N (lb)

Y = vertical fifth wheel force, N (lb)

$Y_1$ = vertical suspension force, forward axle N (lb)

$Y_2$ = vertical suspension frame force, N (lb)

$Y_3$ = vertical suspension force, rearward axle, N (lb)

This system of eleven equations contains 13 unknowns. Two additional equations can be obtained when optimum braking conditions are considered in which the deceleration in g-units equals the tire-road friction coefficient, i.e., a = μ. Then the braking forces $F_{x2RF}$ and $F_{x2RR}$ are replaced by $aF_{z2RF}$ and $aF_{z2RR}$, and a system of eleven equations with eleven unknowns is obtained which can be solved by successive substitution. The results are as follows:

The individual dynamic normal axle loads are:

Tractor front axle:

$$F_{z1F} = W_1(1 - \Psi_1 + a\chi_1) + Y(1 - y + az_1) \quad , \quad \text{N (lb)} \qquad (8\text{-}49)$$

where

$$X = aY \quad , \quad N \text{ (lb)}$$

Tractor rear axle:

$$F_{z1R} = W_1(\Psi_1 - a\chi_1) + Y(y - az_1) \quad , \quad N \text{ (lb)} \tag{8-50}$$

Trailer forward axle:

$$F_{z2RF} = \frac{Y_2 bd}{(c + d)[(b / 2) + av]} + w_F - \frac{w_F ua}{(b / 2) + av} \quad , \quad N \text{ (lb)} \tag{8-51}$$

Trailer rearward axle:

$$F_{z2RR} = \frac{Y_2 bc}{(c + d)[(b / 2) - av]} + w_R + \frac{w_R ua}{(b / 2) - av} \quad , \quad N \text{ (lb)} \tag{8-52}$$

The vertical force Y on the kingpin of the fifth wheel is given by

$$Y = W_{s2} - Y_2 \left\{ \frac{d[(b / 2) - av]}{(c + d)[(b / 2) + av]} + \frac{c[(b / 2) + av]}{(c + d)[(b / 2) - av]} + 1 \right\}$$

$$+ \left[ \frac{w_F u}{(b / 2) + av} - \frac{w_R u}{(b / 2) - av} \right] a \quad , \quad N \text{ (lb)} \tag{8-53}$$

where   $a$ = deceleration, g-units

$\quad$ $w_F$ = weight of tandem forward axle, N (lb)

$\quad$ $w_R$ = weight of tandem rearward axle, N (lb)

The vertical force $Y_2$ on the tandem suspension is determined by

$$Y_2 = \frac{W_{s2} L_2 [\Psi_2 - a(\chi_2 - z_2)] + G_1 a}{H_1} \quad , \quad N \text{ (lb)} \tag{8-54}$$

405

where

$$G_1 = w_F u / [(b / 2) + av][(z_1L_1 - v)a + L_2 - c - b] - w_R u / [(b / 2) - av]$$
$$\times[(z_1L_1 - v)a + L_2 + d + b] \quad , \quad \text{Ncm (lb-in)} \tag{8-55}$$

$$H_1 = d / (c + d)[(z_1L_1 - v)a + L_2 - c - b]\{[(b / 2) - av] / [(b / 2) + av]\}$$
$$+c / (c + d)[(z_1L_1 - v)a + L_2 + d + b]\{[(b / 2) + av] / [(b / 2) - av]\}$$
$$+(z_1L_1 - v)a + L_2 \quad , \quad \text{cm (in)} \tag{8-56}$$

For a typical 2-S2 combination the dynamic axle loads are illustrated in Figure 8-23. Inspection of the lines reveals that the axle load of the forward axle of the trailer suspension will decrease for increasing deceleration.

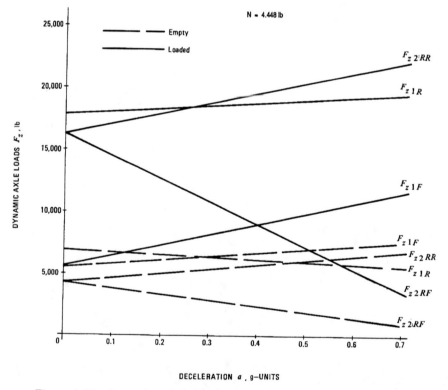

*Figure 8-23. Dynamic axle loads for a tractor-semitrailer combination.*

The braking performance diagram is shown in Figure 8-24. It illustrates the relationship between pedal force, brake line pressure for the empty and laden conditions, and friction utilization. Test data indicated show good correlation between theory and measurements.

## 8.4.2. 2-S2 Combination—Trailer with Walking Beam Suspension

Use the terminology shown in Figure 8-25. The application of the equilibrium conditions to the combination vehicle results in the following dynamic normal axle loads (Ref. 10):

Tractor front axle:

$$F_{z1F} = W_1(1 - \chi_1 + a\chi_1) + Y(1 - y + az_1) \quad , \quad N \text{ (lb)} \qquad (8\text{-}57)$$

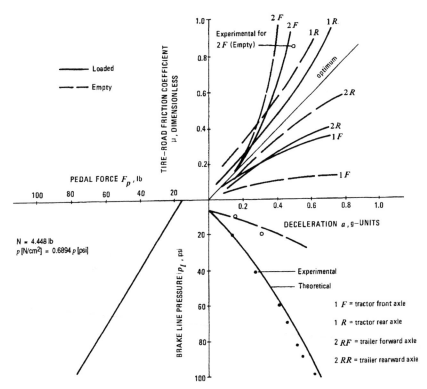

*Figure 8-24. Braking performance diagram for a tractor-semitrailer combination.*

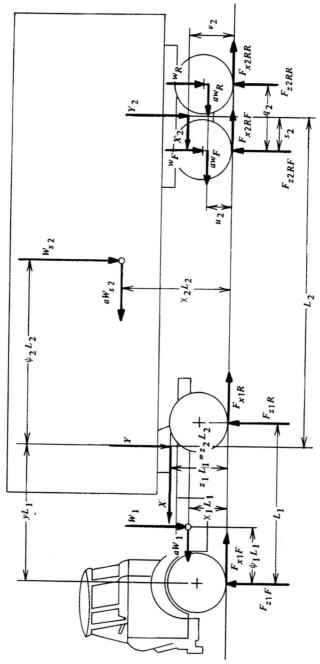

*Figure 8-25. Forces acting on a tractor-semitrailer equipped with walking beam suspension.*

Tractor rear axle:

$$F_{z1R} = W_1 + Y - F_{z1F} \quad , \quad N \text{ (lb)} \qquad (8\text{-}58)$$

Trailer forward axle:

$$F_{z2RR} = [Y_2 s_2 + w_R q_2 - au_2(w_F + w_R)$$
$$-X_2 v_2] / q_2 \quad , \quad N \text{ (lb)} \qquad (8\text{-}59)$$

Trailer rearward axle:

$$F_{z2RF} = Y_2 + w_F + w_R - F_{z2RR} \quad , \quad N \text{ (lb)} \qquad (8\text{-}60)$$

where

$$X_2 = F_{x2RF} + F_{x2RR} - a(w_F + w_R) \quad , \quad N \text{ (lb)} \qquad (8\text{-}61)$$

$$X = W_{s2}a - X_2 \quad , \quad N \text{ (lb)} \qquad (8\text{-}62)$$

$$Y_2 = [W_{s2}\Psi_2 L_2 - W_{s2}a(\chi_2 - z_2)L_2$$
$$-X_2 z_1 L_1 v_2] / L_2 \quad , \quad N \text{ (lb)} \qquad (8\text{-}63)$$

$$Y = W_{s2} - Y_2 \quad , \quad N \text{ (lb)} \qquad (8\text{-}64)$$

and $\quad a$ = deceleration, g-units

$F_{x2RF}$ = actual brake force of semitrailer tandem forward axle, N (lb)

$F_{x2RR}$ = actual brake force of semitrailer tandem rearward axle, N (lb)

$q_2$ = dimension, tandem axle, cm (in.)

$s_2$ = dimension, tandem axle, cm (in.)

$u_2$ = dimension, tandem axle, cm (in.)

$v_2$ = dimension, tandem axle, cm (in.)

$\chi_2$ = horizontal suspension force, N (lb)

### 8.4.3 3-S2 Combination—Tractor with Walking Beam and Trailer with Two-Leaf Spring Suspension

Use the terminology shown in Figure 8-26. The individual dynamic normal axle loads are:

Tractor front axle:

$$F_{z1F} = W_{s1} + Y - Y_4 \quad , \qquad N \text{ (lb)} \tag{8-65}$$

Tractor tandem forward axle:

$$F_{z1RR} = [Y_4 s_1 - X_4 v_1 + (w_{1F} + w_{1R}) a u_1$$
$$+ w_{1R} q_1] / q_1 \quad , \qquad N \text{ (lb)} \tag{8-66}$$

Tractor tandem rearward axle:

$$F_{z1RF} = Y_4 + w_{1F} + w_{1R} - F_{z1RR} \quad , \qquad N \text{ (lb)} \tag{8-67}$$

where

$$X_4 = F_{x1RF} + F_{x1RR} - a(w_{1F} + w_{1R}) \quad , \qquad N \text{ (lb)} \tag{8-68}$$

$$Y_4 = [W_{s1} \Psi_1 L_1 + yY L_1 - (aW_{s2} - F_{x2RF} - F_{x2RR}) z_1 L_1$$
$$- aW_{s1} \chi_1 L_1 + X_4 v_1] / L_1 \quad , \qquad N \text{ (lb)} \tag{8-69}$$

and     $a$ = deceleration, g-units

$q_1$ = dimension, tandem axle, cm (in.)

$s_1$ = dimension, tandem axle, cm (in.)

$w_{1F}$ = unsprung weight of tractor tandem forward axle, N (lb)

$w_{1R}$ = unsprung weight of tractor tandem rearward axle, N (lb)

$W_{s1}$ = tractor weight minus weight of tandem axle, N (lb)

The normal trailer axle loads are identical to those derived in Section 8.4.1 for a two-axle tractor coupled to a tandem axle trailer and may be determined from Eqs. (8-51) and (8-52). The vertical force Y on the fifth wheel kingpin is obtained from Eq. (8-53).

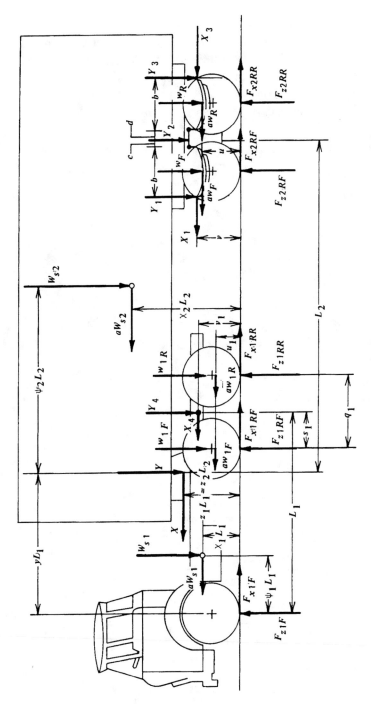

*Figure 8-26. Forces acting on a tandem axle tractor-tandem axle semitrailer combination.*

The application of the braking performance calculations to a vehicle combination consisting of a tractor equipped with a walking beam suspension and a trailer equipped with a two-elliptic leaf spring tandem axle resulted in the dynamic axle loads, braking performance, and braking efficiencies as presented in Figures 8-27 through 8-29. Test data obtained for the vehicle are indicated in the braking performance and braking efficiency diagrams.

The theoretical results demonstrate that considerable dynamic load transfer occurs on tandem axles without equalization among individual axles. Analysis has shown that for tandem axle designs as indicated in Fig. 7-17(B), the load

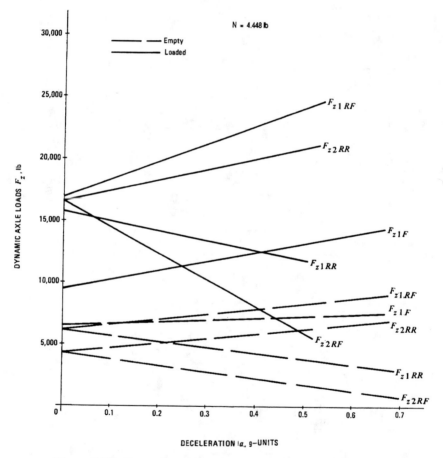

*Figure 8-27. Dynamic axle loads for a tandem axle tractor-tandem axle semitrailer combination.*

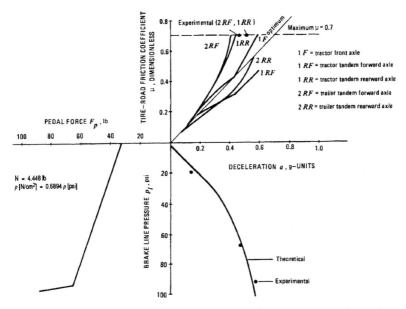

Figure 8-28. Braking performance diagram for a loaded tandem axle tractor-tandem axle semitrailer combination.

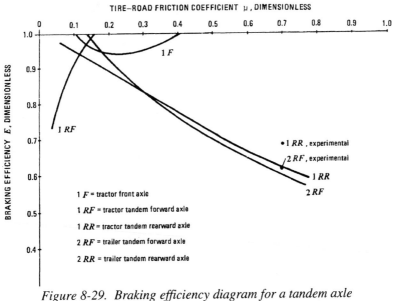

Figure 8-29. Braking efficiency diagram for a tandem axle tractor-tandem axle semitrailer combination.

transfer occurring between the forward and rearward axle can be reduced by decreasing the design measurement "v" as illustrated in Figure 8-30. For example, a change of v from 32 in. to 16.8 in. will decrease the axle load on the forward axle to about 47% of its static value for a deceleration of 0.5 g as compared to approximately 5% for v = 32 in. It also means that the wheels-unlocked deceleration on the forward axle can be increased to about 0.45 g, instead of 0.34 g.

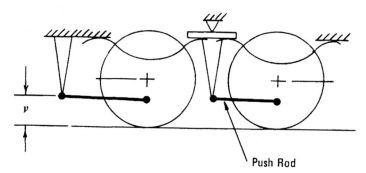

*Figure 8-30. Two-leaf/two-rod suspension.*

### 8.4.4 2S1-2 Combination—Two-Axle Tractor, Single-Axle Semitrailer and Double-Axle Trailer

Use the terminology shown in Figure 8-31. The individual dynamic normal axle loads are:

Tractor front axle:

$$F_{z1F} = W_1(1 - \Psi_1) + W_1 a(\chi_1 - z_1 - z_4 + z_2 + z_4 y - z_2 y) + W_2(1 - \Psi_2)(1 - y)$$
$$-W_2 a(z_4 - \chi_2)(1 - y) + (F_{x1F} + F_{x1R})$$
$$\times (z_1 - z_2 + z_4 + z_2 y - z_4 y) + F_{x2R} z_4(1 - y) \quad , \quad N \text{ (lb)} \qquad (8\text{-}70)$$

Tractor rear axle:

$$F_{z1R} = W_1 \Psi_1 - W_1 a(\chi_1 - z_1 + z_4 y - z_2 y) + W_2(1 - \Psi_2)y - W_2 a(z_4 - \chi_2)y$$
$$-(F_{x1F} + F_{x1R})(z_1 + z_2 y - z_4 y) + F_{x2R} z_4 y \quad , \quad N \text{ (lb)} \qquad (8\text{-}71)$$

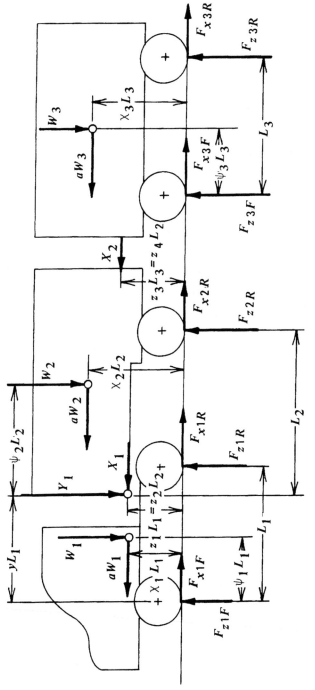

*Figure 8-31. Forces acting on a tractor-semitrailer-double trailer combination.*

Semitrailer axle:

$$F_{z2R} = W_1 a(z_4 - z_2) + W_2 \Psi_2 + W_2 a(z_4 - \chi_2)$$

$$+ (F_{x1F} + F_{x1R})(z_2 - z_4) - F_{x2R} z_4 \quad , \quad N \text{ (lb)} \qquad (8\text{-}72)$$

Double trailer front axle:

$$F_{z3F} = (W_1 + W_2)az_3 + W_3(1 - \Psi_3) + W_3 a \chi_3$$

$$-(F_{x1F} + F_{x1R} + F_{x2R})z_3 \quad , \quad N \text{ (lb)} \qquad (8\text{-}73)$$

Double trailer rear axle:

$$F_{z3R} = -(W_1 + W_2)az_3 + W_3 \Psi_3 - W_3 a \chi_3$$

$$+ (F_{x1F} + F_{x1R} + F_{x2R})z_3 \quad , \quad N \text{ (lb)} \qquad (8\text{-}74)$$

where     a = deceleration, g-units

$F_{x3F}$ = actual brake force of double trailer front axle, N (lb)

$F_{x3R}$ = actual brake force of double trailer rear axle, N (lb)

$F_{z3F}$ = normal force of double trailer front axle, N (lb)

$F_{z3R}$ = normal force of double trailer rear axle, N (lb)

$L_3$ = wheelbase of double trailer, cm (in.)

$W_3$ = double trailer weight, N (lb)

$z_3$ = double trailer hitch height divided by double trailer wheelbase $L_3$

The calculated braking performance data are presented in Figure 8-32 for the laden vehicle combination and are compared to actual road test data.

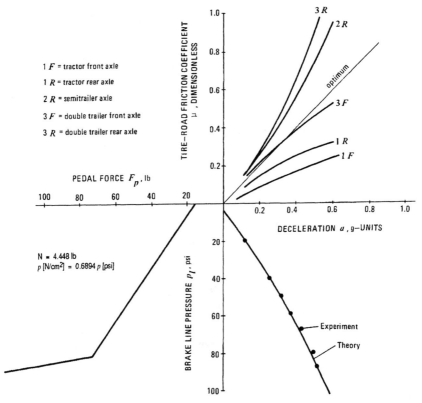

*Figure 8-32. Braking performance diagram for a tractor-semitrailer-double trailer combination*

## 8.4.5 Variable Brake Force Distribution for Tractor-Trailer Combination Equipped with Tandem Axles

The tire-road friction utilization calculations were carried out to determine the improvement in braking performance obtained with a proportioning braking system. The vehicle is a tractor equipped with a walking beam suspension, coupled to a semitrailer equipped with a two-leaf spring tandem axle. Important vehicle data used for the exemplar vehicle are presented in Table 8-2.

The results of the tire-road friction utilization calculations for the vehicle combination equipped with standard brakes (Table 8-2) are presented in Figures 8-33 and 8-34 for the empty and laden vehicle, respectively. Examination of the curves for the empty case shown in Fig. 8-33 indicates that the tandem forward axle of the trailer and the tandem rearward axle of the tractor are heavily overbraked. The tandem forward axle of the tractor (1RF) and the tandem rearward axle of the trailer show good friction utilization values over a wide range of decelerations.

## TABLE 8-2

### Tractor-Semitrailer Data

Tractor:  $W_1$ = 71,969 N (16,180 lb)

$L_1$ = 406 cm (160 in.)

$\chi_1$ = 0.22

$\Psi_1$ = 0.40

Trailer:  $W_2$ = 70,946 N (15,950 lb) (empty)

= 279,423 N (62,820 lb) (laden)

$L_2$ = 993 cm (391 in.)

$\chi_2$ = 0.154 (empty and laden)

$\Psi_2$ = 0.58 (empty), 0.57 (laden)

Brakes:  No brakes on tractor front axle

Tractor tandem axle:

$A_c$ = 194 cm² (30 in.²); BF = 2.3 (unfaded); $\rho$ = 5.5

$r$ = 20.96 cm (8.5 in.); $\eta_m$ = 0.70

Trailer tandem axle:

Identical to tractor brakes except

BF = 1.9 (unfaded)

Tire Radius:  R = 53.34 cm (21 in.)

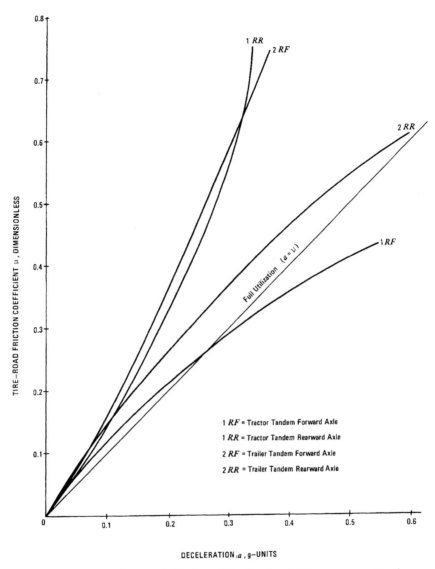

*Figure 8-33. Tire-road friction utilization for 3-S2 tractor-semitrailer combination (empty) with no front brakes.*

Inspection of the tire-road friction utilization curves for the laden case, shown in Fig. 8-34, reveals that the tandem rearward axle of the tractor and the forward axle of the trailer still suffer from low friction utilization, i.e., premature brake lockup. For example, for a tire-road friction coefficient of 0.6, no wheels-unlocked decelerations greater than approximately 0.3 g are possible.

419

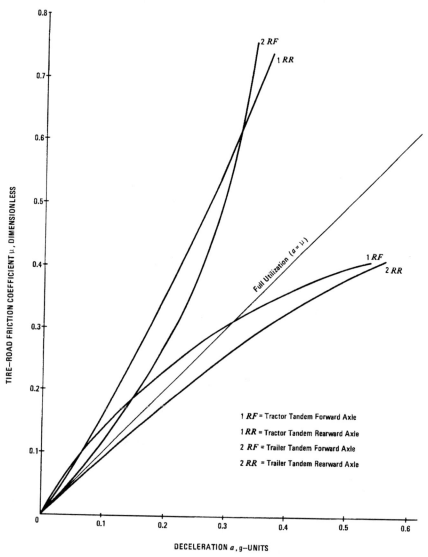

*Figure 8-34. Tire-road friction utilization for 3-S2 tractor-semitrailer combination (loaded) with no front brakes.*

Major portions of braking and tire side forces during turning must be produced by the tandem forward axle of the tractor and the tandem rearward axle of the trailer due to decrease in normal axle loads on the other tandem axles. Comparison of the tire-road friction curves of Fig. 8-34 of the laden vehicle with the curves of Fig. 8-33 of the empty vehicle indicates that the tandem rearward axle of the trailer is overbraked in the empty condition. To avoid possible trailer swing due to premature wheel lockup of the critical trailer rearward axle, load-sensitive proportioning must reduce the brake force concentrated on the tandem rearward axle of the trailer. The relative brake force distribution $\Phi_i$ of the earlier brake system not employing brakes on the tractor front axle, front to rear, is 27%, 27%, 23%, and 23%. The results of the proportioning brake force analysis for the empty vehicle combination are illustrated in Figure 8-35. The proportional brake force distribution, front to rear, is 40%, 20%, 20%, and 20%. Inspection of Fig. 8-35 reveals that the tandem rearward axle of the tractor (1RR) is overbraked for decelerations above 0.17 g while the forward axle of the trailer (2RF) is always overbraked. However, the trailer rearward axle (2RR) is braked near optimum conditions for decelerations up to 0.5 g resulting in sufficient tire side force to minimize the potential for trailer swing. The forward axle of the tractor tandem (1RF) is slightly overbraked for decelerations below about 0.58 g.

No major improvements in tire-road friction utilization may be expected from a different proportional brake force distribution without installing brakes on the front axle of the tractor. The effects of tractor front axle braking for proportional brake force distribution $\Phi_i = 17\%, 25\%, 20\%, 19\%,$ and 19% are illustrated in Figure 8-36 for the empty vehicle. Examination of Fig. 8-36 reveals that the tandem forward axle of the tractor and the rearward axle of the trailer—both critical to vehicle combination stability—are near optimum for decelerations below 0.3 g and are slightly (2RR) and moderately (1RF) underbraked for greater decelerations. The tractor front axle always is underbraked, thus maintaining tractor steerability while rear wheels have locked.

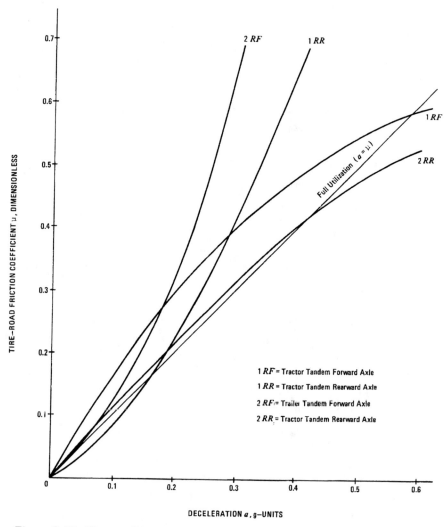

*Figure 8-35. Tire-road friction utilization for 3-S2 tractor-semitrailer (empty) with proportioning (no front brakes).*

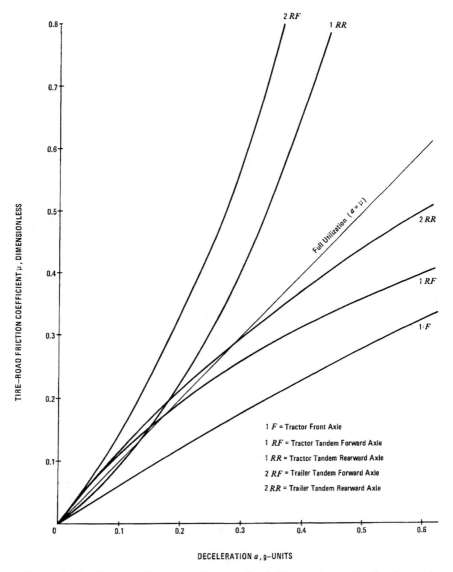

*Figure 8-36. Tire-road friction utilization for 3-S2 tractor-semitrailer (empty) with front brakes and proportioning*

# CHAPTER 9

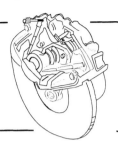

# Anti-Lock
# Brake Systems (ABS)

*In this chapter a historical overview of ABS brake development is presented. Fundamentals of anti-lock brake systems including tire factors are introduced. ABS brakes for hydraulic and air brake systems are discussed.*

## 9.1 Historical Overview

Anti-lock brake systems (ABS) are now used on more and more passenger cars and light trucks (Refs. 29, 30). They are optionally available on some commercial vehicles equipped with air brakes, but will be required in Europe during the early '90s.

Development of ABS brakes, one of the few truly outstanding safety features in the history of the motor vehicle, resulted in the Dunlop "Maxaret" fitted to aircraft in 1952. In 1972 in England, the Jensen Interceptor automobile became the first production car to offer a Maxaret-based ABS using a propeller shaft speed sensor and viscous coupling. In the U.S., vehicle manufacturers worked with various brake manufacturers to develop ABS systems. In 1969, a rear-wheel-only ABS developed by Ford and Kelsey Hayes was offered on the Thunderbird. Chrysler and Bendix produced a four-wheel ABS offered on the '71 Imperial. General Motors likewise offered ABS brakes on some of their luxury models by the mid-'70s. For example, ABS were available on the GM Eldorado, Toronado, Cadillac Deville and Fleetwood between 1976 and 1982.

All manufacturers producing ABS systems in the early '70s used state-of-the-art components including vacuum as energy source and analog electronics. Development of high-pressure energy sources including the use of accumulators was underway in Europe. The basic shortcomings of the early ABS brakes revolved around the low reliability of system electronics, and to some extent slow cycles rates due to limitations associated with the vacuum source. These reasons, and probably low public awareness and additional cost to the buyer, led to their quiet withdrawal from the market in the mid-'70s.

In the early '70s the National Highway and Traffic Safety Administration (NHTSA) of the U.S. Department of Transportation issued a regulation (FMVSS 121) which indirectly required the installation of ABS brakes on air-brake-equipped trucks and trailers by 1975. Early reliability and electronic controller problems caused the government to amend the standard, effectively removing the no-wheels lock requirement of the standard. NHTSA research showed the following breakdown of ABS-system problems: 41% sensors, 16% valves, 8% computers, 3% incorrect installation, 1% electrical connections, 30% electromagnetic wave interference.

In Europe during the early and mid-'70s, brake and electronics suppliers had developed digital electronics changing from analog to integrated circuits and microprocessors which resulted in the introduction of the first Bosch ABS systems on Mercedes passenger vehicles in 1978. This four-wheel system was of the add-on type which is installed in the existing vacuum or hydraulic boost brake system. BMW and others followed shortly. Japanese brake and vehicle manufacturers introduced ABS brakes based on the Bosch system as well as their own designs by the mid-'80s. The Bosch ABS system was used in the '86 Corvette and Cadillac Allante.

In 1984 an integrated ABS system produced by ITT-Teves was introduced on the Lincoln Mark VII in the U.S. and, in 1985, as standard equipment on the Ford Scorpio in Germany. The Teves integrated system combines the ABS actuator, hydraulic booster, and master cylinder into one unit. The Teves ABS system was likewise available on luxury GM cars in the '86 model year.

In the '87 model year, Ford introduced a Kelsey Hayes-developed rear-wheel-only ABS on their pickup trucks, thus effectively eliminating the severe instability problems associated with braking of empty pickup trucks. Since front brakes can still be locked, loss of steering will still occur during severe braking involving front brake lockup.

Since the late '80s and early '90s, ABS systems are found on nearly all top models of every manufacturer including four-wheel systems for four-wheel-drive vehicles developed by Bendix and others.

In the '91 model year, automobile manufacturers offered ABS brakes on approximately one-third of their passenger vehicles. Either as an option or standard equipment, ABS brakes were available on passenger vehicles in the approximate percentages that follow: Chrysler 18%, GM 33%, Ford 43%, Toyota 40%, Nissan 44%, Honda 50%, Mazda 25%, and Mitsubishi 27%. Manufacturers of luxury automobiles like BMW, Mercedes or Porsche have offered ABS brakes for several years prior to the '91 model year. By 1992 more than 15% of all passenger cars produced worldwide were equipped with ABS. The major reason for not offering ABS on mid-size and compact cars is the relatively high option price of approximately $800 to $1300 for a four-wheel system. Changes are made to minimize cost and increase reliability of existing integrated and add-on systems. GM's Delco Moraine Division has developed an effective ABS system (ABS-VI) to be offered for approximately $350 per vehicle, a significant cost saving. GM has introduced the new system on some of their '91 models including Saturn, Oldsmobile Cutlass Calais, Buick Skylark, and Pontiac Grand Am. It is expected that most vehicle manufacturers will offer ABS brakes on nearly all of their models by the mid-'90s. By 1998, practically all passenger cars and light trucks are equipped with four-wheel ABS systems.

Increased use of ABS brakes on passenger cars and particularly on commercial vehicles is expected to improve traffic safety. Some extremely limited German accident data appear to indicate that ABS-equipped automobiles may be overinvolved in certain accidents due to drivers overestimating the safety contribution of the brakes, especially when driving on ice or following too closely. On the other hand, accident data collected during the early '80s with identical model Mercedes Benz vehicles equipped with and without ABS showed 6 to 10% fewer accidents for passenger cars with ABS. US-DOT studies showed that single-vehicle rollover accidents increased for ABS equipped cars. ABS-equipped cars are involved in rear-enders as the bullet vehicle slightly less often than non-ABS cars, and as the target slightly more often. As more complete accident statistics become available over the years, it is hoped that the true safety contribution of ABS brakes will be shown.

Nevertheless, vehicle and brake manufacturers must guard against overstating the safety aspects of ABS brakes in advertisements in order to minimize ABS-induced driver accident causation.

## 9.2 Fundamentals of ABS Analysis

Anti-lock brake systems prevent brakes from locking during braking. Under normal braking conditions the driver operates the brakes as usual; however, on slippery roadways or during severe braking, as the driver causes the wheels to approach lockup, the ABS brakes take over and modulate brake line pressure and, hence, braking force, independent of pedal force.

### 9.2.1 Tire Characteristics

Tire characteristics play an important role in the braking and steering response of a motor vehicle. For ABS-equipped vehicles the tire performance is of critical significance. All braking and steering forces must be generated within the small tire contact or tread patch connecting the vehicle to the road surface. Tire traction forces such as longitudinal or braking forces as well as side forces can only be produced when a difference exists between the speed of the tire circumference and the speed of the vehicle relative to the road surface. It is common to relate tire braking force data to tire braking slip, which is defined as the ratio of the difference of tire circumferential tread speed and absolute tire (or vehicle speed) to absolute speed. Tire side slip is defined in a similar fashion. Since tires are elastic pneumatic structures, the difference in speed consists of elastic tire deformations and tread sliding. Only when the tire is at 100% slip is the braking force produced by complete sliding or skidding of the tire tread patch in contact with the road.

The tire-road contact patch moves in the x-direction with the braking force acting opposite to it, as illustrated in Figure 9-1. The angle $\alpha$ formed by the line of travel of the tire contact patch and the tire plane is commonly called slip angle. The tire side force acts at a right angle to the x-direction.

The absolute braking slip $S_b$ is computed by the expression

$$S_b = (V - V_c) / V \qquad (9\text{-}1)$$

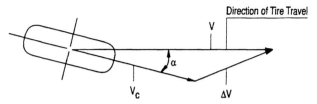

*Figure 9-1. Relative velocity ΔV and tire slip for braking while turning.*

where   V = velocity of tire contact patch, m/s (ft/s)

  $V_c$ = circumferential velocity of tire tread, m/s (ft/s)

Braking slip $S_{b,x}$ in direction of vehicle travel is determined by

$$S_{b,x} = (V - V_c \cos \alpha) \, / \, V \qquad (9\text{-}2)$$

where   x-direction is the direction of travel of the tire

  y-direction is perpendicular to x-direction

  $\alpha$ = slip angle

Similarly, side or lateral slip $S_{b,y}$ is determined by

$$S_{b,y} = V_c \sin \alpha \, / \, V \qquad (9\text{-}3)$$

Typical braking friction-slip curves without any side force are shown in Figure 9-2 (Ref. 1). In general, the μ-slip curve is characterized by a peak friction value obtained at the optimum slip value $S_{b,opt}$, and the sliding friction value obtained for 100% tire slip, i.e., the brake is locked. The shape of the curve is a function of the design of the tire, tread design, and rubber composition, as well as the road surface and possible contamination such as water. The low values of tire slip are mainly related to deformation slip of the tire tread and pneumatic behavior of the tire.

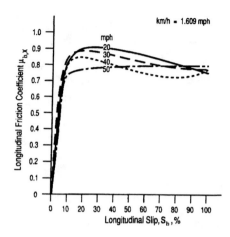

*Figure 9-2. Friction-slip curve for dry concrete as function of speed.*

An important tire parameter significant for ABS analysis is the zero-slip slope of the μ-slip curve. Most tires have $\Delta\mu_b/\Delta S_b$ values ranging between 20 and 30, indicating that a slip value of 1% results in a tire-road friction coefficient of 0.2 to 0.3. The zero-slip slope for most tires is nearly independent of the road surface including wet roads, again indicating that deformation slip dominates friction production at low levels of tire slip.

The actual μ-slip curve illustrated in Fig. 9-2 may be idealized by two linear relationships as shown in Figure 9-3. The zero-slip slope is assumed to remain constant up to the point where it reaches the peak friction value. After the peak value has been reached, increased tire slip causes a straight-line reduction of tire-road friction coefficient until 100% tire slip has been obtained.

Under these conditions the time $t_p$ for the tire to attain optimum slip, that is, peak friction may be computed by (Ref. 13)

$$t_p = (S_p I_w \omega_o \, / \, \mu_p F_z R) + \mu_p F_z R \, / \, k \quad , \quad s \qquad (9\text{-}4)$$

where  $F_z$ = tire normal force, N (lb)

$I_w$ = mass moment of inertia of wheel assembly, $kgm^2$ ($lbfts^2$)

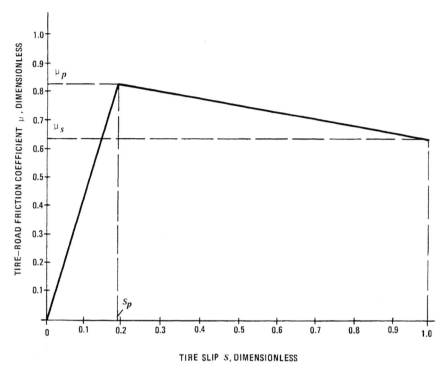

*Figure 9-3. Idealized tire-road friction slip characteristics.*

$k$ = brake torque versus time slope, Nm/s (lb-in./s)

$R$ = tire radius, m (in.)

$S_p$ = tire slip at peak friction

$\mu_p$ = peak tire-road friction coefficient

$\omega_0$ = initial angular velocity of wheel, rad/s

The wheel angular deceleration $\alpha_p$ at which maximum braking traction forces are produced frequently is called threshold deceleration and may be computed by

$$\alpha_p = \omega_0 k S_p \,/\, \mu_p F_z R \quad , \quad \text{rad/s}^2 \tag{9-5}$$

431

For the second linear region of Fig. 9-3 during which the wheel approaches lockup, the tire-road sliding friction coefficient $\mu_s$ affects the time required for the wheel to attain lockup. The time $t_s$ required by the wheel to move from peak friction to lockup is determined approximately by (Ref. 13)

$$t_s = \left\{ \left[ \frac{S_p I_w \omega_p}{(\mu_p - S_p \mu_s) F_z R} \right]^2 + \frac{2\mu_p I_w \omega_p}{k(\mu_p - S_p \mu_s)} \right\}^{1/2}$$

$$- \frac{S_p I_w \omega_p}{(\mu_p - S_p \mu_s) F_z R} \quad , \quad s \tag{9-6}$$

where  $\mu_s$ = sliding tire-road friction coefficient

$\omega_p$ = angular velocity of wheel at peak friction, rad/s

The friction process is stable only for any point on the linearly increasing tire force including peak friction, and at the 100% slip point. For points between peak and sliding friction the process is unstable. ABS control systems must limit the tire slip values to the stable region to prevent wheel lockup because the time required for the tire slip to move through the unstable region and achieve lockup is only a fraction of the time required to achieve peak friction.

Tires with a high peak friction point relative to the sliding friction produce dry road peak friction at approximately 20 to 30% slip. The optimum slip value decreases as tire-road friction decreases. Often, tires exhibiting a ratio of sliding to peak friction of 0.8 or so are rated highly for vehicle handling and steering response, but poorly for braking performance.

Tires showing little or no decrease in friction between peak and sliding condition produce an insignificant effect of slip on braking friction for slip values greater than approximately 30%. Their maximum friction coefficient generally is greater than the sliding friction exhibited by the former tires. These tires generally are judged better relative to their braking than handling performances.

To achieve a directionally stable braking maneuver, tire side forces must be considered along with the braking friction. As indicated earlier, a tire can produce a side force only if it is partially side slipping, i.e., if a slip angle exists between the direction of tire patch motion and the plane of the wheel. A typical tire side force friction coefficient versus slip angle curve is illustrated in Figure 9-4 for a free-rolling tire. The tire side force is measured at a right angle to the direction of motion of the tire contact patch. The side friction coefficient increases to a maximum value between a slip angle range of 8 to 12 degrees for most tires and decreases for higher values of slip angle. The side force production is unstable at and beyond a point where the slope is zero ($\Delta\mu_y$ / $\Delta\alpha = 0$). The cornering stiffness of a tire is expressed by the slope of the curve at the origin, and ranges from 0.25 to 0.4 per degree of slip angle. Consequently, a slip angle of one degree will achieve a tire side friction coefficient of 0.25 to 0.4, depending on tire construction. Operational parameters such as inflation pressure, camber angle, loading, and others will affect side force.

The tire side friction coefficient as a function of slip angle shown in Fig. 9-4 is only valid for a rolling, nonbraked tire. When braking is present during a steering maneuver, the tire side friction coefficient decreases with increasing values

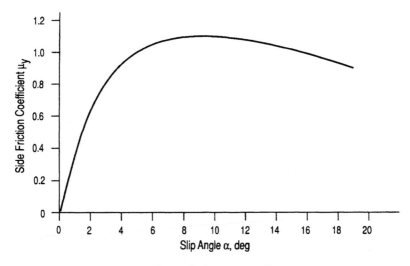

*Figure 9-4. Tire side friction coefficient $\mu_y$ as a function of slip angle for free-rolling tire.*

of tire slip $S_b$ as illustrated in Figure 9-5. Inspection of Fig. 9-5 reveals that braking slip significantly reduces the side friction coefficient for a given slip angle. Braking slip values between 20 and 30% will reduce tire side friction coefficients by 75 to 80% of their free-rolling nonbraked levels. For example, a nonbraking side friction coefficient of 0.6 reduces to 0.15 to 0.18 in the presence of 20% braking slip.

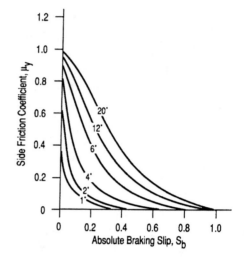

*Figure 9-5. Side friction coefficient $m_y$ as a function of braking slip and slip angle.*

The effects of tire slip angle on braking friction coefficient $\mu$ as a function of tire braking slip are illustrated in Figure 9-6 for a typical passenger-car tire. The diagram shows that braking friction coefficient decreases with increasing values of slip angle. The optimum slip at which tire peak friction occurs increases with higher values of tire slip, indicating the "retarding" component of tire side slipping.

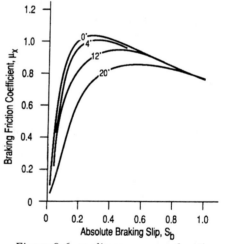

*Figure 9-6. $\mu$-slip curves as a function of slip angle.*

Although the tire characteristics discussed above are elementary, they clearly indicate the complex relationships associated with braking and side force production of a tire. For an ABS system to ensure minimum stopping distances, directional stability, and stable braking while turning, certain performance requirements must be based on the tire characteristics used. Fortunately, certain simplifications in ABS design can be made to achieve an economical and efficient system affordable for vehicles of lower price ranges.

## 9.2.2 ABS Design Concepts

### 9.2.2.a Basic Performance Requirements

The design of an ABS system begins with a complete understanding of the tire-road friction characteristics. The braking process in terms of minimum stopping distance would be optimum if the tire slip of the braked tire could always be kept at values corresponding to peak friction levels. Ideally then, a sensor would detect the magnitude of the coefficient of friction at the tire-road interface under all possible conditions, and the rest of the brake system would use this signal to modulate the brake torque in such a manner that the peak friction coefficient would be used throughout the braking process. In practice, it is not feasible to detect the tire-road friction coefficient directly since this would require a fifth wheel as employed in road friction measuring equipment or complex on-board sensors.

In general, the following methods have been suggested as modulating parameters for the automatic control of brake torque:

1. Angular velocity of the wheel.

2. Braking slip of the tire.

3. Velocity difference between tire and vehicle.

4. Velocity difference between the tire and the other tires of the vehicle.

Practical sensors measure wheel angular velocity from which wheel deceleration is determined by differentiation. The relative tire slip ratio is estimated by comparing a measured wheel velocity with a memory of the wheel velocity before initiation of braking. The memory may consist of a flywheel in the case of purely mechanical ABS systems, a capacitor in the case of the earlier analog systems, or microcomputer memory of currently used integrated circuits. In some ABS systems, longitudinal vehicle deceleration is measured to provide additional data input for brake line pressure modulation. In some rare cases, lateral acceleration is measured to improve pressure modulation.

Performance requirements for wheel anti-lock braking systems include the following:

1. Retention of steering during ABS control for rapid pressure increases up to 1500 bar/s.

2. Retention of vehicle stability and steering ability is generally more important than minimizing stopping distance.

3. Minimum reaction into steering wheel especially on split-coefficient road surfaces.

4. ABS must utilize available tire-road friction optimally.

5. ABS must adapt quickly to changes in tire-road friction levels. For example, a vehicle may be braked at maximum wheels unlocked deceleration on a 0.8 road surface when it suddenly enters onto a slippery 0.2 ice-covered section. A typical passenger car will lock its front brakes in approximately 13 ms for a speed of 5 m/s (16 ft/s) and 50 ms for 20 m/s (66 ft/s). An expression such as front brake lockup time (s) equals 0.0025 velocity (m/s) may be derived as a function of weight and brake force distribution, and wheel size.

6. ABS must minimize the yaw moment effects when braking on a split-coefficient road surface.

7. ABS must recognize hydroplaning and maintain directional stability.

8. ABS must provide stable braking while turning.

9. If an ABS malfunctions, the standard brake system must perform safely, i.e., without loss of directional stability.

10. ABS must perform properly with all tires specified for the vehicle, including mini-spares, if supplied.

11. ABS malfunctioning must be communicated to the driver.

12. Maintenance and repair skills should conform to existing or attainable repair industry practices.

13. ABS is not a substitute for poor brake balance.

14. Owners with trucks towing trailers must be warned against the installation of hydraulically actuated electric trailer brakes which may affect the performance of the truck ABS system by influencing the brake fluid requirement.

These basic requirements can be achieved only with a four-wheel ABS system. The current practice of installing ABS systems on only the rear axles of most light trucks and vans improves directional stability by preventing premature rear brake lockup, but renders the vehicle without steering when the front brakes are locked. Industry efforts should be directed toward offering four-wheel ABS systems on all vehicles.

The ABS system must be designed to respond within the performance characteristics set by the braked tire. Research and experience has shown that vacuum power sources for brake valve actuation nearly always lead to less than optimum performance when braking on dry road surfaces. Reasons for this are related to the slow response time generally exhibited by vacuum-actuated valves. ABS systems in use today in most cases use pressurized fluid as a power source.

The function of the ABS system is to maintain optimum braking performance of all four wheels—or more on a truck—relative to each other while braking under all foreseeable operating conditions. Two basic criteria are used to sense wheel lockup, namely circumferential or rotational wheel speed deceleration (or acceleration during spinup), and relative wheel slip.

### 9.2.2.b  ABS Control Concepts

The control of the different brake/wheels systems can be accomplished in several ways.

In *single-wheel control* the wheel speed sensor of a wheel controls the adjustments made to the brake line pressure of that wheel independent of any other wheels. This control method results in maximum braking on that wheel and, hence, maximum deceleration. On split-coefficient of friction surfaces the different braking forces left and right cause a yaw moment attempting to rotate the vehicle toward the higher traction side. Single-wheel control is generally used on the front wheels of motor vehicles.

In *select-low control* the wheel with the lower traction controls the brake line pressure for both brakes on that axle. The traction force on the higher friction surface is not fully utilized resulting in a lower brake torque and, hence, longer stopping distance. The advantage is a higher side force traction potential and the absence of a yaw moment. Select-low control is typically used on rear wheels of motor vehicles.

In *select-high control* the wheel with the higher traction controls the brake line pressure of both brakes on that axle. The results are higher braking force because all traction is utilized, unbalanced brake forces left and right which causes a yaw moment, and locking of one wheel on the low-friction surface. Honda has used select-high on the front and select-low control on the rear axle. Consequently, one front wheel can lock up on a low-friction surface while the ABS system operates as designed.

Vehicles having a front-to-rear dual hydraulic split system generally use single- or independent-wheel control on the front and select-low on the rear, with the rear wheel speed sensor located at the differential. Since both rear brakes are controlled as one unit, only one hydraulic control valve is required for the modulation of the rear brake line pressure.

Diagonal hydraulic split systems require four wheel-speed sensors, one for each wheel, and two hydraulic control valves for the rear brakes. Although four sensor and hydraulic channels are involved, the system is only a three-channel system. The select-low control of the rear axle, controlled by the electronic module or computer, provides identical brake line pressure modulation to both rear brakes.

In order to minimize production cost, ABS systems with two control channels have been designed and are in use today on several small front-wheel-driven European cars.

A mechanical two-channel system designed by Girling for diagonal split hydraulic systems controls each front wheel separately. As a front wheel approaches lockup, its brake line pressure is modulated. The same brake line pressure modulation is transmitted to the corresponding rear brake. If a proportioning valve is present, the brake line pressure will be reduced before reaching the rear brakes. In recent years the mechanical system has been replaced by two front-wheel speed sensors along with two hydraulic control valves, one for each of the diagonal brake circuits. For small front-wheel-driven vehicles with a relatively low rear axle weight the non-optimum braking of the rear wheels will have an insignificant effect on overall braking. When braking on split-coefficient surfaces the rear wheel on the low-traction side will lock up. The other rear wheel, since it is modulated by the front wheel on the low-traction surface, will continue to rotate, thus providing sufficient directional stability for the vehicle.

For the lightly loaded operating conditions, the two-channel ABS system may cause directional instability when braking on extremely high-traction surfaces. Under these conditions the front wheels have not yet achieved lockup and, hence, remain unmodulated, while the rear brakes are locked. Over the normal range of road surface conditions premature rear wheel lockup should be a rare event, provided the basic brake force distribution and critical deceleration are properly engineered. To ensure a stable brake force distribution front to rear, brakes with a low sensitivity to brake factor changes such as leading-trailing shoe or disc brakes should be used on the rear axle.

Braking tests generally show that two-channel systems will not provide the same deceleration levels as the more expensive three-channel systems. Skilled test drivers operating a vehicle with properly designed standard non-ABS brakes may be able to outperform two-channel systems under certain conditions, particularly on wet roads and when braking from relatively low speeds such as 40 to 48.3 km/h (25 to 30 mph).

The basic tire characteristics important in ABS design considerations are discussed in Section 9.2. The significant physical relationships for ABS control for straight-line braking are illustrated in Figure 9-7 (Ref. 31), and for braking while turning in Figure 9-8. The critical control ranges for ABS function are cross-hatched. Close inspection of Fig. 9-7 reveals that shorter stopping distances are achieved with ABS brake systems when operating on dry (1), wet (2), and icy (4) road surfaces than when all wheels are locked (100% wheel slip). When braking on snow, the snow wedge forming under the locked tire causes an additional retarding effect.

Inspection of the two curves associated with low (2-degree slip angle) and high (10-degree slip angle) lateral acceleration reveals that the ABS control range must cover a wide performance spectrum. During severe braking in a

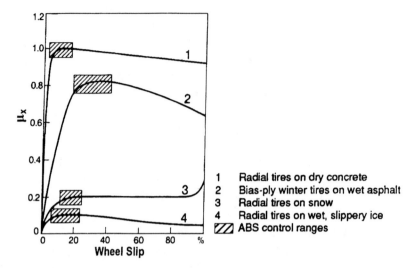

Figure 9-7. *Braking force coefficient with ABS control ranges as a function of brake slip during straight-ahead braking (Bosch).*

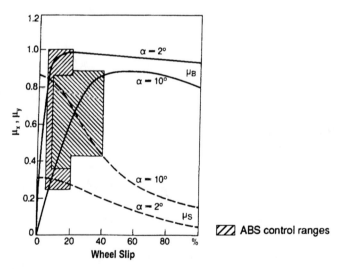

Figure 9-8. *Braking force and lateral force coefficients as a function of brake slip and slip angle α with ABS control ranges (Bosch).*

turn, the ABS system should intervene early on with initially low deceleration values while the lateral acceleration is still near its maximum value permitted by the tire-road friction coefficient. As speed decreases and lateral acceleration drops, the ABS system produces increasing levels of braking slip. For optimally designed ABS systems, the stopping distance while turning is only slightly longer than that associated with a straight stop.

A typical ABS control cycle on a high-traction road surface is illustrated in Figure 9-9. If severe braking occurs on a high-traction road surface, then the modulated pressure increase must occur approximately 5 to 10

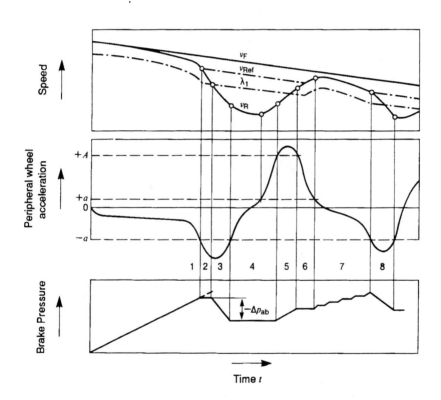

| | |
|---|---|
| $v_F$ | Vehicle speed |
| $v_{Ref}$ | Reference speed |
| $v_R$ | Peripheral wheel speed |
| $\lambda_1$ | Slip switching threshold |
| $+A, +a$ | Thresholds of peripheral wheel acceleration |
| $-a$ | Threshold of peripheral wheel deceleration |
| $-\Delta p_{ab}$ | Brake pressure decrease |

*Figure 9-9. Braking control for high braking force coefficients (Bosch).*

times more slowly compared to the pressure increase prior to ABS control. The braking control characteristics reflected in Fig. 9-9 indicate such requirements.

During initial braking, the brake line pressure in the wheel cylinder and the circumferential tire speed decrease and, hence, wheel deceleration increases. At the end of phase 1, the circumferential speed wheel deceleration exceeds the set threshold level (−a). As a result, the corresponding solenoid valve switches to the pressure holding position. The brake line pressure is kept constant at this level because the threshold wheel deceleration level could be exceeded in the stable range of tire-road friction coefficient/brake slip curve, thus avoiding waste of stopping distance. At the same time the reference speed is reduced along a given ramp function. The value for the slip switching threshold $\lambda_1$ is derived from the reference speed.

At the end of the constant pressure phase 2, the wheel speed has dropped below the threshold $\lambda_1$. The solenoid now switches to the pressure drop position resulting in reduced brake line pressure until the tire circumferential deceleration has exceeded the threshold value (−a). The speed drops below the threshold (−a) again at the end of phase 3 and a pressure holding phase follows. The circumferential tire acceleration increases within this time period until it exceeds the threshold (+a). The brake line pressure still remains constant until the circumferential acceleration has exceeded the relatively high threshold (+a) at the end of phase 4. The brake line pressure then increases as long as the threshold level (+a) is exceeded. During phase 6 the brake line pressure is kept constant since the threshold level (+a) is exceeded. At the end of phase 6 the circumferential acceleration drops below the threshold (+a), indicating that the wheel is operating within the stable range of the friction-slip curve and is slightly underbraked.

The brake line pressure is increased in stages during phase 7 until the circumferential acceleration exceeds the threshold (−a) at the end of phase 7. Here, the brake line pressure is decreased immediately without generation of the $\lambda_1$ signal.

Braking control on a slippery low-friction road surface is illustrated in Figure 9-10. The physical difference between braking on a dry surface is that low brake line pressures may cause ABS brake control and that significantly more time is

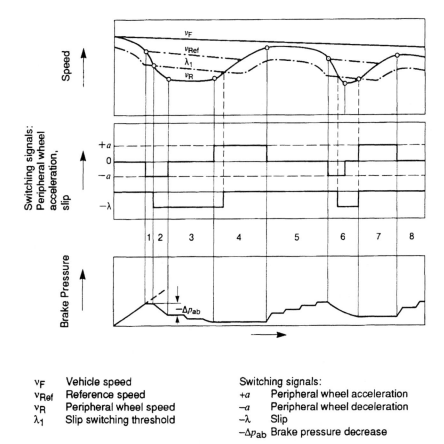

| $v_F$ | Vehicle speed | Switching signals: | |
| $v_{Ref}$ | Reference speed | $+a$ | Peripheral wheel acceleration |
| $v_R$ | Peripheral wheel speed | $-a$ | Peripheral wheel deceleration |
| $\lambda_1$ | Slip switching threshold | $-\lambda$ | Slip |
| | | $-\Delta p_{ab}$ | Brake pressure decrease |

*Figure 9-10. Braking control for low braking force coefficients (Bosch).*

required by the braked wheels to accelerate out of a phase of high slip levels. The logic control unit can recognize these prevailing road conditions and adapt the ABS system characteristics accordingly as shown in Fig. 9-10. In phases 1 and 2, no difference exists between braking on dry and slippery road surfaces. Phase 3 begins with a pressure holding phase of short duration. The wheel speed is then very briefly compared with the slip switching threshold $\lambda_1$. Because the wheel speed is less than the value associated with that of the slip switching level, the brake line pressure is reduced for a short fixed time period. This process is followed by a short constant pressure phase. A next comparison between wheel speed and slip switching threshold $\lambda_1$ is made, leading to a pressure drop during a short fixed time period. The wheel accelerates again in the following pressure holding phase and its circumferential wheel acceleration exceeds the threshold value (+a). This condition leads to further pressure holding until the acceleration is again below the threshold value (+a) at the end of phase 4. In the following phase 5, step-by-step pressure increase occurs similar to the one for dry road surfaces until a pressure reduction is initiated at the beginning of phase 6. The controller logic recognizes when the wheel operates in the range of high slip for a relatively long time. To improve driving stability and steerability, a continuous comparison is made between the wheel speed and slip switching threshold $\lambda_1$. Based on this, brake line pressure is constantly reduced in phase 6 until the circumferential wheel acceleration exceeds the threshold value (+a) in phase 7. Due to the constant decrease in pressure, the wheel runs with high slip for only a short time thereby increasing vehicle stability and steerability compared with the first control cycle.

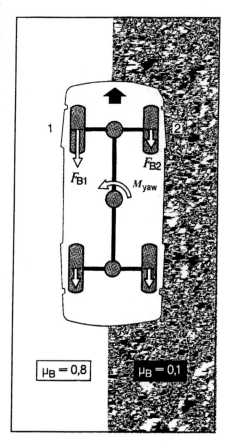

ABS braking control with yaw moment buildup delay is used when a vehicle brakes on split-coefficient of friction surfaces. Extremely different braking forces left-to-right are produced on the front wheels as a result of initial braking on split-coefficient surfaces as illustrated in Figure 9-11.

Heavy passenger vehicles with relatively long wheelbase and large mass moment of inertia about their vertical axis develop a yaw motion relatively slowly, permitting the driver to countersteer. Smaller vehicles with

| | |
|---|---|
| $M_{yaw}$ | Yaw moment |
| $F_B$ | Braking force |
| $\mu_B$ | Braking force coefficient |
| 1 | "high" wheel |
| 2 | "low" wheel |

*Figure 9-11. Yaw moment buildup for greatly differing braking force coefficient (Bosch).*

lower values of mass moment of inertia require a yaw moment buildup delay system in addition to the basic ABS function of the braking system. For vehicles with low critical handling characteristics, brake line pressure is increased at the "high" wheel in stages or steps as soon as the first pressure reduction caused by wheel lockup tendencies takes place at the "low" wheel. When the pressure at the "high" wheel has reached its locking level, it is no longer influenced by the signals of the "low" wheel, that is, it is individually controlled. This feature ensures maximum braking forces at the "high" wheel while retaining steerability for the vehicle. Since the maximum pressure at the "high" wheel is reached in a relatively short time period, the increase in braking distance is small compared to a vehicle without yaw delay.

For vehicles with particularly critical handling characteristics the ABS solenoid at the "high" wheel is controlled with a specific pressure holding and reduction time as soon as the brake line pressure has been reduced at the "low" wheel. The pressure modulation of the "low" and "high" wheel are interrelated depending on vehicle speed. As vehicle speed and the potential for vehicle instability on split-coefficient surfaces increase, the pressure buildup times at the "high" wheel are increasingly reduced while the pressure buildup times at the "low" wheel are increasingly extended.

The optimum yaw moment delay is a compromise between good steering response and minimizing stopping distance. Differences exist between ABS manufacturers. Under extreme conditions directional stability may be lost for ABS designs where minimum stopping distance is a major design objective.

To optimize braking while turning, a lateral acceleration sensor switches off the yaw moment delay feature for lateral acceleration exceeding 0.4 g. Consequently, large braking forces are developed at the outer wheel in the curve which produces a torque directed to the outside of the curve and compensates for the torque of the lateral force. The net result is a slightly understeering vehicle easily controlled by the driver.

The use of ABS systems has resulted in the design of advanced traction controls and vehicle stability systems.

The electronic traction system compares the speed of each wheel. If one wheel tends to spin up, then the control modulates the brake of that wheel. Without pedal force, the hydraulic unit must have its own pressure source.

The advantages include improved traction, increased stability, and relatively low cost. The major disadvantages are thermal loading of brakes, and hence, limitation to lower speeds.

The disadvantages are mostly eliminated when the electronic controls also modulate the engine.

The electronic stability program includes lateral acceleration and yaw sensors, as well as steering wheel sensors in some cases, to compute "intended" dynamic data to actually measured data. When the system "predicts" an instability, individual wheels are braked to correct the path of the vehicle.

The electronic stability program was introduced by Mercedes on the A-model vehicle, after the so-called "Moose" test resulted in rollover in severe turning maneuvers.

## 9.3 Hydraulic Systems

### 9.3.1 Basic Considerations

A schematic of an ABS system is illustrated in Figure 9-12. A wheel-speed sensor transmits a signal of impending wheel lockup to the logic control which, in turn, signals a modulator to reduce brake line pressure, which causes the wheel rotational speed to increase again.

Ideally, an ABS system would modulate all four wheels independently so that maximum longitudinal as well as lateral tire forces would be produced. In the single-wheel control, the wheel speed sensor of the wheel controls the adjustments made to the brake line pressure of the wheel independent of any other wheel. This control results in maximum braking on that wheel and, hence, deceleration. On split-coefficient-of-friction surfaces, the different braking forces left and right will cause a yaw moment attempting to turn the vehicle toward the higher traction side. Such a system would allow panic brake applications even while operating near or at the limit of turning speed of the vehicle. Obviously these specifications require brake system designs with sophisticated and, hence, expensive electronic and hydraulic hardware.

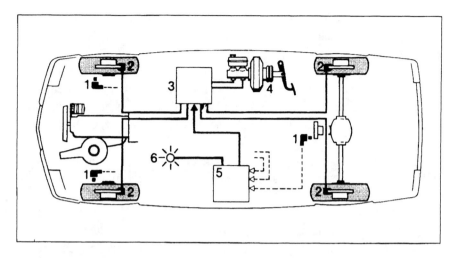

1   Wheel speed sensor
2   Wheel brake cylinder
3   Hydraulic modulator
4   Master cylinder
5   ECU
6   Safety lamp

*Figure 9-12.  Car with ABS 2S (Bosch).*

To reduce cost to a reasonable level relative to the base cost of the vehicle, certain design simplifications can be made while still achieving acceptable ABS performance.

Less expensive than the four-wheel independent system is one that uses independent front wheel modulation and select-low, rear axle control. Here, select-low refers to the fact that the rear wheel operation on the low-coefficient-of-friction road surface controls the modulation of both rear wheels. Some performance degradation from optimum braking on the rear axle occurs when operating on split-coefficient surfaces. This is because the traction force on the higher friction surface is not fully utilized, resulting in a lower brake torque and, thus, longer stopping distance. The advantage is a high side force potential and the absence of a yaw moment. Braking in a turn, as well as straight-line braking performance expressed as vehicle deceleration, approximates that of four-wheel control systems when operating on typical highways including wet and dry surfaces.

A further decrease in costs is obtained through systems modulating only the rear axle, either each wheel independently, or the rear axle by sensing the propeller shaft angular velocity. In a panic brake application or while operating on slippery road surfaces, the front wheels can lock, thus rendering the vehicle unsteerable. Although this provides a stable stop, the accident avoidance characteristics of a vehicle equipped with such a system are not much better than a properly engineered standard brake system when braking on a slippery road while turning. Investigations of accident studies indicate that no significant safety benefits may be expected with stable yet nonsteerable rear ABS vehicles in certain types of accidents, particularly in intersection collisions.

The design of the basic or standard brake system should be such that a stable stop results with the ABS system failed. Consequently, a properly engineered brake force distribution including the use of proportioning valves must be considered to ensure front-before-rear lockup.

## 9.3.2 Vacuum-Powered Systems

Vacuum-powered ABS systems are generally not used on modern vehicles. A detailed evaluation of experimental data obtained with earlier four-wheel ABS systems powered by vacuum revealed no shorter or even longer stopping distances when compared with stops involving some or all wheels locked. Tire-road friction utilizations as low as 60% achieved on dry roads appeared to be unnecessarily low. The major reason for this shortcoming resulted from the significant time delay associated with the rise of vehicle deceleration during ABS operation as compared with the standard brakes.

In 1983 and later, Chrysler and Mitsubishi used vacuum as the power source in some rear-wheel-drive imports equipped with rear ABS. With the front brakes locked the ABS system is deactivated. A pulse generator located at the exit port of the transmission measures the speed of the rear wheels. A decelerometer or G sensor installed on the floor of the luggage compartment measures vehicle deceleration and, hence, the reduction of vehicle speed during braking. The electronic control unit receives signals from the stoplight switch, the pulse generator, and the G sensor. With impending rear wheel skid, the modulator receives appropriate signals from the control unit to modulate rear brake line pressure. It consists of a pressure control section, a vacuum-powered pressure drive unit, and a solenoid valve.

### 9.3.3 Separate ABS Systems (Add-On)

Since the reintroduction of ABS systems between the late '70s and mid-'80s, both separate and integrated systems have been used.

In 1978 the separate ABS 2S design manufactured by Bosch was the first anti-lock brake system to go into mass production with Mercedes Benz automobiles. Other manufacturers followed in the mid-'80s. Developments in digital electronics made it possible to reliably monitor and process the complex performance requirements associated with a braked wheel. The design is very flexible, permitting the addition of the ABS components without modification to the basic hydraulic brake system, as illustrated in Figure 9-13. It operates as follows: During driving, wheel sensors at each front wheel and at the rear axle differential (or at each rear wheel) measure wheel speed. The wheel speed signals are processed by the electronic control unit (ECU), and if a wheel skid is recognized, a solenoid valve is energized in the hydraulic modulator. Each front wheel is controlled by an individual solenoid valve. At the rear axle the wheel with the lower tire-road traction determines the hydraulic brake line pressure modulating both rear brakes (select-low control). In the case of the standard front-to-rear split, a single solenoid valve is used for rear brake control while two solenoid valves are required for a diagonal split. The select-low control employed on the rear axle results in slightly longer stopping distances since the maximum traction is not fully utilized by one rear wheel; however, increased directional stability is achieved.

In the separate ABS design the modulating pressure cannot exceed the brake line pressure produced by the driver's pedal force in the master cylinder. The ECU switches the solenoid valves into three different positions, depending on the wheel speed signal processing results (Fig. 9-13).

The first deenergized condition connects the master cylinder to the wheel cylinders of the individual brakes for standard brake operation; the hydraulic brake line pressure increases as the pedal force produced by the driver increases. In the second position the solenoid valve is excited with half the maximum current resulting in separation of the wheel cylinders from the master cylinder and the return line causing the brake line pressure to remain constant. In the third position the solenoid valve is excited with maximum current, which isolates the master cylinder and simultaneously connects the wheel cylinders with the return line and thus accumulator. In the third position

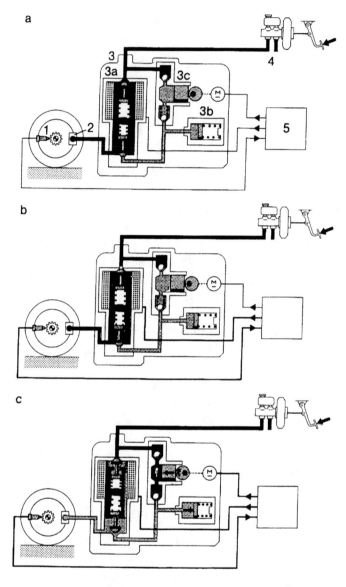

| 1 | Wheel speed sensor | a) | Pressure build-up |
| 2 | Wheel brake cylinder | b) | Pressure holding |
| 3 | Hydraulic modulator | c) | Pressure reduction |
| 3a | Solenoid valve | | |
| 3b | Accumulator | | |
| 3c | Return pump | | |
| 4 | Master cylinder | | |
| 5 | ECU | | |

*Figure 9-13. Brake pressure modulation (Bosch).*

the brake line pressure decreases resulting in a drop of wheel brake torque. The three-position valve control permits pressure increase and decrease in stages by cyclical actuation. Depending on the traction and braking conditions, cycle rates may range from 4 to 10 or more cycles per second.

Incorporated in the ECU are safety checks which detect any electrical failure of the ABS system. At the beginning of a trip, and each time the vehicle has stopped, the controller, safety circuits, and associated equipment are checked for proper performance. In particular, components are checked which are not active when braking without ABS and whose failure would be noticeable only when ABS braking is required. If any fault is recognized by the checking process, the ABS system is switched off, a safety lamp lights up on the dashboard, and the brake system returns to standard braking. Hydraulic leaks or other mechanical failures are not detected by the safety check.

ABS systems used for diagonal split braking designs in many cases are based on the same modular system of the front-to-rear split design. The Bosch ABS 2E diagonal split design illustrated in Figure 9-14 operates as follows: During normal non-ABS braking, brake fluid flows through the rear-axle solenoid valve into the right rear wheel brake marked HR and through the central valve to the left rear wheel brake marked HL. In the case of ABS operation the two left-hand solenoid valves each control one front wheel brake while the rear solenoid acts directly on the right rear wheel; the rear-axle solenoid valve acts directly on the right rear wheel. If the ECU provides the proper signal, the rear-axle solenoid switches to the pressure holding mode.

If the pressure in the lower float-piston chamber connected directly with the brake master cylinder increases with respect to the pressure in the upper float-piston chamber connected with the rear-axle solenoid valve, an imbalance is produced which moves the float piston upward and closes the central valve. If the pressure in the right rear wheel brake (HR) is reduced, the float piston continues to move upward until a state of force equilibrium is achieved, i.e., until the brake line pressures in the rear brakes are nearly identical. This design permits the select-low control of the rear brakes by use of one solenoid valve only.

Add-on ABS system designs have proven to be reliable. They operate with vacuum or hydraulic booster, and are simply installed between the master cylinder and wheel brakes in an existing brake system. Brake force reduction

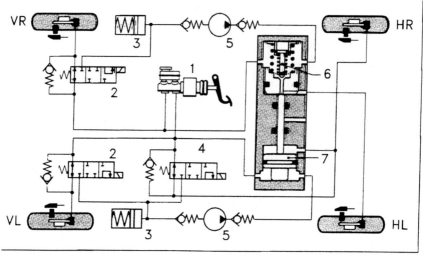

Brake booster      V   Front
Solenoid valves      H   Rear
Accumulator chambers      R   Right
Rear-axle solenoid valve      L   Left
Dual-circuit return pump
Central valve
Float piston

*Figure 9-14. Hydraulic system ABS 2E for diagonal brake circuit
configuration (Bosch).*

is achieved through direct drainage of brake fluid from the wheel cylinder
circuit to be modulated into a reservoir of limited volume. In each brake
circuit a pump driven by a common electric motor returns the brake fluid from
the reservoir to the master cylinder against the effective actuation force of
the driver's foot. This results in brake pedal pulsation during ABS operation,
limits the maximum hydraulic brake line pressure based on the driver's pedal
force, and necessitates the employment of central valves instead of vent ports.

Consumer cost of a four-wheel ABS system may easily range from $800 to
$1300 depending on the type of passenger car involved. Most vehicle and
brake manufacturers have recognized the fact that the high cost of ABS
systems will prevent their introduction in a wide range of low- to medium-cost
vehicles.

Two-channel ABS systems have been designed to reduce cost for use on diagonal split brake systems. The hydraulic modulator has only two solenoid valves, one for each circuit controlled by the front wheels. The diagonally opposite rear wheel is braked with a given delay through a standard proportioning valve. Pressure modulation always acts simultaneously on one front and one rear wheel. The disadvantage of this system still includes the rare potential for premature rear brake lockup as, for example, in the case of extreme high traction on the front wheel and reduced traction on the rear wheels caused by the use of snow tires on the front wheels and excessive tread wear on the rear tires.

GM's Delco Moraine Division has designed a cost-effective ABS system called ABS-VI. ABS-VI is an add-on or separate system, using a conventional master cylinder with vacuum booster. The common ABS solenoid valves, used by other manufacturers similar to the one described in the preceding paragraphs, are replaced by a screw-actuated plunger system referred to as *electromagnetic braking*. The ABS-VI modulator consists of three small screw-operated plungers driven by electric motors. During normal non-ABS braking the plungers are up, with the check valve open, allowing brake fluid to flow from the master cylinder to the wheel brakes. During ABS operation the solenoid closes thereby isolating the particular circuit, and the plunger is turned down to increase volume and reduce brake line pressure to keep the wheel from locking. Depending on conditions, the plunger can be moved up and down to cycle brake line pressure up to seven times per second, which is about half of some of the pump-accumulator ABS systems. The plunger design results in a smooth and gradual pressure modulation, which significantly reduces pedal feedback to the driver. As with other separate designs, the modulation pressure cannot exceed the pressure in the master cylinder. Although the ABS-VI may not perform equally well in all respects when compared with expensive systems, directional stability and steering ability is achieved for all foreseeable operating and road surface conditions. Of course, the significant selling feature is its low cost of approximately $350 per system to the consumer. When compared with the cost range around $1000 for many other systems, GM's new ABS-VI is expected to be a popular safety feature for all vehicles including the price-sensitive small cars.

A system similar to the ABS-VI was developed by FAG Kugelfischer for motorcycles in 1985. In that system the plunger is actuated by a linear electric motor.

### 9.3.4 Integrated ABS Systems

In the integrated ABS design the hydraulic brake booster and the ABS valve block form a closed unit as illustrated in Figure 9-15. Advantages include compact design requiring little space and the ability to select and optimize the performance characteristics of the booster for the particular application at hand (Ref. 33). Unlike conventional brake boosters, the brake master cylinder pistons are decoupled from the brake pedal. This makes it possible to size the diameters of the master cylinder pistons so that, in the event of a pressure supply failure, higher brake line pressures and, hence, greater decelerations can be achieved with normal pedal forces. If a brake circuit fails, the counter-pressure at the brake pedal remains stable as a result of the decoupling process. With conventional brake boosters, a brake line failure may cause sagging of the brake pedal due to the absence of some of the resistance force against the brake pedal.

Teves, in 1985, was one of the first companies to introduce a compact integrated ABS system in 1985 on some Ford passenger vehicles. The four-wheel ABS system uses hydraulic brake fluid for the braking function of the

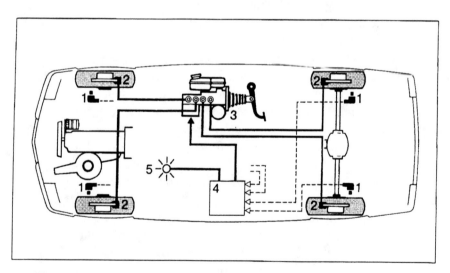

1   Wheel speed sensor
2   Wheel brake cylinders
3   Hydraulic modulator with master cylinder
4   ECU
5   Safety lamp

*Figure 9-15. Car with ABS 3 (integrated ABS) (Bosch).*

wheel brakes and the hydraulic booster. Major system components are master cylinder, hydraulic booster, electric pump, accumulator, electronic controller, reservoir, relays, wheel speed sensors, and warning lights.

The brake booster is located behind the master cylinder in basically a conventional arrangement. The booster control valve is located in a parallel bore above the master cylinder and is operated by a lever mechanism connected to the pushrod of the brake pedal.

The high-pressure electric pump runs at intervals for short periods of time to charge the high-pressure accumulator which supplies the service brakes. The accumulator is a gas-filled pressure chamber mounted to the master cylinder/ booster assembly along with the electric motor and hydraulic pump.

Three solenoid valves are used to operate the ABS system. The hydraulic pump maintains pressure between 1400 and 1800 $N/cm^2$ (2030 and 2610 psi). When the brakes are applied, a scissor-lever arm activates the control valve. Hydraulic pressure, proportional to the pedal travel, enters the booster chamber. This pressure is transmitted through the normally open rear brake solenoid valve to the rear brakes. The same hydraulic pressure moves the booster piston against the master cylinder piston, which shuts off the central valve in the master cylinder. The result is brake line pressure application to the front brakes through the normally open solenoid valves.

During ABS operation the electronic control unit causes the solenoid valves to close and open to properly modulate the respective brake line pressure to prevent wheel lock. Integrated ABS systems afford—among other things— simple installation with small space requirements by replacing the master cylinder/vacuum booster assembly. They employ external lines only leading to the wheel cylinders, provide the shortest possible response time of the hydraulic booster due to the use of high-pressure systems, have monofluidic design, and a single-circuit pump driven by an electric motor independent of engine speed. Neither separate nor integrated fully hydraulic ABS systems can be rated superior in all respects. However, the basic advantages of integrated ABS systems include simple mounting, free choice in the design of pedal characteristics, lower noise potential, and no sagging of pedal height in the event of a hydraulic circuit failure. Major disadvantages of integrated ABS are complex hydraulic modulation and expensive repairs.

Currently three major ABS manufacturers provide four-wheel ABS systems for the U.S. Market. They are Bosch, ITT, and Kelsey Hayes. Bosch and ITT have an excellent safety record, while Kelsey Hayes had difficulties when their four-wheel ABS system was introduced in the early '90s. Their problems were mostly related to manufacturing abnormalities, including system contamination. With a change to new production facilities these problems appear to have been eliminated.

## 9.4 ABS System Components

### 9.4.1 Wheel Speed Sensors

The wheel speed sensors signal the wheel speed to the ECU. The pole pin (5) shown in Figure 9-16 is surrounded by a winding (4) and is located directly over the sensor ring (6), a gear wheel attached to the wheel hub or differential. The pole is connected to a permanent magnet (2), the magnetic field of which extends into the sensor ring. When the ring rotates, the pole is faced alternately by a tooth and by a tooth gap. Consequently, the magnetic field changes repeatedly and induces an AC voltage in the winding which is tapped off at the winding ends. The frequency of the voltage serves as a signal for the

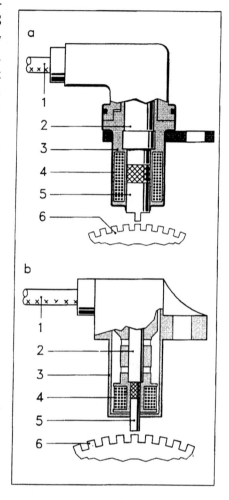

a) Wheel speed sensor DF2
   with chisel-type pole pin
b) Wheel speed sensor DF3
   with round pole pin
1  Electric cable
2  Permanent magnet
3  Housing
4  Winding
5  Pole pin
6  Sensor wheel

*Figure 9-16. Typical wheel speed sensor (section) (Bosch).*

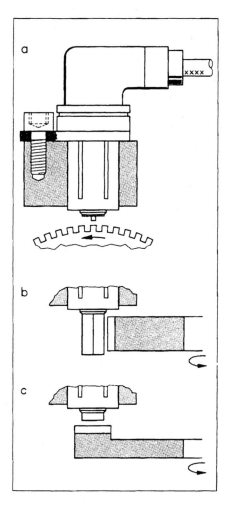

a) Radial installation, radial pick-off with chisel-type pole pin
b) Axial installation, radial pick-off with rhombus-type pole pin
c) Radial installation, axial pick-off with round pole pin

*Figure 9-17. Types of installation and pole pin shapes for wheel speed sensors (Bosch).*

wheel speed. Different pole geometries are illustrated in Figure 9-17. The pole pin must be aligned to the pulse ring. To ensure a faultless signal, wheel speed sensor and sensor ring are separated from each other only by a closely controlled air gap of approximately 1 mm (0.004 in.). Vibration and deflections must be kept at a minimum. Wheel speed sensors are greased before installation to minimize adverse effects from water splash and dirt.

The sensor system described is a passive or variable reluctance system. The output of the system is a function of the speed of the sensor ring. The signal amplitude is very low at low impulse ring speed, e.g., at 3 to 5 km/h (2 to 3 mph). The development of active electronic sensor components powered by the electrical system of the vehicle has greatly expanded the options for sensor design. "Zero" wheel speed is critical for other functions such as traction control, speedometer, and odometer reading.

## 9.4.2 Electronic Control Unit (ECU)

The ECU receives, amplifies and filters sensor signals, as well as measures and differentiates speeds. The input data are used by the ECU to compute circumferential wheel acceleration and braking slip.

The ECU for the ABS 2S Bosch system is illustrated in Figure 9-18. Only 60 components are installed on the board of 140 mm size. The digital controller consisting of two LSI integrated circuits combines 16,000 transistor functions on a chip area of approximately 37 mm² (0.057 in.²). Discrete semiconductor components for filtering, level adaptation, clock generation, and interference suppression, and power transistors for solenoid valve control supplement these components.

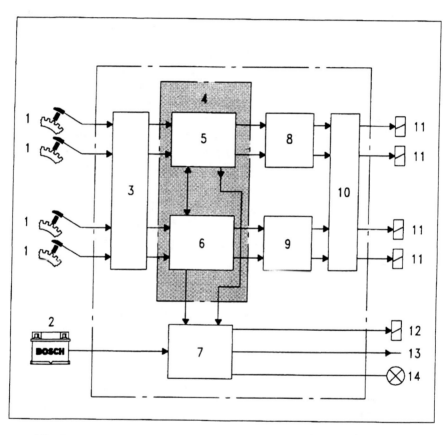

| 1 | Wheel speed sensor (wheel frequencies) | 8 | Output circuit 1 |
|---|---|---|---|
| 2 | Battery | 9 | Output circuit 2 |
| 3 | Input circuit | 10 | Output stage |
| 4 | Digital controller | 11 | Solenoid valves |
| 5 | LSI 1 | 12 | Safety relay |
| 6 | LSI 2 | 13 | Stabilized battery voltage |
| 7 | Voltage stabilizer/fault memory | 14 | Safety lamp |

*Figure 9-18. Control unit for ABS 2S (Bosch).*

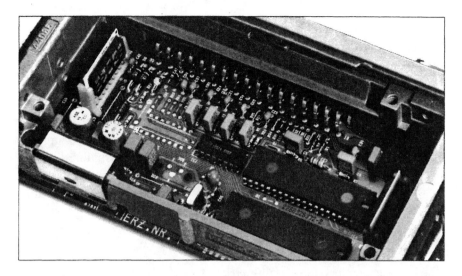

*Figure 9-19. Control unit for ABS 2S (open) for separate installation (Bosch).*

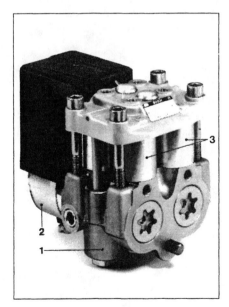

| 1 | Accumulator |
| 2 | Return pump |
| 3 | Solenoid valve |

*Figure 9-20. ABS 2S hydraulic modulator (Bosch).*

The ECU should be installed in a location where high temperatures and splash water are avoided. The basic block diagram of the ECU is shown in Figure 9-19.

### 9.4.3 Hydraulic Modulator

The hydraulic modulator converts the ECU commands for pressure modulation in the wheel brakes by use of the solenoid valves. It acts as the hydraulic link between the master cylinder or pressure accumulator and the wheel cylinders of the individual wheel brakes.

The hydraulic modulator used for the Bosch ABS 2S is illustrated in Figure 9-20. The return pump is used to pump the brake fluid coming from the wheel cylinders

459

during pressure reduction to the master cylinder via the appropriate accumulators. The accumulators temporarily store the brake fluid which is suddenly pumped back as a result of a pressure reduction. The solenoid valves modulate the pressure in the wheel brakes during ABS operation. The basic design of the solenoid valve used in the Bosch ABS 2S system is illustrated in Figure 9-21. Special design features employed to ensure high functional reliability and minimum friction involve guiding the moving armature (6) by nonmagnetic bearing rings (3). Steel balls soldered together with carrier plates (11) act as sealing elements. The steel balls are very small. Their seats are hardened and machined to extremely close tolerances to ensure proper sealing even at pressures as high as 2068 N/cm$^2$ (3000 psi).

## 9.4.4 Electric Circuit

The electrical connections between the individual electrically actuated components are combined in a cable harness designed and routed so that water entry is avoided. The ECU checks the voltage drop across certain critical electric lines and, when found faulty, switches the ABS system off.

## 9.4.5 Drive Train Influence on ABS

A free-rolling wheel provides the best prerequisites for ABS control. Any additional rotational masses will adversely affect speed sensing and data analysis. Ideally, a driver would place a manual transmission into neutral prior to braking. In emergencies, this will not be the case. Automatic transmissions with torque converters are less sensitive to vehicle inertias since they transmit only a small portion of engine drag.

Rear-wheel-drive vehicles have free-rolling front wheels providing optimum ABS control and generation of a reference speed. The additional rear drive inertias are of little consequence, except on slippery roads.

Front-wheel-drive vehicles present less favorable ABS control conditions due to the rotational inertias affecting the front wheels. Two rear-wheel speed sensors are required because the rear wheels are not connected. Rear suspension effects must be minimized through filtering of the speed signals.

Four-wheel-drive vehicles must disconnect front-to-rear coupling as well as differential locks to ensure proper ABS signal analysis and directional stability while braking. Viscous couplings may be stiff enough to adversely affect

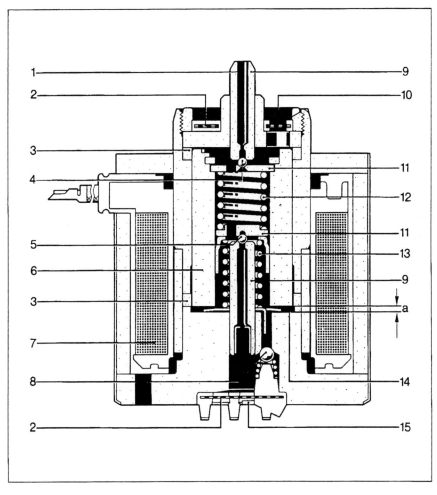

| | | | |
|---|---|---|---|
| 1 | To the return line | 10 | To the wheel brake cylinder |
| 2 | Filter | 11 | Carrier plate |
| 3 | Non-magnetic bearing ring | 12 | Auxiliary spring |
| 4 | Discharge valve | 13 | Main spring |
| 5 | Intake valve | 14 | Recess step |
| 6 | Armature | 15 | From the brake master cylinder |
| 7 | Winding | a | Working air gap |
| 8 | Check valve | | |
| 9 | Valve body | | |

*Figure 9-21. Basic design of ABS 2S solenoid valve (Bosch).*

ABS performance. Provisions should be made in the gear system to automatically engage a "neutral" gear when the brake light is energized by brake pedal movement. The generation of a proper reference speed is made difficult when only one axle is driven, while the second axle is engaged only under adverse road conditions.

## 9.5 ABS Systems for Air Brakes

ABS systems for air brakes use concepts similar to those found in hydraulic brake systems. Major components are: wheel speed sensors, usually one for each wheel on the axle; an electronic control unit, which collects the sensor information, processes it, and sends control signals to the air pressure modulation valve; and an air pressure modulation valve, which accomplishes the air pressure modulating function by the use of electrical solenoids. The basic control process is illustrated in Figure 9-22.

### 9.5.1 Control Analysis

Some details associated with the analysis of the ABS control of an air brake system and, in particular, the brake line pressure modulation as a function of time are reviewed in the paragraphs that follow.

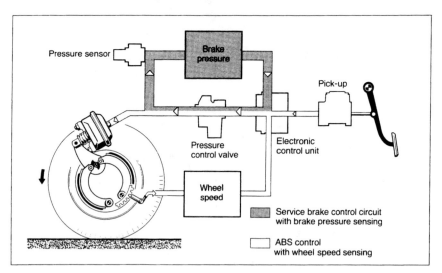

*Figure 9-22. Electronically controlled commercial vehicle braking system (ELB).*

The time-dependent behavior of vehicle speed V and tire circumferential speed $R\omega$ are illustrated schematically for an air brake system in Figure 9-23(A). The angular velocity of the wheel is designated by $\omega$, the tire radius by R measured in ft. Shown in Fig. 9-23(B) are the characteristics of brake line pressure $p_\ell$ as a function of time during the ABS control operation.

As the angular velocity and thus tire circumferential speed $R\omega$ begins to decrease more than the vehicle forward speed V, and reaches a point corresponding to the design threshold angular deceleration $\alpha_p$, the brake line pressure modulating valve receives a signal at time $t_1$ to reduce the pressure. After the response time $\tau_1$ has elapsed, the pressure begins to decrease at time $t_2$ according to the functional relationship

$$p_\ell = p_{max}e^{-c_1(t-t_2)} \quad , \quad N/cm^2 \ (psi) \tag{9-7}$$

where $c$ = time constant indicating pressure decrease characteristics, $s^{-1}$

$p_{max}$ = maximum brake line pressure, $N/cm^2$ (psi)

$t$ = time, s

$t_2$ = time at which maximum brake line pressure is reached and pressure decrease begins, s

The decrease in brake line pressure causes the angular velocity of the wheel to increase again (Fig. 9-23[A]). Parallel to this process, an angular velocity $\omega_c$ is computed from a specified angular deceleration $\alpha_r$ and from the angular velocity $\omega_p$ of the wheel at the instant the threshold value $\alpha_p$ was exceeded as

$$\omega_c = \omega_p + \alpha_r(t - t_1) \quad , \quad rad/s \tag{9-8}$$

where $t_1$ = time at which ABS signal is received by the brake pressure modulator valve, s

$\alpha_r$ = specified angular deceleration, rad/s

$\omega_p$ = wheel angular velocity at peak friction value, rad/s

$\omega_p$ is a function of the ABS stop and not a function of vehicle speed as illustrated in Fig. 9-23(A).

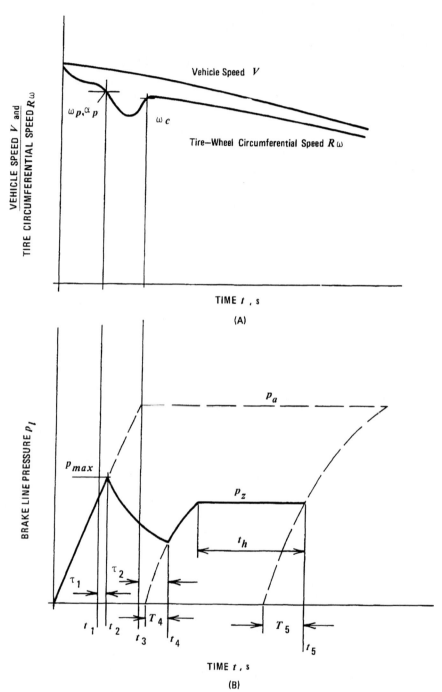

Figure 9-23. Wheel anti-lock control for an air brake system.

When the actual angular speed of the wheel has attained the computed value $\omega_c$ at time $t_3$, the brake line pressure modulating valve receives the signal to increase pressure again. After the response time $\tau_2$ has elapsed, the brake line pressure begins to increase at time $t_4$ according to

$$p_\ell = p_a[1 - e^{-c_2(t-t_4+T_4)}] \quad , \quad N/cm^2 \text{ (psi)} \tag{9-9}$$

where $c_2$ = time constant indicating pressure increase characteristics

$p_a$ = applied pressure, $N/cm^2$ (psi)

$t_4$ = time at which pressure increase begins, s

$T_4$ = difference between time $t_4$ and time associated with pressure increase from zero pressure, s

The brake line pressure $p_\ell$ is increased only to a pressure $p_z$ and always remains below the applied pressure $p_a$. The brake line pressure $p_z$ is generally somewhat smaller than the pressure that causes lockup to occur. However, if the wheel speed tends to lock up again at a pressure equal to or lower than $p_z$, the previous pressure decreasing and increasing processes are repeated until a pressure $p_z$ is produced which does not cause wheel lockup to occur. This brake line pressure is kept constant until the hold time $t_h$ has elapsed, after which the pressure is increased again toward the applied pressure $p_a$ to allow the ABS system to adjust the braking effort to a different tire-road friction situation that might have developed during the hold time $t_h$. The pressure increases toward $p_a$ according to the approximate relationship

$$p_\ell = p_a[1 - e^{-c_3(t-t_5+T_5)}] \quad , \quad N/cm^2 \text{ (psi)} \tag{9-10}$$

where $c_3$ = time constant indicating pressure increase characteristics, s

$t_5$ = time at which pressure increase begins, s

$T_5$ = difference between time $t_5$ and time associated with a pressure increase from zero pressure, s

The use of an adjustable "hold time" $t_h$ has the advantage that the control frequency is adjustable. This allows the prevention of control frequencies near the natural frequencies of suspension and steering components which otherwise may cause undesirable vibrations and damage to suspension components. The air consumption also may be kept low as a result of an adjustable hold time.

### 9.5.2 ABS Air Brake Designs

In Europe, the most commonly encountered ABS system is the four-sensor/four-channel system (4S/4C) as illustrated for a truck or tractor in Figure 9-24. In that system, four wheels are monitored and four channels are individually controlled. The ABS part of the system consists of the wheel speed sensors, the ECU, and the pressure control valves. Pressure control valves can be designed to provide only one modulated pressure output (single-channel), or can be designed to integrate two pressure control valves into one unit (two-channel). The latter reduces cost but requires a larger space for installation.

A single-axle ABS system is illustrated in Figure 9-25. This 2S/2C system was of particular interest to U.S. manufacturers in the past. Due to the low braking effort typically produced by front brakes on tractors in the U.S., this system effectively eliminated the potential for tractor rear axle lockup and, hence, jackknifing.

For commercial vehicles each wheel usually has a speed sensor. The larger rotational masses and air volume result in cycling frequencies between 2 and 3 Hz. On split-coefficient-of-friction road surfaces, and due to the larger scrub radius, large yaw moments will result. Select-low control is generally not used on commercial vehicles due to the significant increase in stopping distance. To reduce adverse effects on steering control when braking on split surfaces, most ABS systems use a modified individual control on the front axle. A pulsed increase in brake torque on the high-traction wheel results in limiting differences in air pressure left and right, which minimizes the yaw and steering moments. The advantages of stability outweigh the increase in stopping distance resulting from the high-traction wheel not operating at its maximum brake force possible.

## 9.6 In-Use Factors and Operation of ABS Systems

The installation of ABS braking systems in passenger cars and light trucks makes it necessary for certain in-use practices to be considered.

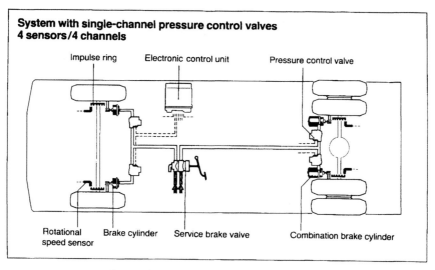

*Figure 9-24. ABS for commercial vehicles*

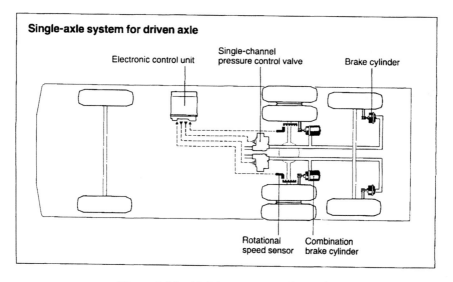

*Figure 9-25. ABS for commercial vehicles.*

### 9.6.1 Brake Fluid

A high-quality, clean, glycol-ether-based brake fluid should be used. Follow the instruction of the manufacturer concerning whether to use DOT 3 or DOT 4. If DOT 3 is required, DOT 4 may be used. Silicone fluid as discussed in Chapter 10 should not be exchanged with regular brake fluid.

### 9.6.2 Tire Size

Changing tire size from the size recommended by the manufacturer may affect the performance of the ABS system. The electronic computer is programmed to analyze the impulses received from the wheel speed sensor and, in conjunction with a given tire size, convert them into speed quantities. If a tire size deviates by more than five percent from the "design" tire size, some ABS systems are programmed to recognize a problem and shut the ABS system down.

Not all ABS systems respond in the same way. Some may store the trouble code for several starts, some have to be reset, and others will operate only after the computer memory has been blanked by removing the fuse. Maintenance manuals provide instruction on how to proceed.

### 9.6.3 Electric Trailer Brakes

Standard electric trailer brakes are actuated by a hydraulic cylinder inserted into the hydraulic braking system of the tow vehicle. Because the actuation system requires brake fluid volume for its performance, care must be taken not to exceed the basic fluid volume capacity of the brake master cylinder of the tow vehicle.

In the case of rear ABS only, found in many pickup trucks, the hydroelectric controller can be inserted into the front brake circuit or into the rear brake line between the master cylinder and the ABS modulator.

Hydroelectric controllers should not be installed in any hydraulic line carrying ABS modulated brake fluid. Consequently, they must not be installed in four-wheel ABS systems. If a trailer with electric brake is required, use a deceleration pendulum-activated trailer brake system. Surge brakes can likewise be used on any type of ABS system.

## 9.6.4 Brake Fluid Level

When checking brake fluid level, the maintenance or owner's manual should always be consulted. Some ABS systems require the accumulator to be depressurized; others require the ignition to be on until the pump has pressurized the system.

# CHAPTER 10

# Analysis of
# Brake Failure

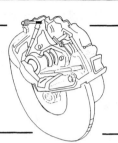

*In this chapter basic relationships presented in previous chapters are applied to the analysis of braking with the system in a partially failed condition. Brake circuit and power boost failure are investigated. Increased pedal travel for various types of brake failure are analyzed. The influence of brake fluid on brake design and brake failure is discussed.*

## 10.1   Basic Considerations

Brake systems for modern motor vehicles are designed to perform safely for a wide range of normal situations and under a variety of extreme conditions. A review of a large number of multidisciplinary accident investigation studies as well as others reveals that brake malfunctioning was noted as an accident causation factor in slightly less than 2% of all accidents (Ref. 1). The brake malfunctioning involved failures of brake lines, wheel cylinders, brake hoses, lining attachment, and lining mismatch. In isolated cases, owners had driven their vehicles without adequate brake maintenance so that ventilated front brake rotors had worn to the cooling vanes, or the swept surface of brake drums had separated from the hub. A review of the individual case reports showed that most brake malfunctionings were caused by faulty maintenance and repair, or lack of maintenance efforts by the owners of the vehicles. Of the 2% brake-related accidents, 89% were associated with brake malfunctioning and 11% with brake imbalance (left-to-right and front-to-rear).

The nature of brake failures for vehicles equipped with ABS brakes are not fully understood at the present time. Driver factors and accident reconstruction limitations must be considered. Drivers frequently concluded that their brakes had failed when the typical ABS hydraulic module noise and brake pedal response were observed. In general, however, most driver complaints can be grouped into extended stopping distance, and soft or pedal to the floor conditions. Normally driver complaints to NHTSA relative to ABS problems number less than 100 to 500 for any one vehicle/brake manufacturer. According to *USA Today*, in case of the Kelsey-Hayes four-wheel ABS (4WAL) used by GM, more than 5000 complaints were collected by 1995, approximately three years after the ABS brake system was placed into production. Many of the accidents reported involved low-speed rear-end collisions. If an ABS brake failure is involved resulting in increased pedal travel, the response time of the brake system may be increased by as much as 0.5 to 0.7 seconds, causing the vehicle to travel a significantly greater distance without braking. In low-speed rear-end crashes this condition may be the brake system induced accident causation factor.

For commercial vehicles equipped with air brakes, the major problems stem from brakes being out of proper adjustment as far as S-cam brakes are concerned. This unsafe condition reveals itself frequently when a runaway truck accident occurs on a downgrade. The brake adjustment may be sufficient for many "normal" brake applications not involving excessive drum temperatures. However, under severe operating conditions, such as on a downgrade, the brake adjustment is such that no adequate brake shoe application force against the drum can be sustained.

Considering the importance of brake adjustment, wheel brakes for hydraulic and air brake systems should be automatically adjusting, and in the case of air brakes, brake adjustment indicators should be provided.

## 10.2 Development of Brake Failure

Failure of braking system components under ordinary driving conditions is likely to occur only if:

1. Parts are defective.

2. Parts become severely worn.

3. Parts become degraded.

4. Parts become mechanically damaged.

Any one of the four conditions indicated can be the result of a design defect, a manufacturing defect, operator abuse, or improper maintenance or repair.

A part becomes degraded (e.g., through oil contamination) as a result of a design or manufacturing defect (e.g., leaking seal) or defective maintenance (e.g., improper repair of a seal, depositing grease on the caliper contaminating the pads, etc.).

A part becomes severely worn through long-time use, abuse, improper installation or wrong part, or other reasons.

A part is defective when it is designed, manufactured, or maintained defectively. Brake design defects could involve one component or the entire system in terms of lack of braking stability caused by premature rear brake lockup, front rotor shudder or vibration causing excessive pad/wheel cylinder piston push back, and other reasons.

Mechanical damage such as a bleeder valve fracture or brake hose leakage caused by road debris can occur suddenly and without warning.

Brake failures generally can be grouped into failures causing (a) insufficient brake force, i.e., stopping distance that is significantly increased over that obtained with normal brakes; (b) excessive component wear; and (c) inconvenience to the driver.

# 10.3  Analysis of Partial Brake Failure

## 10.3.1  Basic Considerations

A partial brake failure exists when certain parts of a brake system fail so that the remaining system can produce only a reduced braking force. The major partial brake failures involve loss of one hydraulic circuit or the loss of power assist. Of lesser significance for passenger cars are brake failures due to excessive brake temperature, simply because they occur only under rare circumstances such as the parking brake remaining partially applied while driving or brake pads not properly returning. Insufficient pad return may be caused by increased friction on disc brake caliper guide pins, or weak or broken return springs in the case of drum brakes.

The purpose of a brake failure analysis is to determine how the intended design effectiveness or deceleration capability of the brake system, i.e., the deceleration/pedal force relationship, is altered if a partial failure should occur within the system.

In the brake failure analysis, some or all of the following questions should be answered:

1. Did the brake failure decrease braking effectiveness?

2. Did the brake failure develop suddenly or slowly over time?

3. Did the brake failure cause the vehicle to lose directional control such as leaving its lane of travel?

4. Did the brake failure cause an increase in pedal travel and/or pedal force requirements?

5. Was the brake failure intermittent, i.e., did it exist at the time of the accident but could not be duplicated after the accident?

6. Were there any physical indications that would allow the driver to notice the brake failure prior to the accident?

7. Is the secondary, emergency or parking brake effectiveness sufficient to avoid the accident, assuming the driver actuated the parking brake?

## 10.3.2 Brake Line Failure

Any brake system contains the mechanisms for pedal force application, pedal force or brake line pressure transmission, and brake force production in the wheel brakes. The mechanism for application of pedal force $F_p$ involves a pedal lever ratio $\ell_p$ such that a pedal effort $F_p\ell_p$ is applied to the master cylinder pushrod. The pedal force transmission involves a dual-circuit master cylinder and the hydraulic brake lines between the master cylinder and wheel brakes. Connected into the brake lines can be special devices such as metering or proportioning valves. The wheel brakes may be divided into those having one or two actuating mechanisms or wheel cylinders. The first category includes leading-trailing and duo-servo-type drum brakes, and single-caliper-type disc brakes with two opposing wheel cylinders. In the event of a circuit failure in a braking system using single actuation mechanisms, no braking action can be developed by this brake since none of the "wetted" surfaces in the failed circuit can be pressurized. Brakes involving two actuation mechanisms or two wheel cylinders per wheel brake may be connected to separate circuits. In the case of one circuit failure the wheel brakes produce a reduced brake force, in most cases a braking action equal to 50% of the non-failed system. Since the two wheel cylinders may be installed at the top and bottom of the drum brake, such a hydraulic split is sometimes referred to as a horizontal split.

The six basic possibilities for installing brake lines between a tandem or dual master cylinder and the wheel brakes to form two independent brake line circuits are shown in Figure 10-1. System 1 shows the standard front-to-rear split generally used in rear-wheel-driven cars. Most large cars of U.S.

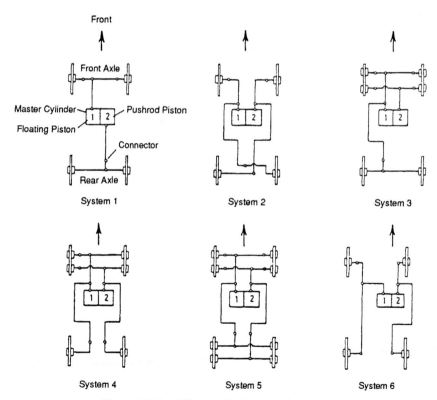

*Figure 10-1. Different dual-circuit brake systems.*

manufacture, as well as others including Mercedes Benz, use this system. Systems 2, 4, 5, and 6 develop equal braking forces for each circuit. In the case of systems 1 and 3, a failure of circuit 1 or 2 will result in different braking forces. The braking forces achievable with systems 2 and 6 are identical in the failed mode for either circuit. The effects on vehicle stability while braking under partial failure mode, i.e., a failure of circuit 1 or 2, will be different, with system 6 showing an undesirable side-to-side unbalance. Since it is obvious that a dual system of type 6 is undesirable, it is not included in any further analysis.

Three measures of braking performance for a dual brake system may be identified.

1. Reduced braking force of the vehicle in the partial failure mode due to a decreased brake system gain between the master cylinder exit and the wheel brakes.

2. Changes in brake force distribution front-to-rear and, hence, reduced braking efficiency, i.e., decreased wheels-unlocked deceleration, or different braking forces left-to-right and, hence, potential for lateral path deviation.

3. Increased pedal application time due to longer brake pedal travel.

All three measures will cause an increase in stopping distance.

## 10.3.2.a Vehicle Deceleration

The wheels-unlocked deceleration of a vehicle with non-failed brakes may be computed by Eq. (5-3). The deceleration a achievable with a complete brake system may be rewritten as

$$a = \frac{1}{W} \sum_{}^{n} [(p_\ell - p_o)A_{wc}BF\eta_c(r / R)]_i \quad , \quad \text{g-units} \qquad (10\text{-}1)$$

where   $A_{wc}$ = wheel cylinder area, cm² (in.²)

BF = brake factor, defined as ratio of brake drag to actuating force of one shoe

i = identifies location of wheel, i.e., front or rear, left or right n = number of wheels braked

$p_\ell$ = brake line pressure, N/cm$^2$ (psi)

$p_o$ = pushout pressure, N/cm$^2$ (psi)

r = effective drum or disc radius, cm (in.)

R = effective tire radius, cm (in.)

W = vehicle weight, N (lb)

$\eta_c$ = wheel cylinder efficiency

Eq. (10-1) may be used to compute the vehicle deceleration with a partially failed system by summing only the brake forces of the wheels not affected by a system failure. In the use of Eq. (10-1) it should be remembered that the brake line pressures front and rear are generally the same in the case of a partial failure since any proportioning valve function is eliminated.

## 10.3.2.b  Pedal Force

The pedal force $F_p$ is computed by Eqs. (5-1) or (5-5), solved for pedal force

$$F_p = p_\ell A_{MC} B / \ell_p \eta_p \quad , \quad N \text{ (lb)} \tag{10-2}$$

where $A_{MC}$ = master cylinder area, cm$^2$ (in.$^2$)

B = power boost ratio

$\ell_p$ = pedal lever ratio

$\eta_p$ = pedal lever efficiency

The pedal force/brake line pressure relationship is not affected by circuit failure as indicated by Eq. (10-2). In other words, a given pedal force still produces the same level of brake line pressure. However, the same brake line pressure is not as effective in producing deceleration, as an inspection of Eq. (10-1) reveals. Consequently, the pedal force/deceleration relationship is affected by a circuit failure.

The power boost ratio B is affected by the performance of the vacuum booster or effectiveness of the hydro-boost. In the case of a complete power assist failure B = 1, and the brake system reverts to manual brakes.

### 10.3.2.c  Braking Efficiency

Braking efficiency is a measure of the capability of the vehicle to use a given tire-road friction coefficient for maximum deceleration prior to wheel lockup (see Chapter 7). The maximum wheels-unlocked deceleration $a_{F,max}$ or $a_{R,max}$ for system 1 that can be obtained with the partially failed system on a road surface with a specified tire-road friction coefficient can be determined from Eqs. (7-17a) or (7-17b) with the front brakes failed ($\Phi = 1$), or rear brakes failed ($\Phi = 0$), respectively, and solved for vehicle deceleration. The results are

Front failed:
$$a_R = \Psi\mu / (1 + \mu\chi) \quad , \quad \text{g-units} \tag{10-3}$$

Rear failed:
$$a_F = \frac{(1 - \Psi)\mu}{1 - \mu\chi} \quad , \quad \text{g-units} \tag{10-4}$$

where  $a_F$ = deceleration with rear brakes failed, g-units

$a_R$ = deceleration with front brakes failed, g-units

$\mu$ = tire-road friction coefficient

$\chi$ = center-of-gravity height divided by wheelbase

$\Psi$ = static rear axle load divided by vehicle weight

Braking efficiency may be computed by dividing the deceleration computed by Eqs. (10-3) and (10-4) by the tire-road friction coefficient $\mu$. Similar relationships may be derived for most of the systems indicated in Fig. 10-1.

### 10.3.2.d  Pedal Travel

The increased pedal travel is a significant factor when braking with one circuit failed. Reasons for this are longer times required to apply the brakes ranging from 0.5 to 0.7 seconds and possible undesirable driver reaction to the unfamiliar pedal position. It is not uncommon for drivers to totally abandon pedal application due to the longer pedal travel (driver says "it went to the floor") and reduced vehicle deceleration (driver says "brakes did not slow the car"). These two factors may confuse a driver sufficiently to not continue to apply hard pedal forces. Manufacturers should properly educate and warn operators about this feature generally unknown to the driving public.

Since pedal travel is determined by the master cylinder piston travel, the functioning of a master cylinder used in dual-circuit brake systems is discussed next for normal and failed operation. A typical dual-circuit or tandem master cylinder is illustrated in Figure 10-2. The operation is in principle the same as a single-circuit master cylinder. When the pushrod piston is moved toward the floating piston, the hole (1) connecting chamber (2) with the reservoir (3) is closed. As pedal travel continues the resulting pressure buildup in chamber (2) is transmitted by the floating piston to chamber (4). The floating piston moves forward and hole (5) connecting chamber (6) closes. At this moment the brake fluid volume in front of each primary cup or seal is closed and pressurization begins. If each master cylinder chamber has the same diameter, equal pressures are developed in each chamber.

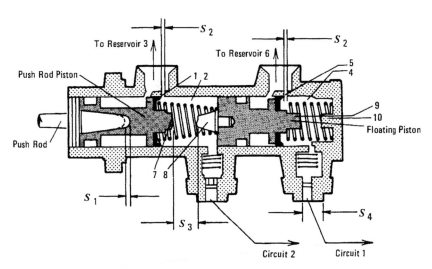

*Figure 10-2. Tandem master cylinder.*

If the circuit connected to chamber (2) fails through a hydraulic leak at any of its "wetted" surfaces (wheel cylinder, brake lines, fittings, master cylinder primary seal), no brake line pressure can be developed in chamber (2). This condition causes the pin (7) of the pushrod piston to contact the pin (8) of the floating piston. The pushrod force is transmitted directly to the floating piston and pressure buildup in chamber (4) results.

Similarly, a leak in the circuit connected to chamber (4) causes the floating piston pin (9) to come in contact with the stop (10). At this instant, brake line pressure can be developed in chamber (2) by the pushrod piston. The pedal travels in either failure mode are longer than in the unfailed condition. Longer pedal travels result in increased time before the brakes are applied and, hence, longer stopping distance. Measurements of pedal displacement indicate that a pedal travel of 5 in. requires approximately 0.25 s for the 90th percentile male driver.

The pedal travel is determined by the travel of the pushrod piston. Since the pushrod piston travel is affected by the travel of the floating piston, the following travels are identified and illustrated in Fig. 10-2:

1. Travel $S_1$ to overcome pushrod play:

   $$S_1 \approx 0.02 S_{av} \quad , \quad \text{mm (in.)}$$

   where $S_{av}$ = average pedal travel required for stop, mm (in.)

2. Travel $S_2$ to overcome hole or compensating port connecting chambers (2) or (4) with reservoir:

   $$S_2 \approx 0.06 S_{av} \quad , \quad \text{mm (in.)}$$

3. Possible travel $S_3$ of pushrod piston available for pressure buildup:

   $$S_3 = k S_{av} \quad , \quad \text{mm (in.)}$$

4. Possible travel $S_A$ of floating piston available for pressure buildup:

   $$S_4 = (1 - k) S_{av} \quad , \quad \text{mm (in.)}$$

where $k$ = ratio of pushrod piston travel $S_3$ to available travel of the pistons of the tandem master cylinder $S_{av}$

$S_{av}$ = $S_3 + S_4$, travel of pistons associated with tandem master cylinder, available for pressure buildup, mm (in.)

The factor $k$ generally assumes values between $0.9\Phi$ and $1.25\Phi$, where $\Phi$ is the ratio of rear brake force divided by total brake force (see Section 7.4).

The maximum pedal travel $S_{p,max}$, determined by the maximum travel of the pistons of the master cylinder, the pushrod play, and the pedal lever ratio is

$$S_{p,max} = \ell_p(S_1 + S_2 + S_3 + S_4) \quad , \quad \text{mm (in.)} \tag{10-5}$$

Eq. (10-5) can be rewritten with the previous expression for the individual travels as

$$S_{p,max} = \ell_p S_{av}(0.02 + 0.06 + k + 1 - k)$$

$$= 1.08\ell_p S_{av} \quad , \quad \text{mm (in.)} \tag{10-6}$$

where $\ell_p$ = pedal lever ratio

The travel of the pushrod piston and floating piston actually used in a normal braking situation is less than the maximum design value $S_3$ and $S_4$. Let $\rho$ be the ratio of the actual travel used for pressure buildup by the pushrod piston and floating piston to the available travel of pushrod piston and floating piston. The brake pedal travel for normal braking for the non-failed and failed brake system may then be represented by the following expressions:

1. Service brake not failed:

$$S_{p,nor} = \ell_p S_{av}[0.08 + \rho k + \rho(1 - k)]$$

$$= \ell_p S_{av}[0.08 + \rho] \quad , \quad \text{mm (in.)} \tag{10-7}$$

where r = ratio of actual piston travel used for pressure buildup to available piston travel

2. Circuit failure, system 1 (Fig. 10-1):

a. Circuit No. 1 failed, i.e., the front brakes are failed and the rear brakes are operative, and the floating piston develops no brake line pressure

$$S_{p,rear} = \ell_p S_{av}(0.08 + \rho k + 1 - k)$$

$$= \ell_p S_{av}[1.08 - k(1 - \rho)] \quad , \quad mm \text{ (in.)} \tag{10-8}$$

Eq. (10-8) is obtained from Eq. (10-7) by substituting the entire travel of the floating piston available into Eq. (10-7), i.e., $\ell_p S_{av}(1 - k)$ and not $\ell_p S_{av}\rho(1 - k)$ used in normal braking without brake failure.

b. Circuit No. 2 failed, i.e., the rear brakes failed and the front brakes are operative, and the pushrod piston develops no brake line pressure

$$S_{p,front} = \ell_p S_{av}[0.08 + k + \rho(1 - k)]$$

$$= \ell_p S_{av}[0.08 + \rho + k(1 - \rho)] \quad , \quad mm \text{ (in.)} \tag{10-9}$$

3. Circuit failure, system 2 (Fig. 10-1):

Any circuit failed:

$$S_{p,failed} = \ell_p S_{av}[0.58 + 0.5\rho] \quad , \quad mm \text{ (in.)} \tag{10-10}$$

4. Circuit failure, system 3 (Fig. 10-1):

a. Circuit No. 1 failed, i.e., the floating piston develops no brake line pressure

$$S_{p,failed} = \ell_p S_{av}[1.08 - k(1 - \rho)] \quad , \quad mm \text{ (in.)} \tag{10-11}$$

b. Circuit No. 2 failed, i.e., the pushrod piston develops no brake line pressure

$$S_{p,failed} = \ell_p S_{av}[0.08 + \rho + k(1 - \rho)] \quad , \quad \text{mm (in.)} \quad\quad (10\text{-}12)$$

5. Circuit failure, system 4 (Fig. 10-1):

Any circuit failed:

$$S_{p,failed} = \ell_p S_{av}[0.58 + 0.5\rho] \quad , \quad \text{mm (in.)} \quad\quad (10\text{-}13)$$

6. Circuit failure, system 5 (Fig. 10-1):

Any circuit failed:

$$S_{p,failed} = \ell_p S_{av}[0.58 + 0.5\rho] \quad , \quad \text{mm (in.)} \quad\quad (10\text{-}14)$$

### 10.3.2.e  Vehicle Stability Analysis

Dual-circuit splits producing different brake forces on the left and right wheels during a circuit failure will generate a rotating moment about the vertical axis of the vehicle. For the diagonal dual circuit, a negative scrub radius on the steered front wheels produces a steering moment during braking which automatically turns the front wheels in such a direction that a rotation of the vehicle is avoided. When properly designed, the vehicle will travel in a straight path without the driver having to turn the steering wheel.

During braking the retarding forces between tire and ground tend to rotate the vehicle in the direction of the non-failed front brake. For the vehicle to continue to travel straight, the side forces generated at the front and rear wheels must be of equal magnitude, but opposite direction. Since the side forces act on different lever arms, the resulting moment counteracts the braking moment. For side forces to be produced, both front and rear tires must generate slip angles, i.e., the vehicle will run under a small slip angle.

A stable braking maneuver will be achieved with a diagonal circuit split failure when the summation of moments due to braking, due to the side forces produced by slip angles, and due to tire realignment moment is equal to zero.

A negative scrub radius will cause the moment summation to be close to zero. If the summation is greater than zero, the driver must countersteer to maintain vehicle control.

### 10.3.3 Performance Calculations

The three performance measures were applied to a particular case. No attempts were made to express longer pedal travels in terms of increased application times. It may be assumed that application time increases linearly with pedal travel.

The deceleration achievable by a passenger vehicle with different circuit failures was computed and is shown in Figure 10-3. A maximum tire-road friction coefficient of $\mu = 1.0$ was assumed for columns 1 and 4. The tire-road friction coefficient for the computations of columns 2 and 3 was assumed large enough to prevent lockup. The column identified by "Wheels Unlocked" (1) represents the maximum deceleration that can be attained by the different systems with either circuit No. 1 or No. 2 failed prior to any brake lockup occurring. The deceleration achievable with the non-failed system is indicated also. The column identified by "No Proportioning" (2) gives the deceleration that can be achieved with a brake system with a fixed brake force distribution and a pedal force of 489 N (110 lb) with either circuit No. 1 or No. 2 failed. The column identified by "Proportioning" (3) represents the deceleration that can be achieved by a vehicle having a brake system with proportioning valve. In this case the proportioning valve is not affected by the failure of the front brakes, i.e., the proportioning valve is not bypassed when the front brakes are failed. The column identified by "Wheels Locked" (4) gives the maximum deceleration achievable on a m = 1.0 friction surface without regard to pedal force and lockup. Column 5 provides information on braking stability.

Inspection of Fig. 10-3 reveals that the performance of system 1 is limited to a deceleration of 0.25 g with circuit No. 1 failed, i.e., the front brakes inoperative, when a proportioning valve is installed which is not bypassed when the front brakes are failed. Proportioning valves generally are bypassed in the event of a front circuit failure. With the proportioning valve bypassed, the deceleration achievable with a pedal force of 489 N (110 lb) is 0.5 g with the front brakes failed. Inspection of the deceleration values reveals whether the pedal force and brake system gain or tire-road friction determines the deceleration

| | 1 | | 2 | | 3 | | 4 | | 5 |
|---|---|---|---|---|---|---|---|---|---|
| | Wheels Unlocked | | No Proportioning Pedal Force 110 lb | | Proportioning Pedal Force 110 lb | | Wheels Locked | | Vehicle Side to-Side Stability |
| No Failure | 0.80 | | 1.23 | | 1.23 | | 1.00 | | Stable |
| Circuit Failed | 1 | 2 | 1 | 2 | 1 | 2 | 1 | 2 | |
| System 1 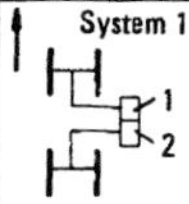 | 0.41 | 0.64 | 0.50 | 0.73 | 0.25 | 0.73 | 0.41 | 0.64 | Stable |
| System 2 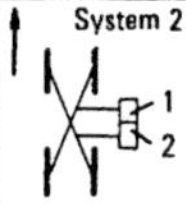 | 0.48 | 0.48 | 0.62 | 0.62 | 0.49 | 0.49 | 0.50 | 0.50 | Unstable |
| System 3 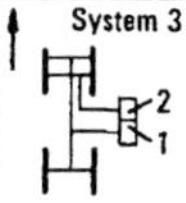 | 0.64 | 0.61 | 0.37 | 0.86 | 0.37 | 0.62 | 0.64 | 1.00 | Stable |
| System 4 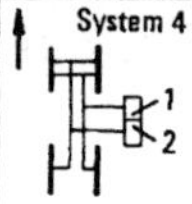 | 0.48 | 0.48 | 0.62 | 0.62 | 0.49 | 0.49 | 0.84 | 0.84 | Unstable |
| System 5 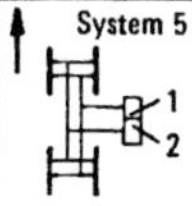 | 0.80 | 0.80 | 0.62 | 0.62 | 0.49 | 0.49 | 1.00 | 1.00 | Stable |

*Figure 10-3. Calculated deceleration in g-units for partial failure.*

available. For example, a system with proportioning and a pedal force of 489 N (110 lb) produces a deceleration of 0.25 g for system 1 with circuit No. 1 failed. If no pedal force limit is set, the tire-road friction coefficient permits deceleration of 0.41 g, which indicates that either the pedal force must exceed 489 N (110 lb), or if that is undesirable, the brake system gain must be increased.

Indicated in Fig. 10-3 are also the effects of side-to-side brake unbalance on vehicle stability. Only systems 2 and 4 exhibit a measure of vehicle stability with circuit failure. The influence of an unbalance on the front and rear wheels such as experienced by system 2 (diagonal brake system split) could, to a large extent, be counteracted by a negative scrub radius. A negative scrub radius forces the wheel to rotate slightly in the direction of the lower brake force, thus producing a tire slip angle sufficiently large to hold the vehicle in a stable controlled stop requiring little driver steering input. No driver steering corrections require that all moments about the vertical vehicle axis be zero. In order to accomplish zero moment, tire slip angles, tire realignment moments, pneumatic trail, and camber must be considered.

The pedal force analysis was applied in detail to the common front-to-rear split identified as system 1. Brake system data of a large domestic automobile were used as a base for computing the pedal travels required to perform a stop under partially failed conditions. The results are shown in Figure 10-4 in terms of pedal travel ratios. The pedal travel ratios shown in Fig. 10-4 are the pedal travel for normal unfailed braking to maximum travels available; the pedal travel with the front brakes failed to the maximum travel available; the pedal travel with the rear brakes failed to maximum travel available. The values were computed for $\rho = 0.5$, i.e., 50% of the available master cylinder piston travel is required for a service brake stop. Inspection of Fig. 10-4 reveals that in the case of a failure of the front brakes the ratio of pedal travel required to apply the rear brakes to maximum pedal travel is nearly twice the ratio for the unfailed or normal brakes for brake force distributions $\Phi$ less than 0.3. In other words, for $\Phi = 0.3$ or less the front brake fluid requirements due to a large wheel cylinder cross-sectional area tend to be significantly greater than those for the rear brakes, and in the case of a front brake failure, cause a significantly greater pedal travel drop toward the floor than in the case of a rear brake failure.

The ratios of pedal travel under failed conditions to those required under normal conditions (not maximum pedal travel as shown in Fig. 10-4) may be a more meaningful pedal feedback indicator to the driver. These ratios are presented in Figure 10-5. Inspection of Fig. 10-5 clearly reveals again that the standard front-to-rear split requires significant pedal travels in the event of a front brake failure and small values of $\Phi$. This condition exists in spite of a rather long master cylinder piston travel as indicated by $\rho = 0.5$, used in the calculations. Fortunately, front-to-rear dual brake systems are used on vehicles having front-to-rear weight distributions not requiring extremely low $\Phi$-values to prevent premature rear wheel lockup, i.e., relatively large $\Psi$-values.

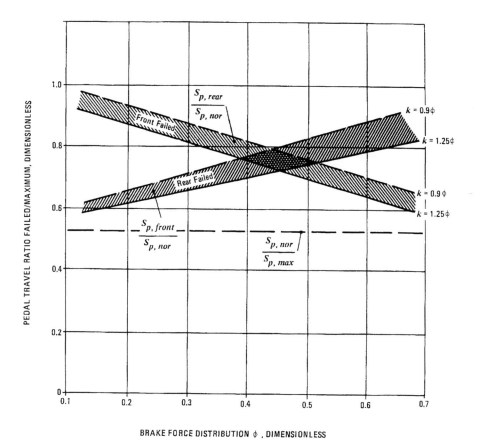

*Figure 10-4. Maximum pedal travel ratios required for partial failure stops, system 1.*

To show the effect of a long versus short master cylinder piston travel for the front-to-rear split dual design, the pedal travel ratios failed/normal are plotted in Figure 10-6 as a function of $\rho$, i.e., versus utilization of effective master cylinder piston travel required for a normal stop. A ratio of floating piston travel to pushrod piston travel of 74:26 and a brake force distribution $\Phi = 0.32$ were used in the calculations. Inspection of the curve representing front brake failure reveals that long master cylinder or large pedal travel reserves, i.e., low values of $\rho$ result in undesirably long pedal travels in the case of front brake failure. This condition exists in spite of the highly acceptable normal-to-maximum pedal travel ratio.

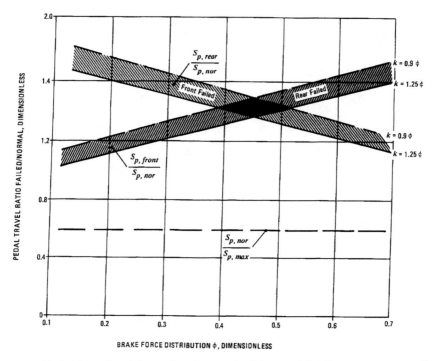

*Figure 10-5. Normal pedal travel ratios required for partial failure stops, system 1.*

Braking systems using duo-servo brakes on the rear and disc brakes on the front generally have relatively small master cylinder piston travels for the rear circuit chamber. Under these conditions a front brake failure can easily cause excessive pedal travels often "interpreted" by drivers as a complete brake failure. The reason for this undesirable characteristic stems from the high brake factor associated with the duo-servo brake.

### 10.3.4  Improved Dual Master Cylinder Design

Significant improvements in minimizing the effects of partial failure on pedal force and pedal travel have been accomplished by means of a special stepped bore master cylinder (Ref. 32). The brake line pressures achieved under failed conditions are double the pressure achieved under normal conditions. This change in effective piston area is accomplished by means of a stepped bore as illustrated in Figure 10-7. Whereas the larger master cylinder feeds both circuits in the case of an intact brake system, the smaller diameter master

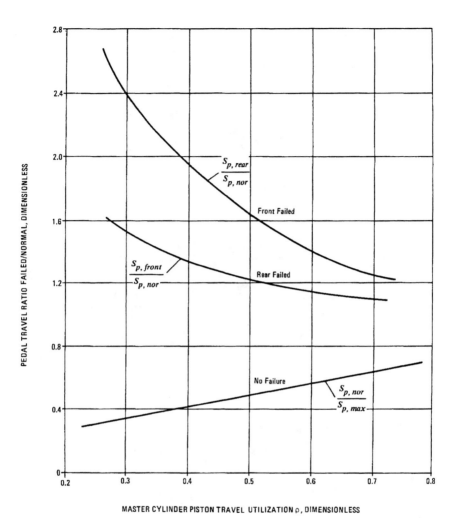

MASTER CYLINDER PISTON TRAVEL UTILIZATION $\rho$, DIMENSIONLESS

*Figure 10-6.   Pedal travel ratio as a function of piston travel utilization.*

cylinder is utilized in the event of circuit No. 1 failure, and double brake line pressure is produced because the smaller piston area is one-half the area of the larger piston. If circuit No. 2 fails, the differential area between the larger and the smaller bore, i.e., one-half of the standard area, becomes the effective brake line pressure producing area. The result is a brake line pressure under failed conditions that is twice as large as under normal conditions. Since under circuit failure conditions not all wheel brakes are actuated and normally

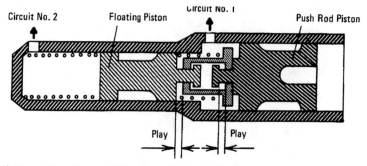

Circuit No. 2     Floating Piston     Circuit No. 1     Push Rod Piston

Play     Play

(A) Stepped Bore Master Cylinder, Normal Brake Application

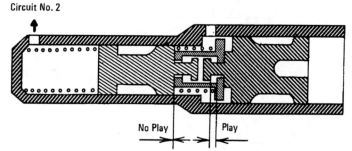

Circuit No. 2

No Play     Play

(B) Stepped Bore Master Cylinder, Brake Application With Leakage in Circuit Number 1

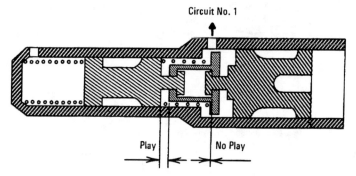

Circuit No. 1

Play     No Play

(C) Stepped Bore Master Cylinder, Brake Application With Leakage in Secondary Circuit Number 2

*Figure 10-7. Stepped bore tandem master cylinder.*

are not locked, a nearly identical wheels-unlocked deceleration results with the same pedal force. The pedal travel in the failed condition exceeds the pedal travel in the unfailed condition by not more than approximately 30%. Thus with this master cylinder, pedal forces and pedal travel are no longer the limiting factors in the case of partial failure braking.

Brake system splits using unequal volume distribution between circuits and significantly different cross-sectional areas for the floating and pushrod piston chambers may lead to excessively high brake line pressures in case of partial failure.

## 10.4 Comparison of Dual Brake Systems

A comparison of the dual brake systems represented in Fig. 10-1 indicates that a different number of connectors and flexible hoses is required for the different systems. For example, system 1 requires 17 connectors compared to 34 for system 5.

A leak is more likely to develop in a hydraulic circuit that contains more removable connections, wheel cylinder seals, and other devices such as valves and control elements. A comparison of the complexity of the different dual-circuit splits is shown in Figure 10-8. All removable connections, such as T-fittings, are included in Fig. 10-8. The data indicate system 5 has a higher failure probability than the remaining systems. Difficulties also may arise in properly installing the flexible hoses near the wheels. A larger number of fittings with machined cavities will also trap a greater amount of residual air in the brake system. Not included in the number of removable connections is the third bleeder screw required for double caliper disc brakes if they are designed as one unit.

A hydraulic brake system may experience a temporary brake failure due to brake fluid boiling and vaporization. This condition may cause an accident although shortly after the accident the brake pedal is found to be firm and the brake system mechanically intact. The maximum temperature of a brake should be kept below certain limits. High brake temperatures will result in: (a) brake fade and increased lining wear, (b) high tire bead temperatures, and (c) increased temperature of the brake fluid in wheel cylinders. Modern brake fluids boil at approximately 705 K (450°F). Consequently, prolonged downhill travel with a fully loaded vehicle (trailer) or school bus, may cause the brake fluid to boil, vapor to develop, and the brake system to fail. When the driver applies light pedal force, the pressurized brake fluid will boil at higher temperatures. It may be possible that the brakes worked without failure on the downhill portion, with the brake fluid just below boiling temperature, only to fail after the driver

| Brake System | Single-Circuit System | System No. | | | | |
|---|---|---|---|---|---|---|
| | | 1 | 2 | 3 | 4 | 5 |
| Wheel Cylinder | 8 | 8 | 8 | 12 | 12 | 16 |
| Removable Connections | 17 | 15 | 16 | 25 | 26 | 34 |

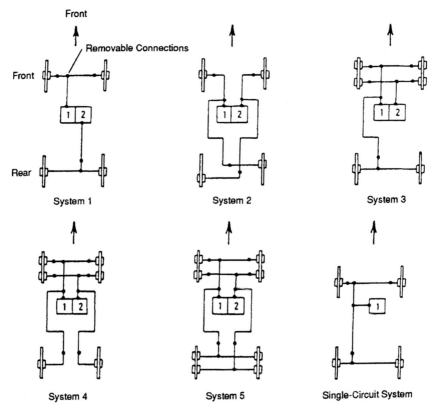

*Figure 10-8. Comparison of system complexity.*

relaxed his or her foot off the brake pedal at the bottom of the hill, where the brake fluid began to boil due to the lower fluid pressure some short distance farther down at the first intersection.

To determine excessive brake temperature for this type of intermittent brake failure, all brake components affected by high temperature must be carefully examined. It is expected that the swept surfaces of rotors and drums would show sign of discoloration; brake pads and linings may be blackened; rubber seal or dust boots may be charred from excessive heat.

One effect of brake fluid vaporization from thermal overloading on circuit failure of the different dual systems is indicated in Figure 10-9. It is assumed that the front brakes are experiencing excessive brake temperature leading to fluid vaporization and, hence, brake failure of the circuits connected to the front brakes. Inspection of Fig. 10-9 reveals that only system 1, the front-to-rear

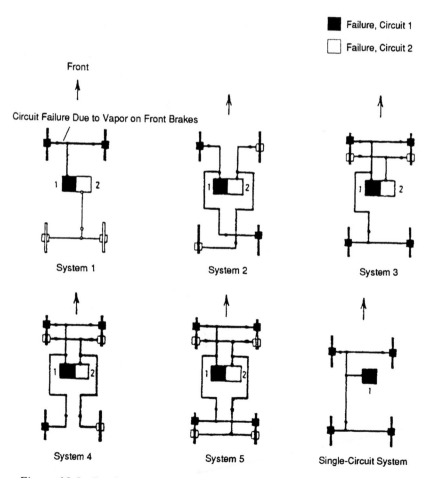

*Figure 10-9. Dual systems, front brake failure due to brake fluid vaporization.*

split, provides a partial braking capability on the rear wheels with the front brakes failed due to vaporization. If, however, the rear brakes exhibit vaporization and circuit failure, all but systems 1 and 3 will fail completely.

## 10.5 Vacuum Assist Failure

Vacuum assist or booster failure exists when the assist function of the power boost unit is degraded partially through insufficient vacuum (engine problems, vacuum pump defect, booster leaking) or complete loss of vacuum (stalled engine, loose hose connection).

With the complete or partial failure of the vacuum assist unit, the reduced brake line pressure may be obtained from Eq. (10-2) with the boost ratio reduced accordingly. The lower brake line pressure may then be used in Eq. (10-1) to compute vehicle deceleration under a booster failure condition. Typical results of such an analysis are presented in Figure 10-10 in the form of a braking performance diagram. The following observations can be made with respect to various levels of power boost failure:

1. <u>No assist</u>. To produce a deceleration of 0.9 g, a pedal force of 1201 N (270 lb) is required. A deceleration of only 0.32 g is produced with a pedal force of 445 N (100 lb).

2. <u>32 % assist</u>. The deceleration, produced by a pedal force of 445 N (100 lb), is 0.52 g. A deceleration of 0.9 g requires a pedal force of about 956 N (215 lb).

3. <u>60 % assist</u>. The deceleration, produced by a pedal force of 448 N (100 lb), is 0.76 g. A deceleration of 0.9 g requires a pedal force of about 667 N (150 lb).

## 10.6 Full Power Brake Failure

Full power hydraulic brake systems use a tandem master cylinder in conjunction with a pressurized accumulator or circulating pump system. Dual-circuit failure analysis is identical to that of the unassisted or vacuum power-assisted brake system. In the event of a power source failure, a sufficient level of energy is stored in the accumulator, or the emergency accumulator (spring or

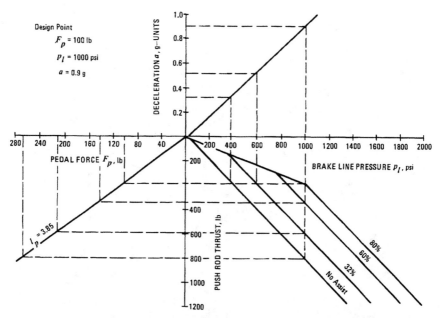

*Figure 10-10. Braking performance diagram for vacuum-assisted brake system.*

gas loaded) for the circulating pump system, to produce a certain number of successive emergency stops. Generally, three stops are required under complete primary energy source failure. Furthermore, most systems provide a manual "push through" capability from the driver's pedal effort.

## 10.7  Degraded Braking Due to Air Inclusion

A detailed analysis of the braking performance based on the brake fluid requirements of various brake system components is presented in Section 5.4.3. An example calculation is shown in which the maximum deceleration is computed as the brake pedal contacts the floor.

In connection with a particular accident involving a four-wheel disc brake sports car, braking tests were conducted to determine the partial braking performance as air was artificially introduced into the brake system. The results are shown in Figure 10-11, where pedal travel, vehicle deceleration, and pedal force are plotted as a function of amount of air inclusion. Percent air identifies the percentage of brake fluid volume of the master cylinder that

495

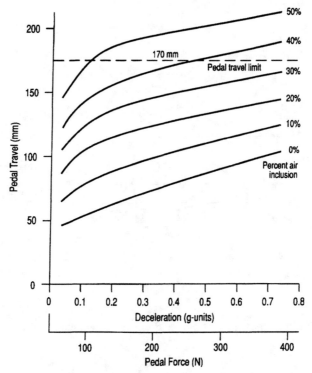

*Figure 10-11. Vehicle deceleration as a function of pedal travel
(1966 Porsche 911).*

had been replaced by air. A maximum pedal travel of 170 mm (6.7 in.) was
available. Inspection of Fig. 10-11 reveals that even when 30% of the
master cylinder volume has been replaced by air, over 0.8 g deceleration
can be achieved before the pedal contacts the floor. With 50% air the pedal
will bottom out at a deceleration of approximately 0.15 g at a pedal force of
120 N (25 lb).

## 10.8  Brake Fluid Considerations in Design and Failure Analysis

Since the characteristics of the different types of brake fluids available are
important in the design of brake systems and in the investigation and analysis
of brake failures, some details are presented in the following sections.

### 10.8.1 Brake Fluid Properties

The most important properties of brake fluids are:

1. High boiling point to avoid the development of vapor in the brake system at elevated temperatures.

2. Small amounts of water in the fluid (less than 0.2%) should not reduce the boiling point significantly.

3. Viscosity should be low when cold, or high when hot.

4. Compressibility should be low; temperature and pressure should have little effect.

5. Corrosion protection against metal parts to ensure long component life.

6. Good lubrication for long component life of moving parts.

7. Good chemical reaction with rubber parts to avoid shrinking and cause only little swelling.

8. Little or no gas production caused by turbulent-type flow processes and low pressures.

9. Rapid elimination of foam.

10. Ability to mix with property-improving additives.

11. Ability to absorb residual air during vacuum filling of the brake system at the factory.

12. Oxidation stability in foreseeable temperature ranges.

13. Low vehicle paint aggressivity.

14. Low toxicity.

15. Proper color to avoid inadvertent use of improper fluids.

### 10.8.2 Different Types of Brake Fluids

In the past, three different types of fluids have been used in automotive applications. No single type satisfies all 15 requirements equally well.

The three types of brake fluids are based on:

a. Polyglycolether, the most commonly used brake fluid such as DOT 3 or DOT 4.

b. Silicone, rarely used except by U.S. military (DOT 5).

c. Mineral oil, rarely used in automotive applications, except, for example, Rolls-Royce using fully hydraulic brake system design without master cylinder, and motorcycles.

*Polyglocolether*-type brake fluids are commonly used in vehicles equipped with hydraulic brake systems. Some differences exist among DOT 3, 4, and 5. For example, DOT 4 has a better reaction ability than DOT 3 and, hence, lower sensitivity to water found in the brake fluid than DOT 3.

*Silicone-based* brake fluids are used less frequently. The U.S. military currently uses silicone fluids. The basic element of the silicone fluids is a silicium-based polymer. Advantages include non-hygroscopic characteristics; good viscosity/temperature behavior; compatible with commonly used elastomers for rubber parts; good corrosion properties; high boiling point. Disadvantages of silicone fluids are high compressibility, especially under higher temperatures; high air absorption which will be released again into the water when vibrations occur; no ability to absorb water (the flip side of being non-hygroscopic); minute amount of water may cause brake failure.

*Mineral oil* brake fluids are polyalcylethylene-based polymers. They are not compatible with commonly used seal or cup materials SRB (styrol-butadiene-rubber) or EPDM (ethylene-propylene-diene-materials). Mineral oil fluids are used in hydro-boost hydraulic brake systems where the pedal effort is used only to control brake line pressure, i.e., in those without master cylinders. The major disadvantage is the inadvertent contamination of the mineral oil with regular brake fluid. Small amounts less than 0.2% of regular brake fluid mixed with mineral brake fluids may deteriorate rubber seals and hoses.

## 10.8.3 Brake Fluid Performance

The performance of brake fluids is regulated by FMVSS 116, which in most parts is based on SAE J1703. Brake fluid classifications are determined by the boiling point (dry), wet boiling point (3.5% water by weight in the fluid), and viscosity at 233 K (−40°F). An additional 12 other measures are evaluated in FMVSS 116. Based on FMVSS 116 the dry/wet boiling point temperatures are: for DOT 3, 477/413 K (400/285°F); DOT 4, 503/428 K (446/311°F); and DOT 5, 555/453 K (500/356°F). On occasion, brake fluid manufacturers market newly formulated brake fluids to improve existing ones. For example, DOT 4 Plus has higher dry and wet boiling point temperatures as well as lower cold temperature viscosity than regular DOT 4. DOT 5.1 is a new brake fluid that is not compatible with regular DOT 5.

Regular brake fluids based on polyglycolether are hygroscopic, which means that they will absorb water from the surrounding atmosphere. More water will be found in the brake fluid with time, provided the fluid has not been changed during maintenance of the vehicle. Entry of water into the brake system through the master cylinder reservoir is generally of little significance. The small breather hole or diaphragm effectively prevents water from entering the master cylinder. In addition, no serious threat of overheating of brake fluid in the master cylinder—assuming proper maintenance, repair, and design—and, hence, no brake fluid boiling occurs due to its location away from the heated friction brakes. However, water will enter the brake system through its flexible hoses, through a diffusion process. The amount of diffusion water is a function of the hygroscopic properties of the fluid and the diffusion resistance of the hose material. The commonly used SRB hose material diffuses more water than EPDM materials.

Higher levels of water contamination of brake fluid were found in the flexible brake hoses near the wheel cylinders than at the master cylinder, which generally has no flexible hoses near it or in the reservoir. Tests have shown that the boiling point of brake fluid taken from the brake system near the brake hoses is up to 311 K (100°F) lower than that of the fluid taken from the master cylinder reservoir. In this respect, DOT 4 and 5 exhibit lower water contamination than DOT 3, which should be changed every year.

## 10.9  Seal and Rubber Materials

Polyglycolether and silicone-based brake fluids are compatible with SRB (styrol-butediene-rubber) and EPDM (ethylene-propylene-diene-rubber). Mineral oil is not compatible with these elastomers and requires seal and rubber component materials such as neoprene. Most seal elastomers begin to deteriorate when exposed to temperatures above the boiling point for long periods of time. Pulsating pressure loadings as may occur at the master cylinder can only be withstood by SRB seals for temperatures less than 352 to 377 K (175 to 220°F). EPDM seal materials have temperature limits of up to 394 K (250°F). As engine compartment temperatures increased in the past, difficulties may arise for heavy high-powered passenger cars since the cup hardness has to be maintained for minimum seal wear at the compensating ports. The change to EPDM cup material required design changes of the master cylinder since excessive wear at the compensating ports occurred. The commonly used compensating port design was replaced by central valves. DOT 4 exhibits excellent performance in connection with EPDM material in the form of slight seal swelling essential for proper operation over extended periods of vehicle life. Neoprene materials used for mineral oil have lower allowable operating temperatures than either SRB or EPDM cups.

## 10.10  Failure of Air Brake Systems

Air brake systems are designed with a dual brake system split; generally the front brakes form one circuit, the rear brakes the other. Consequently, air loss in one circuit will provide full air pressure to the remaining circuit. The reduced vehicle deceleration can be computed by use of the appropriate equations of Chapter 6 (Eq. [6-7]). Modern air brake systems are designed so that the failure of one circuit of the tractor will still provide full braking of the trailer and partial braking of the tractor.

If an air loss occurs, automatic or driver-controlled brake application of the emergency/parking brakes results. The spring brakes generally provide an application force corresponding to an air pressure in the brake chamber of approximately 41 N/cm$^2$ (60 psi). It should be noted that when the brakes are out of adjustment, i.e., the pushrod travel is too long to effectively push the shoes against the drum, then the spring of emergency brakes will be ineffective also.

The most common brake failure found in accidents related to or caused by air brakes results from brakes being out of adjustment or being at a critical level of adjustment so that the driver cannot easily identify how "bad" the brakes are. Adjustment indicators will provide a significant safety benefit for the truck driver as well as the motoring public.

If the glad hands between tractor and trailer are switched, the brakes will operate in one of the following modes:

(a) <u>Trailer reservoir without air pressure</u>: For older trailers without spring brakes the trailer supply valve will stay open when attempting to charge the trailer. However, the trailer will not be charged because the lines are crossed. When the brakes are applied by the driver, the trailer brakes apply at maximum pressure and stay applied.

With spring brakes, the brakes cannot be released when the dash-mounted valve is activated.

(b) <u>Trailer reservoir with air pressure</u>: For older trailers without spring brakes, the brakes apply fully when the dash-mounted valve is activated and the brakes stay on.

# Appendix A

## CONVERSION OF UNITS

| TO CONVERT | INTO | MULTIPLY BY |
|---|---|---|
| bar | psi | 14.5 |
| BTU | foot-lb | 778.3 |
| BTU/hr | horsepower | $3.929 \times 10^{-4}$ |
| cubic feet | cubic yards | 0.03704 |
| °Celsius | °Fahrenheit | $(°C \times 9/5) + 32$ |
| cubic centimeters | gallons (U.S.) | $2.642 \times 10^{-4}$ |
| feet | meters | 0.3048 |
| feet/sec | kilometers/hr | 1.097 |
| feet/sec/sec | meters/sec/sec | 0.3048 |
| foot-lbs | Newton meters | 1.36 |
| foot-lbs/sec | BTU/hr | 4.6263 |
| horsepower | foot-lbs/sec | 550 |
| horsepower | kilowatts | 0.7457 |
| inches | meters | $2.54 \times 10^{-2}$ |
| inches | millimeters | 25.4 |
| kilometers/hr | miles/hr | 0.6214 |
| kilowatt-hrs | BTU | 3413 |
| knots | feet/sec | 1.689 |
| liters | gallons (U.S.) | 0.2642 |
| liters | cubic inches | 61.02 |
| meters | feet | 3.281 |
| meters | millimeters | 1000 |

## CONVERSION OF UNITS (CONTINUED)

| TO CONVERT | INTO | MULTIPLY BY |
|---|---|---|
| miles/hr | feet/sec | 1.4667 |
| miles/hr | kilometers/hr | 1.609 |
| millimeters | inches | 0.03937 |
| moment of inertia: kg-m$^2$ | lb-ft-sec$^2$ | 0.737 |
| Newton | pounds | 0.2248 |
| Newton-meter | foot-lbs | 0.74 |
| pounds | kilograms | 0.4536 |
| psi | kPa | 6.8948 |
| yards | meters | 0.9144 |

# References

1. Limpert, Rudolf, <u>Motor Vehicle Accident Reconstruction and Cause Analysis</u>, The Michie Company, 1994.

2. Burckhardt, Manfred, <u>Fahrwerktechnik: Bremsdynamik und Pkw-Bremsanlagen</u>, Vogel Buchverlag, Germany, 1991.

3. Burckhardt, Manfred, <u>Reaction Times in Emergency Braking Maneuvers</u>, Verlag TUV Rheinland, Germany, 1985.

4. Nieman, G., <u>Machine Elements</u>, Vol. 1, Springer-Verlag, Germany, 1981.

5. *Auto, Motor and Sport Magazine*, No. 9, 1991, pp. 197-201.

6. Kuhlman, A., <u>Introduction into the Safety Sciences</u>, Verlag TUV Rheinland, Germany, 1981.

7. Kuhlman, A., *et al.*, <u>Prognose der Gefahr</u>, Verlag TUV, Germany, 1969.

8. Schaden, Richard F., and Heldman, Victoria C., <u>Product Liability Design</u>, Practicing Law Institute, 1982.

9. Burkman, Albin J., "A Laboratory Method for Testing Moisture Sensitivity of Brake Lining Materials," SAE Paper No. 620128 (458A), Society of Automotive Engineers, Warrendale, Pa., 1962.

10. Limpert, R., <u>Engineering Design Handbook, Analysis and Design of Automotive Brake Systems</u>, US Army Material Development and Readiness Command, DARCOM-P-706-358, 1976.

11. <u>Automotive Handbook</u>, Bosch, 1986.

12. Muller, W., "Contribution to the Analysis and Testing of Motor Vehicle Drum Brakes," <u>Deutsche Kraftfahrtforschung und Strassenverkehrstechnik</u>, No. 207, 1971.

13. Mitschke, M., <u>Motor Vehicle Dynamics</u>, Springer Publisher, 1972.

14. Arpaci, Vedat S., <u>Conduction Heat Transfer</u>, Addison-Wesley Publishing Company, 1966.

15. Limpert, R., "Temperature and Stress Analysis of Solid-Rotor Disc Brakes," Ph.D. Dissertation, University of Michigan, 1972.

16. Kreith, F., <u>Principles of Heat Transfer</u>, International Textbook Company, 1965.

17. Limpert, R., "Cooling Analysis of Disc Brake Rotors," SAE Paper No. 751014, Society of Automotive Engineers, Warrendale, Pa., 1975.

18. *Auto, Motor and Sport Magazine*, No. 25, 1991.

19. Pierce, B.F., "Human Force Considerations in the Failure of Power Assisted Devices," US Department of Transportation No. DOT-HS-800889, July 1973.

20. Klein, H.C. and Strien, H., "Hydraulically Boosted Brakes - An Important Part of Central Hydraulic Systems," SAE Paper No. 750867, Society of Automotive Engineers, Warrendale, Pa., 1975.

21. Limpert, R., "Minicars RSV Brake System," 5th International Automobile Congress, 1977.

22. Radlinski, R., Williams, S., and Machey, J., "The Importance of Maintaining Air Brake Adjustment," SAE Paper No. 821263, Society of Automotive Engineers, Warrendale, Pa., 1982.

23. Murphy, R., *et al.*, "Bus, Truck, Tractor-Trailer Braking System Performance," US Department of Transportation Contract No. FH-11-7290, March 1972.

24. Sido, F., "Analysis of Response Characteristics of Pneumatic Brake Systems Through Model Test," *Automobiltechnische Zeitschrift*, Vol. 71, No. 3, 1969.

25. Flaim, Thomas, "Vehicle Brake Balance Using Objective Brake Factors," SAE Paper No. 890804, Society of Automotive Engineers, Warrendale, Pa., 1989.

26. Warner, Ch., and Limpert, R., "Proportioning Braking of Solid-Frame Vehicles," SAE Paper No. 710047, Society of Automotive Engineers, Warrendale, Pa., 1971.

27. Grandel, Juergen, "Braking of Cars Towing Unbraked Trailers of Different Weight," *Verkehrsunfall und Fahrzeugtechnik*, Oct. 1987.

28. Burckhardt, Manfred, "The Computation of Braking Deceleration of a Car-Trailer Combination with Unbraked Trailers," *Verkehrsunfall und Fahrzeugtechnik*, Oct. 1987.

29. Anti-Lock Braking Systems for Passenger Cars and Light Trucks - A Review, SAE PT-29, Society of Automotive Engineers, Warrendale, Pa., 1986.

30. Ostwald, F., "Development of Wheel-Antilock Systems for Motor Vehicles," *Automobile Revue*, No. 40, 1964.

31. "Braking Systems for Passenger Cars," Bosch Technical Instruction, Stuttgart, Germany.

32. Larson, A., and Larson, L., "Stepped Bore Master Cylinder—A Way of Improving Dual Brake Systems," SAE Paper No. 750385, Society of Automotive Engineers, Warrendale, Pa., 1975.

# Index

# About the Author

Dr. Rudolf Limpert is a consulting engineer on motor vehicles and traffic safety, specializing in accident reconstruction, product liability, and vehicle safety seminars. He has been an automobile mechanic and a brake design engineer. He has also served as safety standards engineer for the National Highway Traffic Safety Administration and professor of mechanical engineering at the University of Utah, where he directed a multidisciplinary accident investigation team for the U.S. Department of Transportation. The author of four other books related to automotive safety, Dr. Limpert received his Ph.D. in mechanical engineering from the University of Michigan, his M.S. and B.E.S. from Brigham Young University, and his B.S. from the Engineering School of Wolfenbuettel.